Average global temperatures have risen approximately 1.2°C since the pre-industrial age.[1]

In the Intergovernmental Panel on Climate Change's (IPCC) 2021 report a group of 234 top scientists from 66 countries concluded that 'it is unequivocal that human influence has warmed the atmosphere, ocean and land. Widespread and rapid changes in the atmosphere, ocean, cryosphere and biosphere have occurred.'

2021

Global temperature anomaly

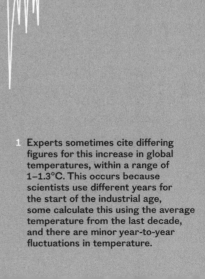

1850

1 Experts sometimes cite differing figures for this increase in global temperatures, within a range of 1–1.3°C. This occurs because scientists use different years for the start of the industrial age, some calculate this using the average temperature from the last decade, and there are minor year-to-year fluctuations in temperature.

Greenhouse gas emissions – which include carbon dioxide, methane, nitrous oxide and fluorinated gases – from human activities have risen to concentrations in the atmosphere that have not been seen in millions of years, since a time when trees grew at the South Pole and the sea level rose by 20 metres.

~420 ppm
2022

Atmospheric CO$_2$ concentration in parts per million

Emergence of *Homo sapiens*

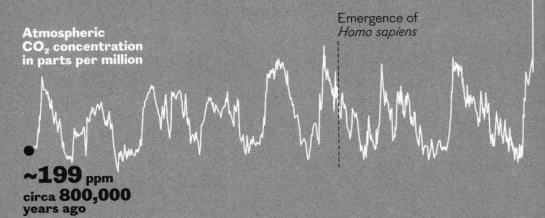

~199 ppm
circa **800,000** years ago

Despite dire warnings in the 1980s and 1990s, we have emitted more CO$_2$ since 1991 than in the rest of human history.

According to the IPCC's estimate, our remaining carbon budget for a 67 per cent chance of limiting warming to 1.5°C at the beginning of 2020 was 400 gigatonnes.[2] At the current rate of emissions, we will exceed this carbon budget before 2030.

2021

1990

Annual global CO$_2$ emissions from burning fossil fuels

785 GtCO$_2$ emitted

948 GtCO$_2$ emitted

1750

Some countries are vastly more historically responsible for emissions than others; the largest emitters released hundreds of billions of tonnes of CO_2 into the atmosphere between 1850 and 2021.

420.0 $GtCO_2$	**US**
241.8	**China**
117.3	**Russia**
93.1	**Germany**
74.9	**UK**
66.7	**Japan**
57.1	**India**
38.5	**France**
34.2	**Canada**
30.0	**Ukraine**

In 2015, nearly every country in the world – 195 in total – committed to the Paris Agreement. The goal of the Paris Agreement is to limit global warming to well below 2°C, and ideally below 1.5°C, compared to pre-industrial levels.

The world is not on track to meet these goals. There is a vast gap between the promises governments have made and the actions they have taken. Many emissions – such as those from international transport and shipping, as well as many of those associated with the military – go unrecorded or are unaccounted for.

Based on current policies, the IPCC estimates that global warming will reach 3.2°C by 2100.

2 The carbon budget is the maximum amount of CO_2 that humanity can emit while still having a chance at limiting warming to 1.5°C or 2°C.

THE CLIMATE BOOK

CREATED BY

GRETA THUNBERG

PENGUIN PRESS

New York

2023

PENGUIN PRESS
An imprint of Penguin Random House LLC
penguinrandomhouse.com

First published in Great Britain by Allen Lane, an imprint of Penguin Random House UK, 2022.

Pages 444–446 constitute an extension of this copyright page.

LIBRARY OF CONGRESS CATALOGING-IN-PUBLICATION DATA IS AVAILABLE
ISBN 9780593492307 (hardcover)
ISBN 9780593492314 (ebook)

Printed in the United States of America
1st Printing

Designed by Jim Stoddart

PART THREE /

How It Affects Us

PART FOUR /

What We've Done About It

PART FIVE /

What We Must Do Now

320

The contributors have assembled thousands of references and citations for the chapters in *The Climate Book*. While these notes are too numerous to be printed within this book, they can be found at **theclimatebook.org**

Next pages
Frozen bubbles of
methane in Lake
Baikal, Russia.

PART ONE

How Climate Works

'Listen to the science. Before it's too late'

To solve this problem, we need to understand it

Greta Thunberg

The climate and ecological crisis is the greatest threat that humanity has ever faced. It will no doubt be the issue that will define and shape our future everyday life like no other. This is painfully clear. In the last few years, the way we see and talk about the crisis has started to shift. But since we have wasted so many decades ignoring and downplaying this escalating emergency, our societies are still in a state of denial. This is, after all, the age of communication, where what you say can easily outweigh what you do. That is how we have ended up with such a great number of major fossil-fuel-producing – and high-emitting – nations calling themselves climate leaders, despite not having any credible climate mitigation policies in place. This is the age of the great greenwashing machine.

There are no black-and-white issues in life. No categorical answers. Everything is a subject for endless debate and compromise. This is one of the core principles of our current society. A society which, when it comes to sustainability, has a lot to answer for. Because that core principle is wrong. There *are* some issues that are black and white. There are indeed planetary and societal boundaries that must not be crossed. For instance, we think our societies can be a little bit more or a little bit less sustainable. But in the long run you cannot be a little bit sustainable – either you are sustainable or you are unsustainable. It is like walking on thin ice – either it carries your weight, or it does not. Either you make it to the shore, or you fall into the deep, dark, cold waters. And if that should happen to us, there will not be any nearby planet coming to our rescue. We are completely on our own.

It is my genuine belief that the only way we will be able to avoid the worst consequences of this emerging existential crisis is if we create a critical mass of people who demand the changes required. For that to happen, we need to rapidly spread awareness, because the general public still lacks much of the basic knowledge that is necessary to understand the dire situation we are in. My wish is to be part of the effort to change that.

I have decided to use my platform to create a book based on the current best available science – a book that covers the climate, ecological and sustainability crisis holistically. Because the climate crisis is, of course, only a symptom of a much larger sustainability crisis. My hope is that this book might be some kind of go-to source for understanding these different, closely interconnected crises.

In 2021, I invited a great number of leading scientists and experts, and activists, authors and storytellers to contribute with their individual expertise. This book is the result of their work: a comprehensive collection of facts, stories, graphs and photographs showing some of the different faces of the sustainability crisis with a clear focus on climate and ecology.

It covers everything from melting ice shelves to economics, from fast fashion to the loss of species, from pandemics to vanishing islands, from deforestation to the loss of fertile soils, from water shortages to Indigenous sovereignty, from future food production to carbon budgets – and it lays bare the actions of those responsible and the failures of those who should have already shared this information with the citizens of the world.

There is still time for us to avoid the worst outcomes. There is still hope, but not if we continue as we are today. To solve this problem, we first need to understand it – and to understand the fact that the problem itself is by definition a series of interconnected problems. We need to lay out the facts and tell it like it is. Science is a tool, and we all need to learn how to use it.

We also need to answer some fundamental questions. Like, what is it, exactly, we want to solve in the first place? What is our goal? Is it to lower emissions, or to be able to go on living as we are today? Is our goal to safeguard present and future living conditions, or is it to maintain a high-consumption way of life? Is there such a thing as green growth? And can we have eternal economic growth on a finite planet?

Right now, many of us are in need of hope. But what is hope? And hope for whom? Hope for those of us who have created the problem, or for those who are already suffering its consequences? And can our desire to deliver this hope get in the way of taking action and therefore risk doing more harm than good?

The richest 1 per cent of the world's population are responsible for more than twice as much carbon pollution as the people who make up the poorest half of humanity.

Perhaps, if you are one of the 19 million US citizens or the 4 million citizens of China who belong to that top 1 per cent – along with everyone else who has a net worth of $1,055,337 or more – then hope is perhaps not what you need the most. At least not from an objective perspective.

Of course, we hear, some progress is being made. Some nations and regions report quite astonishing reductions in CO_2 emissions – or at least in the years since the world first started negotiating the frameworks for how we manage our statistics. But how do all those reductions hold up once we include our total emissions, rather than carefully managed territorial statistics? In other words, all those emissions that we so successfully negotiated out of these figures. For instance, outsourcing factories to distant parts of the world and negotiating emissions from international aviation and shipping out of our statistics – which means that we not only manufacture our products by using cheap labour and exploiting people, we also erase the associated emissions – emissions that have, in reality, increased. Is that progress?

To stay in line with our international climate targets we need to get our individual per capita emissions down to somewhere around 1 tonne of carbon dioxide a year. In Sweden, that figure currently stands at around 9 tonnes, once you include consumption of imported goods. In the US that figure is 17.1 tonnes, in Canada 15.4 tonnes, in Australia 14.9 tonnes and in China 6.6 tonnes. When you add biogenic emissions – such as emissions from the burning of wood and vegetation – those figures will in many

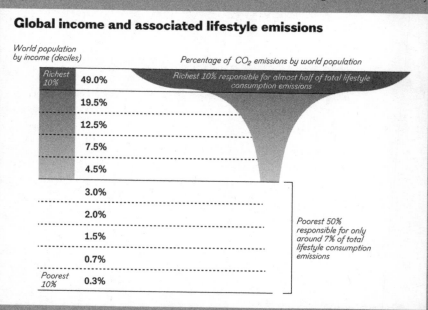

Global income and associated lifestyle emissions

World population by income (deciles)

Percentage of CO_2 emissions by world population

Richest 10%	49.0%	Richest 10% responsible for almost half of total lifestyle consumption emissions
	19.5%	
	12.5%	
	7.5%	
	4.5%	
	3.0%	
	2.0%	Poorest 50% responsible for only around 7% of total lifestyle consumption emissions
	1.5%	
	0.7%	
Poorest 10%	0.3%	

cases be even higher. And in forestry nations such as Sweden and Canada, significantly higher.

Keeping emissions below 1 tonne per person a year will not be a problem for the majority of the world's population, since they will only need to make modest reductions – if any – in order to live inside the planetary boundaries. In many cases, they would even be able to increase their emissions quite substantially.

But the idea that countries such as Germany, Italy, Switzerland, New Zealand, Norway, and so on will be able to achieve such enormous reductions within a couple of decades without major systemic transformations is naive. And still this is what the leaders of the so-called Global North are suggesting will happen. In Part Four of this book we will be looking at how that progress is coming along.

Some people believe that if they were to join the climate movement now, they would be among the last. But that is very far from true. In fact, if you do decide to take action now, you would still be a pioneer. The final part of this book focuses on solutions and things we can actually do to make a real difference, from small, individual actions to a planetary system change.

This book is intended to be democratic, because democracy is our best tool to solve this crisis. There may be subtle disagreements between the people writing from the front lines. Each person in this book is speaking from their own point of view and may arrive at different conclusions. However, we need all of their collective wisdom if we are to create the enormous public pressure required to make change. And rather than having one or two 'communication experts' or individual scientists drawing all the conclusions for you as a reader, the idea behind this book is that, taken together, their knowledge in their respective areas of expertise will lead you to a point where you can start to connect the dots yourself. At least, this is my hope. Because I believe the most important conclusions are yet to be drawn – and hopefully they will be drawn by you.

The Deep History of Carbon Dioxide

Peter Brannen

All life is conjured from CO_2. This is the original magic trick, from which everything else in the living world follows. At Earth's surface, with mere sunlight and water, it is transformed into living matter through photosynthesis, leaving oxygen in its wake. This plant carbon then flows through animal bodies and ecosystems and back out into the oceans and air as CO_2 once again. But some of this carbon slips the churn of the surface world altogether and passes into the Earth – as limestone, or as carbon-rich sludge, slumbering deep in the planet's crust for hundreds of millions of years. If it isn't buried, this plant stuff is quickly burned on Earth's surface in the fires of metabolism, by animals, fungi, bacteria. In this way, life uses up 99.99 per cent of the oxygen produced by photosynthesis – and would use it all, if it weren't for that infinitesimal leak of plant matter into the rocks. But it is from this leak into the rocks that the planet has been gifted its strange surplus of oxygen. In other words, the Earth's breathable atmosphere is the legacy not of forests and swirls of plankton alive today but of the CO_2 captured by life over all of our planet's history and commended to Earth's crust as fossil fuels.

If this was the end of the story, and CO_2 was *merely* the fundamental substrate of all living things on Earth and the indirect source of its life-sustaining oxygen, that would be interesting enough. But it just so happens that this same unassuming molecule also critically modulates the temperature of the entire planet and the chemistry of the entire ocean. When this carbon chemistry goes awry, the living world is warped, the thermostat breaks, the oceans acidify and things die. This astounding significance of carbon dioxide to every component of the Earth system is why it's not just another noisome industrial pollutant to regulate, like chlorofluorocarbons or lead. It is rather, as the oceanographer Roger Revelle wrote in 1985, 'the most important substance in the biosphere'.

The most important substance in the biosphere is not one to be treated cavalierly. The movement of CO_2 – as it billows from volcanoes, stirs into the air and oceans, swirls through eddies of life and soaks back into the rocks again – is what makes the Earth the *Earth*. This is called the carbon cycle, and life on Earth crucially depends on this global cycle maintaining a kind of delicate, if dynamic, balance. While CO_2 perennially issues from volcanoes (at a hundredth the rate of human emissions) and living organisms exchange it in a ceaseless frenzy at the Earth's surface, the planet is meanwhile constantly scrubbing it from the system at the same time, preventing climate catastrophe. Feedbacks that draw down CO_2 – from the erosion of whole mountain chains to the sinking of blizzards of carbon-rich plankton to the bottom of the sea – serve to maintain a kind of planetary equilibrium. Most of the time. This is an unlikely, miraculous world we live on, and one that we recklessly take for granted.

Sometimes in the geologic record, though, the planet has been pushed beyond a threshold. The Earth system can bend, but it can also break. And sometimes – in exceedingly rare, exceedingly catastrophic episodes buried deep in Earth history – the carbon cycle has been completely overwhelmed, undone, spun out of control. And the reliable consequence has been mass extinction.

What would happen if, say, continent-scale volcanoes, burning through kingdoms of carbon-rich limestone and igniting massive coal and natural gas deposits underground, injected thousands of gigatonnes of CO_2 into the air – from exploding calderas and from steaming, incandescent expanses of basalt lava? This was the predicament for the hapless creatures alive 251.9 million years ago, in the moments before the greatest mass extinction in the history of life on Earth. At the end of the Permian period, 90 per cent of this life would learn the fatal cost of a carbon cycle completely deranged by too much carbon dioxide.

In the End-Permian mass extinction, carbon dioxide blasted out of Siberian volcanoes for thousands of years and nearly ended the project of complex life. All the normal guardrails in the carbon cycle buckled and failed in this, the single worst moment in the entire geologic record. The temperature soared by 10°C, the planet convulsed with lethally hot, acidifying oceans which pulsed with lurid blooms of algal slime that robbed their ancient waters of oxygen. This anoxic ocean instead filled with poisonous hydrogen sulphide as hurricanes roared overhead, taking on an unearthly intensity. In the aftermath, when the fever finally broke, one could travel the world without seeing a tree, the world's coral reefs had been replaced by bacterial slime, the fossil record went silent and the planet took nearly 10 million years to pull itself back from oblivion. Thanks, in large part, to burning fossil fuels.

Every single mass extinction in Earth history is similarly marked by massive disruptions of the global carbon cycle, the signals of which have been teased out of the rocks by geochemists. Given the central importance of CO_2 to the biosphere, perhaps we shouldn't be surprised to find that pushing this system so far from equilibrium can so reliably result in planetary devastation.

Now, what if one lineage of the primate Homo tried to do the exact same thing as those ancient volcanoes hundreds of millions of years ago? What if they immolated those same massive reservoirs of underground carbon – buried by photosynthetic life over all of Earth history – not by mindlessly exploding it all through the crust like a supervolcano but in a rather more mannered fashion, retrieving it from the deep and burning it all at the surface in a more diffuse eruption, in the pistons and forges of modernity . . . and at a rate ten times that of the ancient mass extinctions? That is the absurd question we now demand the planet answer for us.

The climate is not responsive to political sloganeering; it is not accountable to economic models. It is accountable only to physics. It doesn't know, or care, whether the excess CO_2 in the atmosphere comes from a once-in-a-100-million-year volcanic event or from a once-in-the-history-of-life industrial civilization. It will react the same way. And we have in the rocks an unmistakable warning – a fossil record littered with the tombstones of ancient apocalypses. The good news is that we're still a long way from matching the gruesome crescendos of those cataclysms past. And it could even be the case that the planet is more resilient to carbon cycle shocks today than in those very bad old days. There is no reason we need to etch our names on this ignominious roster of the worst events ever in Earth history. But if the rocks tell us anything, it is that we are pulling the most powerful levers of the Earth system. And we pull them at our peril. /

This is an unlikely, miraculous world we live on, and one that we recklessly take for granted.

1.3

Our Evolutionary Impact

Beth Shapiro

The earliest evidence of humans as an evolutionary force comes from the fossil remains recovered at the earliest sites of human occupation on the planet's continents and islands. As people dispersed out of Africa more than 50,000 years ago and spread around the globe, the communities they joined began to change. Animal species, and particularly megafauna, including giant wombats, woolly rhinoceroses and giant sloths, started to go extinct. Our ancestors were efficient predators armed with uniquely human technologies – tools that improved the chance of a successful hunt and an ability to communicate and quickly refine these tools. The coincidence in timing of the megafaunal extinctions and the first appearance of people is recorded in the fossil records of every continent other than Africa. But coincidence does not necessarily prove causality. In Europe, Asia and the Americas, human arrival and the extinctions of local megafauna occurred during periods of climatic upheaval, leading to decades of debate about the relative culpability of these two forces in causing the megafaunal extinctions. Proof of our culpability comes, however, from Australia, where the earliest extinctions tied to humans are recorded, and from islands, where some of the most recent human-caused extinctions – the moa of Aotearoa (New Zealand) and the Mauritian dodo both became extinct within the last several hundred years – have taken place. The Australian and more recent island extinctions did not occur during periods of major climate change, and neither are extinctions recorded during more ancient climate events. Instead, these extinctions, like those on other continents, are the consequence of changes to the local habitat brought about by the appearance of people. In our earliest phase of interacting with wildlife, we had already begun to determine other species' evolutionary fate.

By 15,000 years ago, humans had entered a new phase of interactions with other species. Grey wolves that had been attracted to human settlements as sources of food had transformed into domestic dogs, and both dogs and humans were benefitting from their increasingly close relationship. The last ice age ended and the climate improved, and expanding human settlements demanded reliable sources of food, clothing and shelter. Around 10,000 years ago, people began to adopt hunting strategies that sustained prey populations rather than driving them towards extinction. Some hunters took only males or non-reproductive females, and later started to corral prey species and keep them close to their settlements. Soon, people began to choose which animals would be the parents to the next generation, and those animals that could not be tamed were taken for food. Their experiments were not limited to animals. They also planted seeds, choosing to propagate those that produced more food per plant or were ripe for harvest at the same time as others. They created irrigation networks and trained animals to clear land for farms. As our ancestors transitioned from hunters to herders and from gatherers to farmers, they transformed the land on which they lived and the species on which they increasingly relied.

By the turn of the twentieth century, the successes of our ancestors as herders and farmers were threatening the stability of the societies that they created. Wildlands had been replaced by farmland or rangeland and degraded by continuous use. Air and water quality had begun to decline. Extinction rates were again on the rise. This time, however, the devastation was more obvious, people were wealthier and technology was more advanced. As once-widespread species became scarce, an appetite emerged to protect what wild species and spaces remained. Our ancestors once again entered a new phase of interactions with other species: they became protectors, guarding endangered species and habitats from the dangers of the natural and increasingly human world. With this transition, humans became the evolutionary force that would decide the fate of every species, as well as the habitats in which these species live. /

We are the evolutionary force that will decide the fate of every species, as well as the habitats in which those species live.

Civilization and Extinction

Elizabeth Kolbert

The beginning of this story is shrouded in mystery.

Around 200,000 years ago, in Africa, a new species of hominin evolved. No one knows exactly where, or who its immediate ancestors were. Members of this species, which we now call 'anatomically modern humans', or *Homo sapiens*, or, simply, ourselves, were distinguished by their rounded skulls and pointy chins. They were lighter in build than their relatives and had smaller teeth. Though physically not very prepossessing, they were, it seems, unusually clever. They produced tools that were at first rudimentary and gradually grew more sophisticated. They could communicate not just across space but across time. They were able to live in very different climates and, perhaps just as importantly, to adjust to different diets. Where game was abundant, they hunted it; where shellfish were available, they consumed those instead.

This was the Pleistocene, a time of recurring glaciations, and much of the world was covered in vast sheets of ice. Nevertheless, around 120,000 years ago – perhaps even earlier – our species, no longer so new, began to press north. Humans reached the Middle East by 100,000 years ago, Australia by around 60,000 years ago, Europe by 40,000 years ago and the Americas by 20,000 years ago. Somewhere along the way – probably in the Middle East – *Homo sapiens* encountered their stockier cousins, *Homo neanderthalensis*, better known as Neanderthals. Humans and Neanderthals had sex – whether consensual or forced is impossible to say – and produced children. At least some of these children must have survived long enough to have children of their own, and so on through the generations, because today most people on Earth possess a smattering of Neanderthal genes. Then something happened, and the Neanderthals disappeared. Perhaps humans actively did them in. Or perhaps they just outcompeted them. Or perhaps, as a group of researchers at Stanford University recently theorized, humans carried with them tropical diseases their more cold-adapted cousins couldn't cope with. In any event, almost certainly the 'something' that happened to Neanderthals involved humans. As Svante Pääbo, a Swedish researcher who led the team that deciphered the Neanderthal genome once put it to me, 'their bad luck was us'.

The Neanderthals' experience would prove to be unremarkable. When humans arrived in Australia, the continent was home to an assemblage of extraordinarily large beasts. These included marsupial lions, which, pound for pound, had the strongest bite of any known mammal; Megalania, the world's largest monitor lizards; and diprotodons, also sometimes referred to as rhinoceros wombats. Over the course of the next several thousand years, all of these giant creatures disappeared. When humans arrived in North America, it hosted its own menagerie of oversized animals, including mastodons, mammoths, and beavers that grew to be 8 feet long and weigh 200 pounds. They, too, died off. Ditto for the giants of South America – massive sloths, giant armadillo-like creatures known as glyptodonts, and a genus of rhinoceros-sized herbivores known as Toxodon. The loss of so many large species in such a (geologically speaking) short amount of time was so dramatic it was noted back in Darwin's day. 'We live in a zoologically impoverished world, from which all the hugest, and fiercest, and strangest forms have recently disappeared,' Darwin's rival Alfred Russel Wallace observed in 1876.

Scientists have been debating the cause of the so-called megafauna extinction ever since. It's now known that the extinction took place at different times on different continents, and that the order in which species became extinct corresponds to that in which human settlers showed up. In other words, 'their bad luck was us'. Researchers who have modelled human–megafauna encounters have found that even if bands of hunters picked off a mammoth or a giant ground sloth only once a year or so, this would have been enough, over the course of several centuries, to drive such slow-reproducing species over the brink. John Alroy, a biology professor at Australia's Macquarie University, has described the megafauna extinction as 'a geologically instantaneous ecological catastrophe too gradual to be perceived by the people who unleashed it'.

Meanwhile, people continued to spread. The last large landmass to be settled by humans was New Zealand; Polynesians arrived there sometime around the year 1300, probably from the Society Islands. At that point, New Zealand's North and South islands were home to nine species of moa – ostrich-like birds that grew to be almost the size of giraffes. Within a few centuries, all the moas were gone. In this case, the cause of their demise is clear: they were butchered. A Māori saying, *Kua ngaro I te ngaro o te moa*, translates as 'Lost as the moa was lost.'

When Europeans began to colonize the world, in the late fifteenth century, the pace of extinction increased. The dodo, native to the island of Mauritius, was first noted by Dutch sailors in 1598; by the 1670s, it was gone. This was probably partly the result of slaughter and partly the result of introduced species. Wherever the Europeans went, they brought with them rats, in their case ship rats. The Europeans also, often purposely, introduced other predators, like cats and foxes, which pursued many species the rats left alone. Since the first European colonists arrived in Australia, in 1788, dozens of animals have been exterminated by introduced species, including the big-eared hopping mouse, which was decimated by cats, and the eastern hare-wallaby, which may also have been killed off by cats. Since the British started settling in New Zealand, around the year 1800, another twenty species of birds have become extinct, including the Chatham Islands penguin, the Dieffenbach's rail and the Lyall's wren. A recent study published in the journal *Current Biology* estimated it would take 50 million years of evolution for New Zealand's avian diversity to return to pre-human-settlement levels.

All this damage was done with relatively simple tools – clubs, sailing boats, muskets – and a few highly prolific introduced species. Then came the mechanized killing. Towards the end of the nineteenth century hunters armed with punt guns, which could fire nearly a pound of birdshot at once, managed to do in the passenger pigeon, a North American bird that once numbered in the billions. Around the same time, hunters shooting from trains managed to nearly wipe out the American bison, a species once so plentiful its herds were described as 'thicker than . . . stars in the firmament'.

Our most dangerous weapon would prove to be modernity and its trusty sidekick, late capitalism. In the twentieth century human impacts began to increase not just linearly but exponentially. The decades following the Second World War were a time of unprecedented growth in population on the one hand and consumption on the other. Between 1945 and 2000 the number of people in the world tripled. During the same period water use quadrupled, the marine fish catch increased sevenfold and fertilizer consumption rose tenfold. Most of the population growth occurred in the Global South. Most of the consumption was driven by the US and Europe.

The 'Great Acceleration', as it's often called, radically transformed the planet. As the environmental historian J. R. McNeill has observed, this wasn't because people were doing anything new, exactly; it's just that they were doing so much more of it. 'Sometimes differences in quantity can become difference in quality,' McNeill writes. 'So it was with twentieth-century environmental change.' At the start of the century agriculture occupied about 8 million square kilometres around the globe. By this point, people had been farming for some 10,000 years. Most of the great forests of Europe had long ago been felled, and the US's forests and prairies, too, were largely gone. By the century's close, more than 15 million square kilometres were under cultivation, meaning that in just ten decades people ploughed up as much land as they had in the previous ten millennia. The expansion entailed mowing down great stretches of the Amazon and Indonesian rainforests, areas high on the list of 'biodiversity hotspots'. How many species were lost in the process is unknown; many probably vanished before they were even identified. Among the animals that are known to have disappeared are the Javan tiger, now extinct, and the Spix's macaw, now extinct in the wild.

People didn't start using fossil fuels in the twentieth century – the Chinese were already burning coal in the Bronze Age – but, for all intents and purposes, this is when the problem of climate change was invented. In 1900 cumulative carbon dioxide emissions totalled around 45 billion tonnes. By 2000 that figure was 1,000 gigatonnes, and since then it has – horrifyingly – increased to 1,900 gigatonnes. What proportion of the world's flora and fauna can survive in a rapidly warming world is one of the great questions – perhaps *the* great question – of our time.

Most species alive today have persisted through multiple ice ages; clearly they were able to survive colder global temperatures. Whether they can handle warmer ones, though, is unclear; the world hasn't been much hotter than it is today for millions of years. During the Pleistocene, even very small creatures, like beetles, migrated hundreds of miles to track the climate. Today, countless species are once again on the move, but unlike in the ice ages, their way is often blocked by cities, highways or soy plantations. 'Certainly, our knowledge of their past response may be of little value in predicting any future reactions to climate change, since we have imposed totally new restrictions on [species'] mobility,' Russell Coope, a British paleo-climatologist, has written. 'We have inconveniently moved the goal posts and set up a ball game with totally new rules.'

Of course, there are also many species that just can't move. In 2014 Australian researchers conducted a detailed survey of Bramble Cay, a tiny atoll in the Torres Strait. The cay had its own species of rodent, a rat-like creature known as the Bramble Cay melomys, which was the only mammal known to be endemic to the Great Barrier Reef. Owing to rising sea levels, the cay was shrinking, and the researchers wanted to know if the melomys was still there. It wasn't, and in 2019 the Australian government declared the creature extinct. It was the first documented extinction to be attributed

Next pages:
The Hardy
Reef Lagoon,
Queensland.
The Great Barrier
Reef is the largest
living structure on
Earth, providing
a habitat for nearly
9,000 species
of marine life.

to climate change, though almost certainly many undocumented ones had preceded it.

Coral reefs themselves are highly vulnerable to climate change. Reef-building corals are tiny gelatinous animals; what lends corals their colour are the even tinier symbiotic algae that live inside their cells. When water temperatures spike, the symbiotic relationship between the corals and the algae breaks down. The corals expel the algae and turn white; this is what's known as coral bleaching. Without their symbionts, the corals go hungry. If the episode doesn't last too long, they can recover, but ocean temperatures are warming fast and bleaching events are growing both longer and more frequent. A 2020 study by a team of Australian researchers found that coral cover on the Great Barrier Reef has declined by half since 1995. Another 2020 study, by a team of American scientists, reported that, over the last fifty years, the majority of Caribbean reefs have been transformed into habitats dominated by algae and sponges. A 2021 study warned that the reefs of the western Indian Ocean are 'vulnerable to ecosystem collapse'. It's estimated that if reefs collapse, they could take with them species numbering in the millions.

The end of this story is, of course, also unknown. Over the last half a billion years, there have been five mass extinctions, each of which wiped out something like three quarters of the planet's species. Scientists warn that we are now sliding towards another, the Sixth Extinction. This event has the distinction of being the first to be caused by a biological agent – us. Will we act in time to prevent it? /

Most species alive today have persisted through multiple ice ages; clearly they were able to survive colder global temperatures. Whether they can handle warmer ones, though, is unclear.

1.5

The science is as solid as it gets

Greta Thunberg

The remarkable climatological stability of the Holocene era enabled our species – *Homo sapiens* – to make the move from being hunter gatherers to being farmers who cultivated the land. The Holocene began about 11,700 years ago, as the last ice age came to an end. In this relatively brief period of time we have completely transformed our world – 'our', as in the world of humans. 'Our world', as in a world belonging to one specific species – and that species is us.

We developed farming, we built houses, we created languages, writing, mathematics, tools, currencies, religions, weapons, arts and hierarchical structures. Human society expanded at what was, from a geological perspective, an unbelievable speed. Then came the Industrial Revolution, which marked the beginning of the 'Great Acceleration'. We went from going through an incredibly fast development to something else – something mind-blowing.

If the world's history was translated into the time span of a single year, the Industrial Revolution would have occurred at roughly one and a half seconds to midnight, on New Year's Eve. Since the rise of human civilization, we have cut down half of the trees in the world, wiped out more than two thirds of the wildlife and filled the oceans with plastics, as well as initiating a potential mass extinction and a climate catastrophe. We have begun destabilizing the very life-supporting systems we all depend on. We are, in other words, sawing off the branch we're living on.

Yet the vast majority of us are still not fully aware of what is happening, and many simply do not seem to care. This is due to various factors, many of which will be explored in this book. One of them goes by the name of 'shifting baseline syndrome' or 'generational amnesia', which refers to the way we get used to new things and begin to see the world from a different perspective. An eight-lane motorway junction would probably have been unimaginable to my great-grandparents, but for my generation it is completely normal.

To some of us, it even seems natural, safe and reassuring, depending on the circumstances. The distant lights of a megacity, an oil refinery glittering by the side of a dark freeway and bright airport runways lighting up the night skies are sights we are so used to that to many of us their absence would seem odd.

The same goes for the comfort some people find in overconsumption, among other things. The once unthinkable can very quickly become a natural – and even irreplaceable – part of our daily lives. And as we distance ourselves further and further from nature, the harder it becomes to remind ourselves that we are a part of it. We are, after all, an animal species among other animal species. We do not stand above the other elements that make up the Earth. We are dependent on them. We do not own this planet, no more than the frogs or the beetles, the deer or the rhinoceros own it. This is not our world, as Peter Brannen's chapter reminds us.

The rapidly escalating climate and ecological crisis is a global crisis: it affects all living plants and beings. But to say that all of humankind is responsible for it is very, very far from the truth. Most people today are living well within the planetary boundaries. It is only a minority of us who have caused this crisis and who keep driving it forward. This is why the popular argument that 'there are too many people' is a very misleading one. Population does matter, but it is not people who are causing emissions and depleting the Earth, it is what some people do – it is some people's habits and behaviour, in combination with our economic structures, that are causing the catastrophe.

The Industrial Revolution, which was fuelled by slavery and colonization, brought unimaginable wealth to the Global North, and in particular to a small minority of people living there. That extreme injustice is the foundation that our modern societies are built upon. This is the very heart of the problem. *It is the sufferings of the many that have paid for the benefits of the few.* Their fortune came at a price – namely oppression, genocide, ecological destruction and climatological instability. There is a bill for all this destruction that has not yet been paid. In fact, it hasn't even been added up; it is still waiting to be invoiced.

So why does this matter? Why not, in an emergency like this, just let bygones be bygones and get on with finding solutions to our current problems? Why make things more difficult by bringing up some of the most complicated issues in the history of humanity? The answer is that this is not just a crisis happening here and now. The climate and ecological crisis is a cumulative crisis that ultimately dates back to the colonial era and beyond. It is a crisis based on the idea that some people are worth more than others and that they therefore have the right to steal other people's land, resources, future living conditions – even their lives. And this is still going on.

Around 90 per cent of the CO_2 emissions that make up our entire carbon budget have already been emitted – the carbon budget being the maximum amount of carbon dioxide we can collectively emit to give the world a 67 per cent chance of staying below 1.5°C of global temperature rise. This carbon dioxide has already been pumped into the atmosphere or into the oceans, where it will stay, disrupting the delicate balance in the biosphere for many centuries to come – not to mention the risk of passing many tipping points and triggering feedback loops over that same period of time. The budget of remaining CO_2 we can emit while staying below the agreed targets is almost used up – but many low- and middle-income countries are yet to build the infrastructure that the wealth and welfare of higher-income countries are based upon, and for them to do so will require significant CO_2 emissions. It would seem obvious that the 90 per cent of CO_2 already emitted should be at the centre of our climate negotiations, or at least have some effect on the global climate discourse. What is happening, however, is the opposite. Our historical debt – among many other crucial aspects – is being completely ignored by the nations of the Global North.

Some argue that this all happened such a long time ago, that the people in power weren't aware of the problems when they were building our energy systems and started mass-producing all the stuff that we consume. But they were aware, as Naomi Oreskes demonstrates in her essay. Evidence clearly shows that major oil companies such as Shell and ExxonMobil have known about the consequences of their actions for at least the past four decades. So did the nations of the world, as Michael Oppenheimer explains. Still, it remains true that over 50 per cent of all the anthropogenic (human-caused) carbon dioxide ever emitted has been emitted since the Intergovernmental Panel on Climate Change (IPCC) was founded and since the UN held its 1992 Earth Summit in Rio de Janeiro. So they knew. The world knew.

It goes back to those black-and-white issues. Some say that there are many shades in between, that things are complicated and the answers are never simple. But I say again, there are lots of black-and-white issues. Either you fall off a cliff, or you do not. Either we are alive, or we are dead. Either all citizens are allowed to vote, or they are not. Either women are given equal rights to men, or they are not. Either we stay below the climate targets set in the Paris Agreement and thus avoid the worst risks of setting off irreversible changes beyond human control, or we do not.

These issues are as black or white as it gets. When it comes to the climate and ecological crisis, we have solid unequivocal scientific evidence of the need for change. The problem is, all that evidence puts the current best available science on a collision course with our current economic system and with the way of life many people in the Global North now consider their right. Limitations and restrictions are not exactly synonymous with neoliberalism or modern western culture. Just look at how some parts of the world reacted to restrictions during the Covid-19 pandemic.

You can of course argue that there are different scientific views and opinions; that not every scientist agrees with every other scientist. And this is true: scientists spend huge amounts of time debating different aspects of their results – that is how science works. This argument can be used in countless topics of discussion, but it can no longer be used in connection with the climate crisis. That ship has sailed. The science is as solid as it gets.

What largely remains is tactics. How to package, frame and convey the information. How disruptive do scientists dare to be? Should scientists applaud the politicians' inadequate proposals, because they are better than nothing, and because doing so might also help them to gain – or keep – a seat at the table? Or should the scientists risk being dismissed as alarmists and tell it like it is, even though that might lead to an increase in the number of people surrendering to defeat and apathy? Should they maintain a positive, hopeful, 'glass is half full' approach or should they set communication tactics aside and just focus on delivering the facts? Or perhaps a bit of both?

A big divider today is whether to include equity and historical emissions in discussions of the actions needed to tackle the environmental crisis. Since those figures have been negotiated out of our international frameworks, it is no doubt tempting to ignore them, as they will make a bleak message appear far bleaker. It does, however, make those who try to be holistic and include them appear far more alarmist than their colleagues, and that is a big problem. For instance, the prospect of Global North nations such as Spain, the US or France reaching net zero emissions by the year 2050 seems completely inadequate if you include the aspect of equity and historical emissions. But if you are, say, an American scientist aiming to reach a big domestic audience, you will probably not be too keen on dismissing the whole idea of net zero 2050 as totally insufficient. The idea of reaching net zero emissions in three decades is already considered extremely radical in the US discourse. And this tactic makes perfect sense. The problem, however, is that to make the Paris Agreement work on a global scale we have to include equity and historical emissions. There is no way around it. And it's not as if we have the time to move the conversation along slowly.

We have come a long way from our hunter-gatherer ancestors. But our instincts have not been given enough time to keep pace. They still operate largely as they did 50,000 years ago, in another world, long before we developed farming, houses, Netflix and supermarkets. We are built for another reality altogether, and our brains find it hard to react to threats that aren't immediate and sudden for many of us, threats like the climate and ecological crisis. Threats that we can't see clearly because they are too complex, too slow-moving and too far away.

The evolution of *Homo sapiens* has, from a larger geological perspective, happened at the speed of light. Is this what is coming back to haunt us? Was our foundation built on unstable ground right from the start, tens of thousands of years before the beginning of the Industrial Revolution? Were we too gifted as a species? Too superior for our own good? Or can we change? Will we be able to use our skills, our knowledge and our technology to create a cultural shift that will make us change in time to avert a climate and environmental catastrophe? We are clearly capable of doing so. The question of whether we will is entirely up to us. /

If the world's history was translated into the time span of a single year, the Industrial Revolution would have occurred at roughly one and a half seconds to midnight, on New Year's Eve.

1.6

The Discovery of Climate Change

Michael Oppenheimer

In the beginning, it was a scientific curiosity rather than a problem. Svante Arrhenius, a Swedish chemist, evinced no concern when, in 1896, he published his now-famous prediction that by releasing carbon dioxide into the atmosphere through burning coal, humankind would gradually warm Earth by several degrees. His findings were almost universally ignored until the 1950s, when a handful of scientists pointed out that this warming might have catastrophic consequences. A decade later, a young meteorologist, Syukuro Manabe, developed the first modern computer simulations of the climate;[1] his prediction of how hot Earth would become showed that Arrhenius was not far off base. In Manabe's wake there came a new wave of scientific research that started to sketch a picture of progressively worsening impacts, and by the late 1970s a scientific consensus had emerged on how much Earth might warm once carbon dioxide levels in the atmosphere doubled. I was a graduate student in chemical physics when I first heard of 'the greenhouse effect' from a 1969 issue of *Technology Review*, and the idea that humans could come to control Earth's climate scared the hell out of me. Gradually it dawned on me that I could constructively channel this worry and contribute to fixing the problem by combining my interest in politics with my expertise on Earth's atmosphere. I joined a growing chorus of scientists raising the alarm through the 1980s. Only a handful of policymakers were listening, but today warming is impossible to ignore.

The basic physics behind the greenhouse effect, and the reasons that global warming is occurring, are even clearer now than they were a century ago. The gases that make up Earth's atmosphere, primarily nitrogen and oxygen, are by and large transparent to sunlight. As a result, most sunlight passes through the atmosphere and warms Earth's surface.

---

1 In 2021 Manabe won the Nobel Prize in Physics for this achievement.

As Earth warms, it sheds heat back into space in the form of infrared radiation. However, water vapour and some other gases that are present in our atmosphere in trace amounts, particularly carbon dioxide, absorb or trap much of this infrared radiation, sending some of it back towards the surface and increasing Earth's temperature.

These are the greenhouse gases, so-called because the process of trapping heat is analogous to the way the glass of a greenhouse keeps the interior warm even on a frigid day, allowing plants inside to thrive. Without these gases, the heat radiated from Earth's surface would be lost into space and the planet would be about 33°C colder. The atmosphere's greenhouse effect has kept our planet's temperature within a life-supporting range that allowed humans and other species to evolve.

This process had remained stable for thousands of years until the onset of widespread industrialization during the nineteenth century. The fossil fuels that came to power industrial society – coal, oil and natural gas – are remnants of carbon-based plant matter buried millions of years ago. These have been extracted by mining and drilling to fuel our factories, power plants, automobiles, tractors, boats and aeroplanes, and also to heat our homes and workplaces. Fossil fuel combustion releases tens of billions of tonnes of carbon dioxide every year.

Farming, including raising cattle, has also resulted in rising emissions of methane and nitrous oxide, greenhouse gases that exert an even greater warming effect per molecule than carbon dioxide. The drilling and transport of natural gas has leaked yet more methane into the air. Rampant deforestation and other land-use changes have become another large source of carbon dioxide and other greenhouse gases. As a result of these various human influences, carbon dioxide levels in the air are now 50 per cent higher than in pre-industrial times.

The hundreds of billions of tonnes of greenhouse gases already added to the atmosphere would still have had a relatively modest effect on Earth's temperature were it not for the impact of feedback loops which have made it even warmer. Warming has increased evaporation from the ocean surface, putting more of the greenhouse gas water vapour in the air, which has in turn accelerated the warming. Arctic sea ice has melted, allowing more sunlight to be absorbed at the ocean surface rather than reflected back into space by the ice, speeding warming even more. Clouds both trap heat and reflect sunlight, and the net effect of changes in cloudiness due to warming is yet another feedback that makes Earth warmer. Taken together, these feedbacks are causing the Earth to heat up three times as fast as would be the case otherwise.

What makes the atmospheric build-up of carbon dioxide a special concern is that this excess can be permanently removed from the atmosphere only by a very very slow, centuries-long process of dissolving in the oceans. While some experts are exploring ways to artificially speed the removal, no such technology is currently available that is efficient and affordable.

Much like the basic physics, the scope of effort required to tackle warming, and the need for early action, were crystal clear over thirty years ago. So why, for decades, did we do almost nothing? At the heart of the problem was the fact that, even as the scientific community became aware of what was about to unfold, it was extremely difficult to awaken politicians to the dangerous nature of the situation we were in.

In 1981, as a scientist at the Environmental Defense Fund, I began to work with others in the environmental community, along with scientists and a few interested governments, to try to bring this issue before the public and our elected leaders. Yet at that time most governments thought that since the impact of warming was not yet evident no action should be taken – even as the science and the potential cost of inaction were becoming clear.

In 1986 I testified before a committee of the US Senate and watched as a string of officials from various government agencies spoke first – most were uninformed, unconcerned and uninterested in any concerted action to slow the greenhouse gas build-up. I tried to make the stakes very clear to politicians and the public: this was 'a problem which if left unchecked, will come to dominate all others in its effect on the environment . . . the viability of many ecosystems is at stake, as is, perhaps, the viability of civilization as we know it'. Reflecting on the persistence of carbon dioxide, I noted that this was a different type of problem than ordinary air pollution, and that we just could not afford to sit back and wait to see the consequences before implementing policies to stem emissions as then it would be too late to avoid serious impacts.

Two years later, in the midst of a heatwave afflicting the eastern US, I was invited to testify before another Senate committee alongside Professor Manabe and NASA's James Hansen, who that day delivered his famous testimony proclaiming that 'the greenhouse effect has been detected and is changing our climate now'. My testimony covered the report of an international scientific conference I had co-organized under United Nations auspices which concluded that the problem of human-caused climate change must be addressed and which made specific policy recommendations aimed at limiting future greenhouse gas emissions.

Among the stark findings I highlighted that day were that slowing warming to an acceptable rate, and ultimately stabilizing the atmosphere, would require reductions in fossil fuel emissions 'by 60% from current levels, along with similar reductions in emissions of other greenhouse gases. Given the projected doubling in emissions over the next 40 years in "business-as-usual" scenarios,' I noted, 'we have a daunting task ahead.'

The numbers above from the conference report are now outdated because little has been done to rein in emissions, and thus they are far smaller than the reductions now required. If countries worldwide, especially in the Global North, had taken concerted actions back then, we would be in a far better position today to stem the climate crisis, rather than confronting the myriad disasters that now afflict us.

In the same year, 1988, the IPCC was formed through the United Nations, which harnessed the efforts of thousands of scientists across the globe to assess the climate issue and come up with solutions. This was an unprecedented effort by the world's leaders to engage the scientific community to look into the future and project the looming environmental damage to human society and ecosystems. I became involved in the IPCC's First Assessment Report, published in 1990, and have been an IPCC author ever since, during all six assessment cycles.

A race had begun between the irreversible build-up of carbon dioxide and the on-again, off-again efforts of governments to transform their countries to carbon-free economies. I and many of my scientific and environmentalist colleagues understood that we were facing a near future in which countries would be battered by extreme weather triggered or exacerbated by climate change, including worsening droughts, hurricanes and heatwaves. Our goal was to move nations to action before they experienced widespread death and destruction from the ever-increasing extremity of the climate that science foresaw. We have clearly lost that race.

The mitigating steps we took were too slow and too small. Countries did come together to sign the UN Framework Convention on Climate Change at the Earth Summit in Rio de Janeiro in 1992. The aim of the treaty was to reduce greenhouse gas emissions back to 1990 levels by the year 2000. However, the agreement was toothless because its emissions reduction obligations were unenforceable. The participation of the US was important and a cause for hope, given that it had thus far contributed the most to global carbon dioxide emissions. The US Congress ratified the agreement and Bill Clinton's election to the presidency that same year seemed to bode well for climate action. But when the new president tried to implement an energy tax as a first mandatory measure to restrain emissions, he encountered strong opposition in Congress and withdrew his proposal. Taxes are the 'third rail' of US politics and, to this day, carbon taxes face a difficult path to adoption.

Recognizing that progress towards the Framework Convention's goals was falling short, countries came together again at Kyoto in 1997 to agree on binding emissions commitments for developed countries. However, like the Framework Convention, the Kyoto Protocol did not require emissions reductions from developing countries – a serious limitation on its effectiveness, since China's emissions were about to balloon and some other developing countries would eventually follow suit.

The United States never ratified the Kyoto Protocol and, in 2001, newly elected President George W. Bush withdrew the US's initial signature of the document. Science lost the battle because of the political influence of corporations that produce fossil fuels as well as of those firms that heavily consumed them. Many of these firms and their various trade associations had established effective disinformation campaigns involving so-called think tanks, while some politicians from regions that produced fossil fuels promoted distortions and outright lies about the science. In a situation where private interests were creating a public miasma of falsehoods and deception it was all too easy for the general public to discount the risks.

Europe was less distracted and divided by disinformation campaigns from fossil fuel companies and emerged early on as a global leader on the climate issue. UK Prime Minister Margaret Thatcher, a former research chemist, respected the scientific warnings and, also driven by her determination to break the power of the coal-mining unions, lent her support in 1989 to the concept of negotiating the UN Framework Convention. In Germany – another of Europe's major greenhouse-gas-emitting nations – the Green Party had been growing in influence since the mid-1980s, causing the two major political parties there to adopt environmental and energy goals, which Angela Merkel, also a former chemist, continued to pursue after her ascension to Chancellor in 2005. Thus, when the US stepped back from leadership on the climate issue, the European Union, led by the UK and Germany as well as the Netherlands and its Scandinavian member states, partly filled the void and pushed for global action to address the problem. Benefitting from German reunification and the collapse of the former East Germany's emissions, and those from other former Soviet states, the EU achieved the target it agreed to at Kyoto.

Other developed countries, particularly Canada and Australia, swayed by their fossil-fuel-exploiting regions, paid lip service to the Kyoto Protocol's commitments but made little or no effort to actually rein in their emissions.

In 2014 China and the US joined together and offered national emission goals that paved the way for the following year's Paris Agreement. The Paris Agreement was, in some ways, a landmark, but it has proven only modestly effective, as China – and, more recently, India – has fast-rising emissions and an economy that still relies heavily on coal. Nevertheless, China has ample reasons to keep pressing ahead with its climate commitments: it urgently needs to reduce air pollution and stands to gain massively from selling solar photovoltaic modules, wind generators and electric cars to the rest of the world. Yet China's leaders oppose full transparency in the monitoring, reporting and verification of its Paris commitments, and until they change this stance they cannot be relied upon as a model of responsible leadership.

We lost one race – the race to prevent harmful impacts – but now, as warming accelerates, we find ourselves at the start of another: the race to mitigate a climate crisis and maintain a habitable planet. Winning this race will require emerging leaders to face down fossil fuel interests and public myopia in a way that my generation never did. Advances in energy technology, combined with the now incontrovertible understanding of the crisis we face and the admirable combination of determination and purposeful pressure exerted by the younger generation, lead me to be hopeful. It won't be easy, but now the stakes are crystal clear and, this time, no one can say they didn't see it coming. /

Figure 1:
Trends in global atmospheric CO₂ vs. year. Both the carbon dioxide concentration in our atmosphere and average global temperatures (with a darker bar indicating a higher temperature) have soared despite global climate conferences and international agreements to curb emissions.

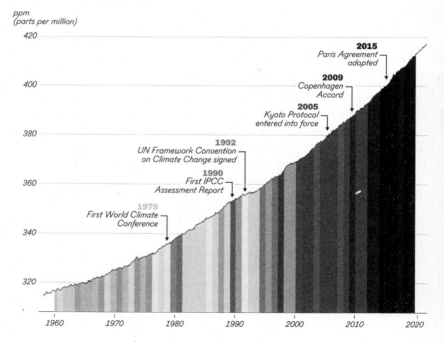

Why Didn't They Act?

Naomi Oreskes

When future historians ask, 'Why didn't people take action to stop the climate crisis when they had known about it for decades,' a prominent part of the answer will be the history of denial and obfuscation by the fossil fuel industry, and the ways in which people in positions of power and privilege refused to acknowledge that climate change was a manifestation of a broken economic system.

Scientists, journalists and activists have documented the many ways the fossil fuel industry spread disinformation about climate change to prevent action. Much of this work has focused on the industry Goliath ExxonMobil. In the 1970s and '80s, Exxon's own scientists informed them of the threat of climate change caused by their company's products. However, from the 1990s onwards the company promoted a public message of high scientific uncertainty, insisting that policy action was at best premature and perhaps unnecessary. ExxonMobil was a key node in a network – sometimes called the 'carbon-combustion complex' – that included coal corporations, automobile manufacturers, aluminium producers and others who profited from cheap fossil fuel energy.

Through advertisements, public relations campaigns, reports commissioned from 'experts for hire', and more, the carbon-combustion complex deliberately created confusion about the climate crisis. Many of the strategies and tactics were taken directly from the tobacco industry, including cherry-picking and misrepresenting scientific evidence; promoting outlier scientists to create the impression of scientific debate where there was little or none; funding research intended to deflect attention from the primary causes of climate change; impugning the credibility of climate scientists; and falsely portraying the fossil fuel industry as supporting 'sound science' rather than protecting profits. They also deflected attention from their role by insisting citizens should take 'personal responsibility' by lowering their 'carbon footprints'.

The fossil fuel industry worked in tandem with a network of politically conservative, libertarian and neo-liberal think tanks who echoed and amplified the message of climate doubt. Some were independent think tanks, such as the CATO Institute in the United States and the Institute for Economic Affairs in the United Kingdom, whose ideological commitments to laissez-faire economic policies made them hostile to government-led action. (Often, these groups borrowed from the tobacco playbook by asserting that acting on the climate crisis would threaten freedom.) Others were front groups, such as the Global Climate Coalition, spearheaded by Mobil Corporation, and the 'Informed Citizens for the Environment', created by a group of US-based coal producers. In 2006 the UK Royal Society – one of the world's oldest and most venerable scientific honour societies – identified thirty-nine organizations funded by ExxonMobil that denied or misrepresented the state of climate science.

The fossil fuel industry and its allies acted indirectly to prevent climate action by poisoning the well of public debate, but they also acted directly when government action appeared imminent. One well-documented case is the American Clean Energy and Security Act of 2009, which would have created an emissions trading system to reduce greenhouse gas emissions. It seemed destined for success, until the US Chamber of Commerce, electric utilities, oil and gas companies, trade associations and think tanks lobbied fiercely against it, and it failed. Between 2000 and 2016, fossil fuel interests in the US alone likely spent close to $2 billion blocking climate action.

Industry disinformation, misdirection and lobbying were abetted by the wishful thinking of people who accepted industry arguments about natural gas as a 'bridge fuel', resisted acknowledging industry malfeasance and insisted on the power of 'corporate engagement'. One prominent example involves Harvard University. In 2021 the university announced that it would divest its endowment of fossil fuel holdings. But for many years Harvard's leaders had declined to criticize the industry, contending that they could not 'risk alienating and demonizing possible partners'. Yet many of these 'partners' had demonized climate scientists and activists and harmed billions of people around the globe.

Most economists now recognize climate change as a market failure, but only a few understand it as part of the larger pattern of environmental destruction that scientists have labelled the 'Great Acceleration'. Capitalism as currently practised has imperilled the existence of millions of planetary species, as well as the health and well-being of billions of humans. It also threatens the prosperity that it was intended to create. Challenging 250 years of dominant economic thinking, the climate crisis has shown that the unrestrained pursuit of self-interest does not serve the common good. It has shown, in the words of economist Joseph Stiglitz, that Adam Smith's invisible hand – the idea that free markets lead to efficiency as if consciously guided – is invisible 'because it is not there'. And it has proved, in the words of Pope Francis, that 'technological products are not neutral, for they create a framework which ends up conditioning lifestyles and shaping social possibilities along the lines dictated by the interests of certain powerful groups'.

These are heavy conclusions for people to accept. No one wants to admit to being duped by disinformation or blinded by a myth, and people in positions of privilege rarely examine the basis for that privilege. Perhaps, most deeply, the climate crisis breaks the promise of progress. And so, even today, many people who are not necessarily climate-change 'deniers' resist meaningful action, refuse to acknowledge just how broken our economic systems are, and deny how much damage industry disinformation has done. /

People in positions of power and privilege refused to acknowledge that climate change was a manifestation of a broken economic system.

1.8

Tipping Points and Feedback Loops

Johan Rockström

Scientifically, it is now well established that Earth has entered a new geological epoch, the Anthropocene, where our globalized world constitutes the largest driver of change on Earth. The amount of CO_2 emitted so far from our fossil fuel burning (some 500 billion tonnes of carbon) and the environmental destruction caused by us is sufficient to affect our planet's future during the next half a million years. We are in the driver's seat, determining the future state of our home, planet Earth. We triggered the Anthropocene some seventy years ago when our fossil-fuel-driven industrialized world economy went truly global, causing multiple 'hockey sticks' of rising human pressures; the 'Great Acceleration' is a fact, manifested in an accelerated rise in greenhouse gas emissions, fertilizer consumption, water use, marine fish catching and terrestrial biosphere degradation, to name just a few (Fig. 1).

The drama, though, is much larger than this almost mind-boggling insight. We haven't just triggered an entirely new geological epoch. We are deep into the Anthropocene, and our planet is showing the first signs of an inability to absorb more human abuse. Merely seventy years after its onset, we are forced to conclude that the Earth system seems to be running out of resilience, losing its biophysical capacity to buffer and dampen the pressure, stress and pollution we are exposing it to.

The scientific community now must explore whether we are at risk of destabilizing the entire Earth system, which means pushing biophysical systems and processes – like the ice sheets, forests and the ocean's circulation of heat – past their tipping point, where feedbacks shift from cooling and dampening to warming and self-reinforcing, which could culminate in an irreversible drift of the entire planet away from the stable interglacial state of the planet, the Holocene, that we have benefitted from since the emergence of human civilizations some 10,000 years ago, and still completely depend on.

This means that we have reached an existential fork in the road. We are in the Anthropocene and are seeing rising signs of the approach of irreversible tipping points. Still, the Earth system, while showing worrying signs of destabilization, remains in an interglacial Holocene-like state. This may seem odd, but it is the reason we can still talk of hope. While the Holocene is a state of the planet (an interglacial with two permanent ice caps in the Arctic and Antarctica), the Anthropocene, so far, is 'only' a trajectory – a movement away from a Holocene state and not, yet, a new state.

But the risk is that there is only limited hope. At 1.1°C of global warming (as of 2021), we have exceeded the warmest global mean surface temperature (GMST) on Earth since we left the last ice age. We have reached the ceiling of the comfortable interglacial state where temperatures never left a 'corridor of life' of plus or minus 1°C. Our grand challenge is to stop our current trajectory and to prevent the Anthropocene from becoming a new, self-reinforcing hot state. The only way to succeed in this human quest is to avoid crossing tipping points in the Earth system that regulate the state of the climate and the living biosphere. This in turn requires that we govern and manage the global commons – all biophysical systems that are critical in regulating the state of the planet – within planetary boundaries that provide a scientifically defined safe operating space on Earth.

We have built our economies, our societies and our civilizations on two assumptions about the natural world: first, that change happens in an incremental, linear way (allowing for regret and simple repair); second, that the biosphere has essentially infinite space and capacity to absorb human impacts (our waste) and cope with our extraction of resources (our consumption).

The science of resilience and complex systems debunks both of these assumptions. Earth's biophysical systems – from ice sheets to forests – ultimately determine how inhabitable the planet is. They do this not only by delivering immediate services to us humans (like food and clean water) but also by having a built-in resilience – the capacity to absorb shocks and stress (like warming due to greenhouse gas emissions and deforestation) and thereby cool and hold the planet within a narrow temperature range. But only up to a certain point. Pass this threshold and the system – be it a coral reef, a frozen tundra or a temperate forest – will irreversibly tip over from one state to a qualitatively different state.

Importantly, tipping points are reached when a small change – for example, a small rise in global temperatures due to fossil fuel burning – triggers a big and irreversible change – like a rainforest becoming a dry savannah. This change is propelled by self-perpetuating feedback loops – so the change

The 'Great Acceleration'

Earth system trends since 1750

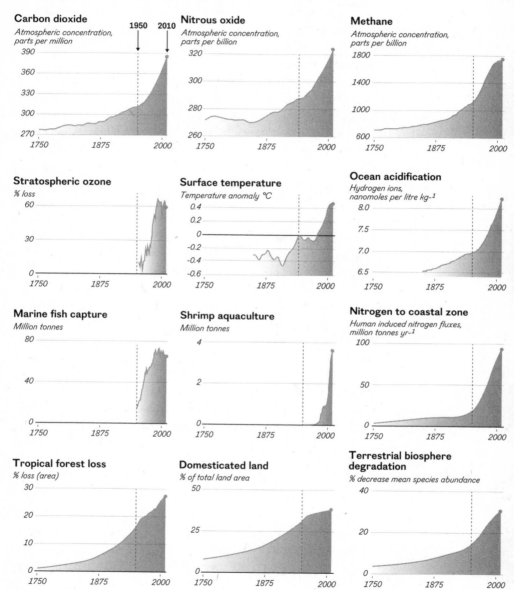

Figure 1

Socio-economic trends since 1750

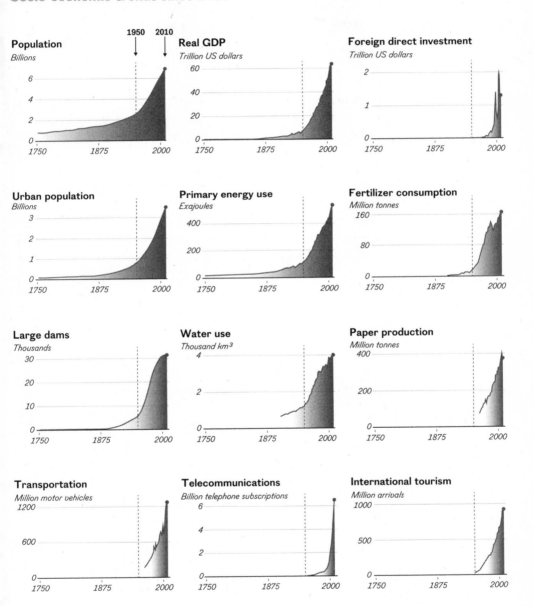

Population
Billions

1950   2010

Real GDP
Trillion US dollars

Foreign direct investment
Trillion US dollars

Urban population
Billions

Primary energy use
Exajoules

Fertilizer consumption
Million tonnes

Large dams
Thousands

Water use
Thousand km³

Paper production
Million tonnes

Transportation
Million motor vehicles

Telecommunications
Billion telephone subscriptions

International tourism
Million arrivals

can continue, even if the pressure (global warming) has stopped. Therefore the system would remain 'tipped' even if the background climate falls back below the threshold. It does not generally happen overnight: it may take decades or centuries before a system finds a new, stable state. The key, though, is that crossing the tipping point is like pushing the 'on' button that causes new biophysical machinery to crank up, with destabilizing feedbacks taking over, shifting a system gradually yet unavoidably towards a new state (Fig. 2), with severe impacts on the environment and the livelihoods of many people.

The fact that crossing tipping points does not have to be abrupt is one of the great challenges we face. If we cross tipping points now or within the next few decades, their full impact might only become apparent, unstoppable, after hundreds or even thousands of years. Sea-level rise from land ice melt is one such example: it will continue for centuries and millennia, and then stay at high levels for thousands of years. As the IPCC now shows, even at 1.5°C of warming we might commit all future generations to a sea-level rise of at least 2 metres, although it may take 2,000 years to reach that level. This introduces a new ethical time dimension. It is *now* that we are determining whether we leave to our children and their children

How can we think of tipping points?

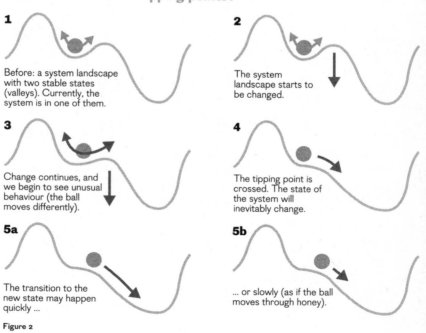

1

Before: a system landscape with two stable states (valleys). Currently, the system is in one of them.

2

The system landscape starts to be changed.

3

Change continues, and we begin to see unusual behaviour (the ball moves differently).

4

The tipping point is crossed. The state of the system will inevitably change.

5a

The transition to the new state may happen quickly ...

5b

... or slowly (as if the ball moves through honey).

Figure 2

a planet that will continue drifting towards less and less inhabitable states in the future. It may take hundreds or thousands of years, but it would be unstoppable.

It is absolutely fundamental to understand the interactions between systems on Earth and Earth system feedbacks in order to assess the risks of pushing the planet too far. Interactions reinforce changes. For example, when warmer oceans accelerate ice melt, a feedback shift is triggered when the white ice surface, which usually reflects 80–90 per cent of the incoming heat from the sun back to space, crosses an albedo (or reflectivity) threshold, because the ice surface gets darker when it melts and becomes flowing, liquid water. At a certain point the system feedback shifts from negative (net cooling) to positive (net warming) and the whole system moves towards a new, ice-free equilibrium as a result of the feedback shift.

As far as we know, not all biophysical systems on Earth have different stable states separated by thresholds which can cause tipping behaviour. Some systems do; others don't. Common to all of the biological, physical and chemical systems and processes (like global cycles of carbon, nitrogen and phosphorus), though, is that they are interconnected, and the biosphere, hydrosphere and cryosphere all interact with each other. And they have feedbacks that determine how they operate (their state), and the dominating feedbacks can shift from mathematically negative (dampening) to positive (reinforcing).

The large components of the Earth system that are defined as tipping elements are those that are characterized by threshold behaviour (i.e. they can trigger tipping points) and that at the same time play a role in regulating the state of the planet. We all depend on the tipping elements remaining stable and resilient. They are global commons, which we now need to manage and govern due to the risks we are taking in the Anthropocene.

In 2008, a range of climate tipping elements were identified (Fig. 3, top). Since then, science has advanced tremendously and we know much more about tipping-point behaviour and the interactions among tipping-element systems; we have also identified over 200 cases and about 25 generic types of regime shift (that is, large, abrupt and persistent critical transitions in the function and structure of ecosystems, beyond climate tipping points). In 2019, a study providing a ten-year science update on climate tipping-point risks was published, and the conclusion was very troubling. Nine of the original climate-tipping elements are showing signs that they may be approaching tipping points (Fig. 3, bottom). This assessment was to a large extent confirmed in the IPCC's Sixth Assessment Report, which raises concerns about six of those nine

Climate tipping elements first identified in 2008

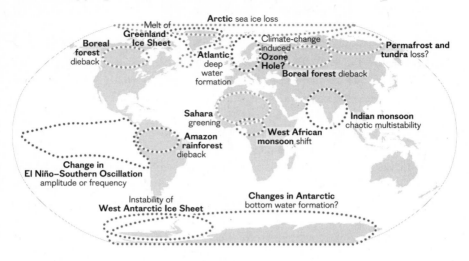

2019 assessment of tipping elements showing signs of instability, and indicated links between different elements

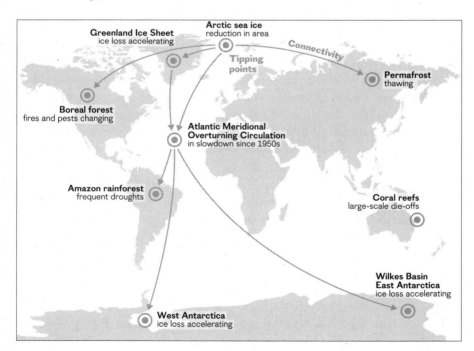

Figure 3

unstable elements: the Ice Sheet, the Greenland Ice Sheet, Arctic sea ice, permafrost, the Atlantic Meridional Overturning Circulation (AMOC) and the Amazon rainforest.

On top of all that, the interactions between tipping-element systems raise particular concern – tipping elements might trigger each other and unleash a domino-like chain reaction. A so-called 'tipping cascade' could push the Earth system towards a new hothouse Earth pathway. At 1.1°C of warming, the Arctic is warming two to three times faster, accelerating ice melt from the Greenland Ice Sheet (and the melting of Arctic sea ice). This in turn slows down the ocean's circulation of heat, the AMOC, which in turn impacts the monsoon system over South America, which can partly explain the rising frequency of droughts over the Amazon rainforest and the subsequent increased severity of fires and abrupt pulses of CO_2 back into the atmosphere, which intensifies the warming. Furthermore, the slowdown of the Atlantic heat conveyor leads to more warm surface waters being stuck in the Southern Ocean, which can explain the accelerated melt of the West Antarctic Ice Sheet.

Admittedly, these complex dynamics are at the scientific frontier, and their precise workings are not yet fully confirmed, but they do raise concerns and provide even stronger scientific support for precaution and rapid action to solve the climate crisis.

We are looking at a rising risk landscape. We can no longer rule out the risk of crossing tipping points, leading to changes that cannot be stopped, or the overall trend in risk assessments as science has advanced over the past twenty years. As we can see in the 'Burning Embers' graph (Fig. 4), the more we learn about the way the climate system works, the greater the cause for concern. As recently as 2001, in the IPCC Third Assessment Report (TAR),

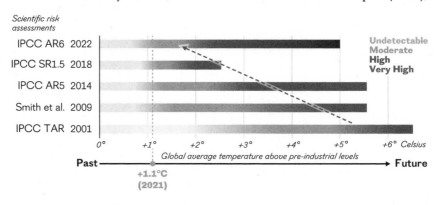

Figure 4: 'Burning Embers', a visualization first used in the IPCC Third Assessment Report in 2001 shows that risk level at a given temperature has increased with each assessment.

we still thought that the risk of irreversible changes with large impacts was very low, and that there was only a serious risk at 5–6°C of warming. This came down to essentially no risk at all, as nobody was or is suggesting that we would reach such disastrous levels of average global heating. With every new IPCC assessment, while the global mean temperature has gone up due to our greenhouse gas emissions, our scientific understanding of the temperature threshold of high risk has gone down further and further. Today, our best understanding is that even at 1.5°C, and certainly between 1.5 and 2°C, we are taking enormous risks. /

We are determining whether we leave to our children and their children a planet that will continue drifting towards less and less inhabitable states.

This is the biggest story in the world

Greta Thunberg

There are currently about 7.9 billion of us living on this beautiful blue planet, peacefully circling around its sun in our tiny little corner of the great cosmos. We are all connected. Just like all other living things and beings, we can trace our origins back through deep time to the sources of life and, therefore, no matter how far from nature we distance ourselves, we are inseparable from it.

All the facts and stories in this book are unnerving enough individually. But they, too, are closely joined together – just like all of us. And once you start to connect them, understanding them as part of a web of inter-linked events, they quickly gain another, far more alarming meaning. Who is responsible for piecing together that greater, holistic tale? Who do we call upon when it comes to addressing the whole picture? Some superior university of universities? Our governments? World leaders? The business world? The United Nations? The answer is no one – or rather, everyone.

We are at the beginning of a rapidly escalating climate and ecological crisis. A sustainability crisis. Technology alone will not be enough to save us, and there are unfortunately no laws or binding resolutions that put us on a path towards a safe future for life on Earth as we know it.

The transition we need in order to guarantee this safe future will not appear out of nowhere. It will come from a change in public opinion, and that change has to be created by us, using whatever effective means we can come up with. It will be driven by how we choose to communicate this story. There won't be a one-size-fits-all message that will work for everyone. Thousands – or even millions – of different approaches will be needed, but right now our resources are quite limited, to say the least. We have to *koka soppa på en spik*, as we say in Swedish. To make do with what we have. And what we do have is morality, empathy, science, media and – in some fortunate parts of the world – democracy. These are some of the best tools we have right now and we all need to start using them.

Some say that we shouldn't involve morality because it can induce feelings of guilt and guilt is not an ideal way to create change. But what else can we do? How do we address this uncomfortable subject without upsetting anyone? How can we talk about an existential human crisis created by inequality, the exploitation of workers and nature, land theft, genocide and overconsumption without any mention of morality? Should we just pretend that the greatest threat we've ever faced is mainly an opportunity to create 'new green jobs' and a better future for all without any major changes for anyone?

There are others – a very small number of people – who think that some kind of dictatorship would be better suited to handle this huge global crisis. But there are no good dictatorships, just look at China or Putin's Russia. The idea of a non-democratic rule that would somehow seek the best for its citizens is nothing less than absurd. Justice and equal rights are essential for solving this crisis – this automatically rules out any form of dictatorship.

Democracy is the most precious thing we have, but as we have been reminded far too many times it is a fragile system and, unless the citizens are well informed and well educated in the matters that fundamentally shape their lives, democracy is easily manipulated.

That is why the contents of this book – science, knowledge and stories – are literally a matter of life and death. Not just to us, but to all future generations and to all living things and beings. There are countless issues that deserve our full attention and that have to be focused on, but the climate and ecological crisis differs from many others as it cannot be undone in the future. And the answers to all the other crises rely on us resolving this one. The climate and ecological crisis cannot be fixed later on. It cannot be left to anyone else to sort out. It has to be us, and it has to be now.

We have to start learning. We have to understand the basic facts. We have to learn to read between the lines. We need to teach each other to tell it like it is. There is no need for exaggeration; the story is bad enough already. There is no need for sugarcoating; we have to be adult enough to handle the truth. And there is no time for despair; it is never too late to start saving as much as we can possibly save. This is the biggest story in the world, and it must be spoken as far and wide as our voices can carry, and much further still. It must be told in books and articles, in movies and songs, at breakfast tables, lunch meetings and family gatherings, in lifts, at bus stops and in rural shops. In schools, boardrooms and marketplaces. At airports, in gyms and in bars. In the fields, in the warehouses and on the factory floors. At union meetings, political workshops and football games. In kindergartens and in old people's homes. In hospitals and car-repair shops. On Instagram,

TikTok and the evening news. On dusty country roads and in the streets and alleys of our towns and cities. Everywhere, all the time.

It has been estimated that we humans who are alive today make up 7 per cent of all *Homo sapiens* that have ever lived. We are all related, in time and space. Together, we stretch back through time and forward into our common future. Thanks to our ability to observe, study, remember, evolve, adapt, learn, change and tell stories, we have gained enough information and knowledge to begin safeguarding our living conditions and our well-being. This has given us an unprecedented possibility to create a fair and affluent world. But that enormous collective achievement – perhaps unique in the entire cosmos – is slipping through our fingers. Up until now, we have been failing. We have allowed greed and selfishness – the opportunity for a very small number of people to make unimaginable amounts of money – to stand in the way of our common well-being.

But now you and I have been given the historic responsibility to set things right. We have the unfathomably great opportunity to be alive at the most decisive time in the history of humanity. The time has come for us to tell this story, and perhaps even change the ending. Together, we can still avoid the worst consequences. We can still avoid catastrophe and start to heal the wounds that we have inflicted. Together, we can do the seemingly impossible. But make no mistake – no one else is going to do it for us. This is up to us, here and now. You and me. /

All the facts and stories in this book are unnerving enough individually. But they, too, are closely joined together – just like all of us.

PART TWO /

How Our Planet Is Changing

'Science doesn't lie'

2.1

The weather seems to be on steroids

Greta Thunberg

'This is the new normal' is a phrase we often hear when the rapid changes in our daily weather patterns – wildfires, hurricanes, heatwaves, floods, storms, droughts, and so on – are being discussed. These weather events aren't just increasing in frequency, they are becoming more and more extreme. The weather seems to be on steroids and natural disasters increasingly appear less and less natural. But this is not the 'new normal'. What we are seeing now is only the very beginning of a changing climate, caused by human emissions of greenhouse gases. Until now, Earth's natural systems have been acting as a shock absorber, smoothing out the dramatic transformations that are taking place. But the planetary resilience that has been so vital to us will not last forever, and the evidence seems to suggest more and more clearly that we are entering a new era of more dramatic change.

Climate change has become a crisis sooner than expected. So many of the researchers I've spoken to have said that they were shocked to witness how quickly it is escalating. But since science is very cautious when it comes to making predictions, maybe this should not come as a big surprise. One result of this, however, is that very few people actually knew how to react when the signs in recent years started becoming obvious. And even fewer still had planned how to communicate what is happening. It seems like the vast majority of people were preparing for a different, less urgent scenario. A crisis that would take place many decades into the future.

And yet here we are. The climate and ecological crisis is not happening in some faraway future. It's happening right here and right now. In the pages that follow we will be looking at some of the major changes taking place as the climate – and the entire planet – starts to destabilize. Each one of these case studies is serious enough in itself, but since they are all interconnected we cannot 'fix' one problem without 'fixing' the others. Holistic problems require holistic solutions. Our main challenge, however, is that all these events are happening at the same time, and at maximum speed.

I realize that reading the chapters that follow might be depressing for some, but we should not be surprised by what is going on. After decades and centuries of distancing ourselves from nature and sustainability, this is what we should expect. There are planetary limits and boundaries. Our resources are not infinite.

Some say that we are not doing enough to halt and address this crisis. But that is a lie, since 'not doing enough' indicates that you are doing something, and the inconvenient truth is that we are basically doing nothing. Or, to be fair, we are doing very, very little – far from what is needed. And, perhaps more importantly, we aren't doing anything to improve or turn things around, we are – at best – playing defence. The forces of greed, profit and planetary destruction are so powerful that our fight for the natural world is limited to a desperate struggle to avoid a total natural catastrophe. We should be fighting for nature but instead we are fighting against those who are set on destroying it.

Imagine where we would have been today were it not for the environmentalists, the activists, the scientists and the Indigenous land defenders. They have fought for us, and in many cases they have risked their lives and their freedom. Imagine if all those millions of people who are trying to improve the conditions for the living planet were given an opportunity to actually start turning this around, instead of just trying to push back against the ongoing destruction or the constant opening of new pipelines, new oil fields and new coal mines, the new sites of deforestation. Then we might start to see improvements, positive feedback loops and positive tipping points. But that is not where we are. Instead it seems like we are stuck in a spiral of negative events – a spiral that is accelerating and becoming more and more difficult to stop the longer we let it continue. And no, unfortunately, this is not 'the new normal'. This crisis will continue to get worse until we manage to halt the constant destruction of our life-supporting systems – until we prioritize people and the planet over profit and greed. /

We are entering a new era of more dramatic change.

Heat

Katharine Hayhoe

Since the dawn of the Industrial Revolution, humans have been producing increasing amounts of carbon dioxide and other powerful heat-trapping gases. As these gases build up in the atmosphere, they essentially wrap an artificial blanket around the planet, trapping more and more of the Earth's heat which would otherwise escape to space. That's why the planet's average temperature is increasing and why climate change is often called global warming.

In our everyday lives, though, what most of us experience isn't so much global warming as global weirding. Imagine weather like a set of dice. There's always a natural chance of rolling a double six: experiencing an extreme weather event like a heatwave, a flood, a storm or a drought. But as the mercury has ticked up, decade by decade, double sixes are showing up more often. Now, we're even rolling some double sevens. How could this be happening? The answer is global weirding.

Heatwaves are one of the most obvious ways climate change is loading the weather dice against us. Extreme heat now begins earlier in the year and stretches later. Heatwaves have grown hotter and more intense, and scientists can even put numbers on how much worse climate change is making them. In 2003, a record-breaking heatwave baked Western Europe with temperatures more than 10°C hotter than average. This heatwave caused flash floods from melting glaciers in Switzerland, triggered forest fires that burned down 10 per cent of Portugal's forests and led to more than 70,000 premature deaths. Scientists found that climate change doubled the risk of that heatwave occurring.

Now, two decades later, the situation is far more dire. In the summer of 2021 a blistering heatwave enveloped western Canada and the US. During that heatwave, the village of Lytton, British Columbia, broke the extreme heat record for Canada three days in a row, with temperatures reaching 49.6°C. Then on the fourth day, a wildfire – exacerbated by the record hot, dry conditions – destroyed most of the town. Scientists estimate that climate change made this heatwave at least 150 times more likely.

Why are heatwaves getting worse? The simple answer is that high temperature extremes become more common as the planet's average temperature increases. However, warmer temperatures also affect weather patterns. During warm weather, it's normal for a dome or a ridge of high pressure to stall over an area for a few days or even weeks. This high-pressure system, also known as a heat dome, is like a 'mountain of warm air' in the sky. Skies under a heat dome are typically clear, so the sun beats unrelentingly all day, every day. The dome also deflects cooler air masses and storms away from the region and suppresses convection, which would normally lead to clouds and rain. So the longer it sits over a region, the drier and hotter it gets. Where does climate change come in? If temperatures were already higher than average to begin with, then a heat dome starts off stronger than it would otherwise be. That's global weirding: in a warmer world, many weather extremes are becoming more frequent, more intense, longer and/or more dangerous.

High temperature extremes are already more common, and the more heat-trapping gases we pump into the atmosphere, the worse they will get. A person born in 1960 will experience only four major heatwaves in their lifetime. A child born in 2020, even if we meet the Paris target of 1.5°C, will experience eighteen such events. And for each additional half-degree of global warming, this number doubles.

What's at risk from more frequent and dangerous heatwaves? It's not the planet itself, but rather many of the living things that call Earth home. In the ocean, eight out of the ten most extreme marine heatwaves on record have occurred since 2010. Marine heatwaves bleach coral reefs, the nurseries of the ocean; kill off billions of shellfish and other marine creatures; and melt the sea ice in the Arctic which polar bears depend on to hunt their prey. On land, extreme heat stresses and kills plants and animals. It can lead to mass die-offs, such as when young birds leap from their nests to cool off before they are able to fly. And extreme heat fuels wildfire outbreaks such as the 2020 Australian wildfires that killed or displaced nearly 3 billion animals. Unchecked, human-caused climate change could lead to the extinction of a third of the plant and animal species on the planet by 2050.

We humans are also living things who call this planet home, and we're also at risk. Extreme heat affects us physically by increasing our risk of heat-related illness and even death, impacting our mental health and even heightening the risk of interpersonal violence and, with other climate impacts, political instability. Air pollution from fossil fuels is already responsible for nearly 10 million premature deaths around the world every year, and hotter air temperatures exacerbate the problem by speeding up the chemical reactions that turn tailpipe emissions into dangerous pollutants. Heatwaves also wither crops, decimate our water supplies, lead to power blackouts, and break down our infrastructure.

We're all affected, but the poorest and most marginalized among us bear the brunt of the impacts. That includes people who already live in highly polluted areas or who have no choice but to work outdoors in extreme heat. They might already lack access to sufficient food and water, or rely on the crops they grow to feed their families. They often don't have access to basic health care or air-conditioning; or they do, but can't afford to pay the bills when the heat spikes. Global weirding primarily affects those who've done the least to contribute to the problem, and that's not fair.

What can we do about it? As the IPCC says, every bit of warming matters, and every action matters. The first step we can each take is simple: use our voices to catalyse action by telling everyone we know how climate change affects us all and what we can do, together, to make a difference. /

In our everyday lives, what most of us experience isn't so much global warming as global weirding.

Methane and Other Gases

Zeke Hausfather

Much of the discussion around climate change focuses on carbon dioxide. There are good reasons for this: CO_2 lasts in the atmosphere for extremely long timeframes, is responsible for around half of the warming the world has experienced to date, and will be the cause of most future warming we project in our climate models.

However, there are other greenhouse gases that also contribute meaningfully to global warming. About a third of historical warming is attributed to methane (CH_4), while the remainder is from a combination of nitrous oxide (N_2O), halocarbons, chlorofluorocarbons, hydrochlorofluorocarbons, and other industrial chemicals), volatile organic compounds and carbon monoxide and black carbon. Major sources of non-CO_2 greenhouse gases include agriculture and waste (nitrous oxide, methane), fossil fuel production and use (methane, volatile organic compounds, carbon monoxide, black carbon), and industrial processes and appliances (halocarbons). Some of these – methane, some halocarbons, black carbon – have a relatively short atmospheric lifetime and are often called short-lived climate forcers.

Methane is the non-CO_2 greenhouse gas that gets the most attention – and for good reason. It is a powerful warming agent – around eighty-three times more potent than CO_2 over a twenty-year timeframe and thirty times more potent over a hundred-year timeframe. However, methane behaves very differently from CO_2 in the atmosphere. In short, methane is temporary, while CO_2 is forever.

When we emit a tonne of methane, more than 80 per cent is removed from the atmosphere via chemical reactions with hydroxyl radicals within twenty years. CO_2, on the other hand, is not removed by chemical reactions; it has to be absorbed by land and ocean sinks. Forty years after methane has been emitted, virtually all of it is gone, while nearly 50 per cent of the CO_2 remains in the atmosphere. Some of the CO_2 we emit today – around 20 per cent – will still remain in the atmosphere 10,000 years into the future.

In practice, this means that the long-term atmospheric CO_2 concentration is a function of cumulative emissions, while atmospheric CH_4 reflects the rate of emissions. To put it another way, if we stop increasing our methane emissions, atmospheric CH_4 will stop increasing. But if we stop increasing our carbon dioxide emissions, atmospheric CO_2 will continue to accumulate until we reduce CO_2 emissions close to zero. This raises a few important points:

- **First,** CO_2 is the primary driver of longer-term warming. In future baseline emissions scenarios (for example, where we don't reduce our emissions) CO_2 drives around 90 per cent of the additional twenty-first-century warming.

- **Second,** reducing warming by cutting methane is a lot easier than it is by cutting CO_2. Cutting methane results in near-immediate temperature declines, while cutting CO_2 only slows the rate of warming until we get to net zero.

- **Third,** methane can be cut at any point and have a large effect on temperatures. CO_2, on the other hand, is cumulative; waiting to cut CO_2 emissions locks in warming in a way that is not the case for methane.

- **Finally,** the degree to which we should focus on CO_2 and methane mitigation depends on short-term versus long-term prioritization. If we think we are close to climate tipping points, methane cuts are a way to quickly reduce warming. If we care more about temperatures in 2050 or 2070, then CO_2 cuts today matter more. When possible, however, we should strive to reduce emissions of both.

To understand the difference between CO_2 and methane, it is helpful to tell a story about why cows are like closed power stations. Let's say there is a rancher named Jane whose family has had a herd of a thousand cows for the past thirty years. Each day these cows happily graze, eating grass and burping out methane that mixes with the atmosphere.

However, the methane in the atmosphere is constantly oxidizing and breaking down. The average lifetime of the methane emitted from her cows is around ten years. This means that while Jane's herd produces around 100 tonnes of methane per year (0.1 tonnes per cow), a similar amount of methane emitted by their predecessors is breaking down, and the amount of atmospheric methane remains unchanged as long as the herd size stays

The drivers of global warming since 1850

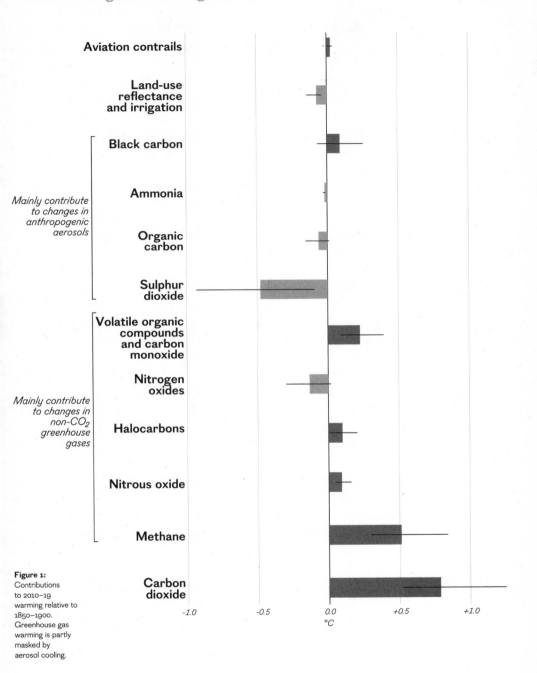

Figure 1:
Contributions to 2010–19 warming relative to 1850–1900. Greenhouse gas warming is partly masked by aerosol cooling.

constant (though methane breaking down does result in a small amount of additional CO_2 in the atmosphere).

Meanwhile, Jane's town has a small coal power plant that powers 500 or so homes. This coal power plant emits around 10,000 tonnes of CO_2 each year. It turns out that 10,000 tonnes of CO_2 has the same warming effect as 100 tonnes of methane if both remain in the atmosphere. So are Jane's cows as bad for the climate as the coal-fired power plant? Not quite.

As long as Jane's herd isn't growing, the methane emitted is balanced out by previously emitted methane breaking down in the atmosphere. The same is not true for the CO_2 from the coal power plant; each year, about half the CO_2 emitted by the coal power plant remains in the atmosphere – with about half being absorbed by land and ocean sinks. So while Jane's cows add no additional methane to the atmosphere, the coal plant adds 5,000 tonnes of CO_2 every year. In fact, the warming effect of the coal power plant is the same as if Jane were adding a further fifty cows to her herd each year.

The next year the town decides that it would be cheaper to generate its electricity with solar panels and battery storage and closes down the old coal power plant. However, the carbon that was previously emitted by the coal-fired power plant remains in the atmosphere. While it will slowly decline over the next few centuries, for the time being the closed coal power plant is still warming the planet just as much as Jane's herd is – even though it is no longer emitting any CO_2.

On the other hand, if Jane were to decide to get out of the ranching business, the methane emissions would fall to zero and most of the methane her cows had ever emitted would be gone from the atmosphere in a decade or two.

This story highlights the critical distinction between CO_2 and methane: once we emit CO_2, we're stuck with it (barring actively sucking it out of the atmosphere). Methane, on the other hand, does not accumulate over the long term; the amount of methane in the atmosphere depends on the rate of emissions rather than the total amount that has ever been emitted. Both are important greenhouse gases, but they behave in very different ways which we need to account for when planning how to reduce our emissions of each. /

2.4

Air Pollution and Aerosols

Bjørn H. Samset

If you light a bonfire and look towards the sky, you will see a rising column of smoke. Spreading upwards and outwards, it twirls and thins until it fades to invisibility. But it's not gone. Smoke particles – one example of what we call aerosols – can remain airborne for days, and in that time they can travel both far afield and high up in the atmosphere. And while there, they have a strong effect on both the weather and the climate. Today, aerosol emissions from our industrial activities effectively mask a good portion of the heating from increased levels of CO_2 and other greenhouse gases. But what happens when we clean up our air and our skies?

It's important to remember that greenhouse gases are not all that we emit into the atmosphere. Aerosols – or minute, airborne particles like those that make up smoke – have been a by-product of our activities for as long as we have had fire and industry. Today, they come from all kinds of combustion, from road traffic and industry, from coal-fired power plants, from ships and aeroplanes, and many other sources. Some are also born in the atmosphere, as the result of emissions of gases such as sulphur dioxide.

Aerosols are hazardous to humans and animals. They are a main component of air pollution, and a notable cause of premature death around the world. When it comes to climate change, however, they play an equally important role, but with a very different effect to greenhouse gases. When airborne, aerosols function as a thin, wispy cloud. They reflect some of the incoming sunlight back to space and therefore act to cool the planet. Furthermore, if aerosols are present in the air when a cloud forms, the cloud droplets will become smaller and more numerous. This makes the cloud whiter and more reflective, which also means a cooler Earth. Aerosols are therefore doubly efficient at cooling the planet's surface.

And since we emit a lot of aerosols every year, they do a lot of cooling. Scientists have measured global warming of around 1.1°C since the period 1850–1900. However, as the latest report from the IPCC shows, if greenhouse gases were the only substances we'd released into the atmosphere, this figure would have been at least 1.5°C. The reason for this difference is mainly our aerosol emissions, which cool the climate by about 0.5°C, while also affecting geographical patterns of rain, monsoon systems, extreme weather, and more.

It is therefore crucial for us to understand aerosols and their effects on the climate as we try to meet the challenge of global warming. Unfortunately, they are also difficult to get a handle on. Today we have a good grasp of where they come from, and in what volume. Where they are transported from, however, is less well known. We also don't quite know what chemical reactions they undergo in the atmosphere, the details of how they interact with clouds and rainfall, or where they ultimately end up. And some aerosols even go against our expectations and heat the climate instead of cooling it. These are dark in colour, like the smoke from bonfires. Dark aerosols like these don't just reflect sunlight but can also capture it, heating the air around them. This, in turn, hinders the formation of rain and can affect both clouds and wind patterns. And if dark aerosols land on snow, they can warm the surface, reducing its reflectivity and speeding up melting.

All these details are important if we want to understand the total effect our emissions have on today's climate – and that of the future. Aerosols are therefore very intensively studied, and new and exciting discoveries are frequent. But we still don't know what will happen to the weather if the amount of human-caused aerosols in the atmosphere changes – and that is a problem, because this is exactly what we expect to happen.

Most scientists predict that the amount of aerosols in Earth's atmosphere will decrease over the coming years. Since the main environmental threat we face today is global warming, it might sound tempting to just leave the cooling aerosols as they are – or even to emit more of them. This is, however, not an option. Not only is air pollution a major health hazard, but many sources of greenhouse gas emissions are also sources of aerosols, such as coal-fired power plants, older diesel cars and container ships. Our path towards net zero CO_2 emissions therefore also inevitably takes us towards cleaner skies.

The clean-up of aerosol emissions is already underway in different regions around the world. China, which was until recently a major emitter of sulphur dioxide, has already made massive clean-up efforts, mirroring what happened in Europe and the US some decades earlier. This is good news for the environment, and ultimately for the climate. However, while this clean-up happens, we may temporarily see the consequences of climate change unfolding even more rapidly in particular regions. Losing the artificial aerosol cooling may speed up surface warming, both globally and close to the emission sources, and therefore make heatwaves stronger and more frequent. The same is true for extreme episodes of heavy rain. And in some parts of the world the opposite could happen. Some countries are still undergoing rapid industrialization, meaning that aerosol emissions – and levels of air pollution – are likely to increase in these areas unless care is taken to employ cleaner technologies than those used in the past.

The range of possible changes in our global aerosol emissions are all included in the scenarios that scientists use to study how climate change will progress. However, just as we don't yet know precisely how much CO_2, methane and other greenhouse gases we will emit in the coming decades, we also don't know what level our future aerosols emissions will reach. Small as they are, aerosols account for a major part of the uncertainty that remains about the future of the climate.

We humans influence the climate in broad and complex ways. Global warming from greenhouse gas emissions is first among these, but in many parts of the world aerosols are just as important. Until now, aerosols have held some global warming in check – but this influence will likely be dramatically reduced as we transition to a climate-neutral society. We are actively studying the details of what this will mean for temperatures, rainfall, extreme weather, and more, but there is no doubt that aerosols must be taken into account as we prepare for the full set of consequences the climate crisis will have for us – and for the rest of nature. /

Until now, aerosols have held some global warming in check – but this influence will likely be dramatically reduced as we transition to a climate-neutral society.

2.5

Clouds

Paulo Ceppi

One of the central aims of climate science is to predict the future amount of global warming for a given level of greenhouse gas emissions. But while we have long known that an increase in greenhouse gas concentration causes warming, exactly how much warming will occur depends to a large extent on clouds.

Why are clouds so critical for climate change? To understand this, we should first consider how clouds affect the present-day climate. Their impact is twofold: on the one hand, clouds reflect sunlight away into space, acting like a parasol to shield the Earth's surface from solar energy; on the other hand, clouds have a greenhouse effect of their own and trap heat radiating from the Earth's surface, similar to an insulating blanket, limiting the loss of heat to space.

Which of the two effects dominates – the cooling parasol or the insulating blanket – depends on the type of cloud: for example, the higher the cloud, the stronger its blanket effect. However, on average globally and considering all cloud types, the cooling parasol effect is nearly twice as large as the insulating blanket effect; our planet would therefore be considerably hotter if clouds didn't exist.

The climate impact of removing all clouds would be about five times as large as the effect of doubling the concentration of carbon dioxide in the atmosphere. This implies that even subtle changes in cloudiness could substantially enhance or mitigate future global warming. As the climate warms, we expect the properties of clouds – their number, thickness and altitude – to change, which will modify their parasol and blanket effects. The resulting knock-on effect on global warming is known as cloud feedback.

Cloud feedback has been a major uncertainty in climate change projections for a long time. Global climate models are not fully up to the task: they cannot accurately simulate the small-scale processes involved in the formation and dissipation of cloud droplets and ice crystals. Meanwhile, directly observing cloud feedback is far from straightforward. Clouds respond to a large variety of meteorological drivers, including temperature, humidity, wind, and air particles known as aerosols. Because all of these drivers vary naturally over time, it is challenging to quantify the component of observed cloud changes that is associated with global warming.

Nevertheless, recent scientific advances have led climate scientists to the conclusion that clouds are amplifying global warming. Observations and modelling suggest this occurs in two main ways: a decrease in the number of low clouds over the oceans in the tropics, meaning a reduced parasol effect and hence greater absorption of sunlight at the ocean surface; and a rise in the altitude of high clouds globally, implying an enhanced blanket effect.

It is important to stress that this amplifying cloud feedback does not mean that climate change will be even worse than we imagined: the possibility of clouds amplifying global warming has been accounted for in climate change projections for a long time. Nevertheless, the newest scientific evidence does confirm that we cannot count on clouds to suppress global warming. Furthermore, it is possible that the amplifying effect of clouds becomes stronger as the climate warms, or worse, that clouds act as a tipping point beyond a certain level of carbon dioxide concentration. To avoid such low-likelihood, high-risk outcomes, the safest path is to rapidly cut carbon emissions now. /

It is possible that the amplifying effect of clouds becomes stronger as the climate warms, or worse, that clouds act as a tipping point beyond a certain level of carbon dioxide concentration.

2.6

Arctic Warming and the Jet Stream

Jennifer Francis

Mother Nature has been on a rampage lately: extreme weather of just about every flavour has wreaked havoc around the northern hemisphere. In 2021 alone, a devastating frigid spell invaded south-central US; severe flooding struck Germany, China and Tennessee; prolonged drought parched western US and Middle Eastern countries; unprecedented heatwaves scorched the Pacific Northwest, Turkey, Japan and Middle East; and deadly hurricanes ravaged the Gulf of Mexico and north-eastern US. And this list is far from complete. Climate change is worsening many types of extreme events in both straightforward and complex ways, and the contributions to this rampage by the melting Arctic are becoming clearer.

The Arctic is warming rapidly and all three of its once-permanent ice types are disappearing: sea ice (ice formed from seawater that floats on the Arctic Ocean), land ice (glaciers and ice sheets) and permafrost (year-round frozen soils). Spring snowcover on high-latitude land areas is also in steep decline. As bright, white surfaces like sea ice and snow shrink, less of the sun's energy is reflected back into space; instead it is absorbed into the climate system, which in turn melts even more ice and snow. This vicious cycle is known as the ice-albedo feedback, and is the main reason the Arctic has been warming at least three times faster than the globe as a whole since the mid-1990s (Fig. 1). Changes of these magnitudes in such a key component of the Earth system cannot help but have enormous impacts on both local and distant weather.

The local effects are relatively straightforward – overall warming favours hotter, drier summers in the far north, which leads to conditions conducive to wildfires, even in boggy regions of the tundra. Connections with weather patterns farther south – where billions of people live – are more complicated, however, and still have researchers scrambling for answers. It all comes down to how disproportionate Arctic warming will affect the jet stream, a river of strong west-to-east winds at altitudes where jets fly that encircles

Near-surface air temperature change in the Arctic and the entire globe since 1995

Figure 1:
The Arctic is now warming over three times faster than the globe as a whole: the trend line for the Arctic is 0.99°C per decade; for the globe it is 0.24°C per decade.

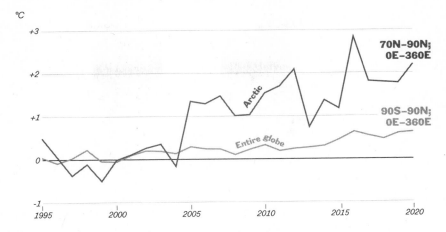

°C

+3

+2

+1

0

-1

1995   2000   2005   2010   2015   2020

70N–90N; 0E–360E

90S–90N; 0E–360E

Arctic

Entire globe

the northern hemisphere (a jet stream exists in the southern hemisphere, as well) (Fig. 2).

The jet stream creates and steers most of the weather systems in temperate latitudes (the zone between the Arctic and the tropics), so anything that affects its strength or path will, in turn, affect the weather that many of us experience. Jet streams exist because of differences in air temperature, such as cold Arctic air butting against warmer air to the south. When the difference is large, jets are strong and tend to flow in a relatively straight path. When the temperature difference is relatively small, jets are weaker and more likely to take bigger swings to the north and south – meanders known as

Figure 2:
The jet stream: strong west-to-east winds that encircle the northern hemisphere.

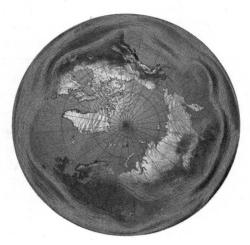

Comparison of conditions with a warm and cold Arctic

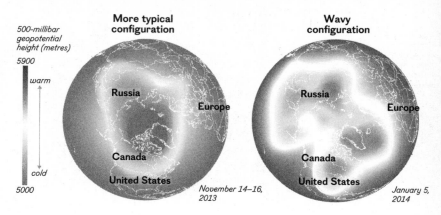

Figure 3:
A cold Arctic and relatively straight jet stream (left) and a relatively warm Arctic and wavier jet stream (right). Warmer regions are represented by shading around the edges of the globe.

Rossby waves. Because the Arctic is warming so much faster than elsewhere, the north–south temperature difference is getting smaller, which is weakening the west-to-east winds in the jet and increasing the likelihood of wavy patterns (Fig. 3). We know that when the Arctic is abnormally warm, pockets of cold air tend to migrate southwards over the continents, creating the so-called warm Arctic/cold continents pattern. Moreover, when the jet stream's waves are large, they tend to progress eastwards more slowly, which means the weather regimes they generate also move more slowly. The upshot is that we notice weather conditions becoming more persistent – be they hot, dry, wet, cold, or even drizzly.

At least, that's the theory. Proving it is not easy because the atmosphere is a chaotic creature and other changes in the climate system are happening simultaneously. For example, shifting ocean temperatures and intensifying bouts of tropical thunderstorms can also affect the behaviour of the jet streams. Recent studies tell us that jet streams in the northern hemisphere are, in fact, getting wavier, but the challenge is determining which factors are in play. The diagnosis varies depending on region, season and the state of fluctuating natural conditions, such as whether an El Niño or La Niña temperature pattern in the tropical Pacific Ocean is present.

Since 2012, when my colleague Dr. Steve Vavrus and I first proposed and documented a connection between rapid Arctic warming and a higher likelihood of extreme weather conditions in temperate regions, a flurry of new research has focused on this topic. While the story is still murky, some aspects of this conundrum are becoming clearer. For example, studies focused on winter weather have identified likely causes of the warm

Arctic/cold continents pattern. When late-autumn sea ice is greatly reduced in the Barents–Kara Seas, a region north of western Russia, intense cold spells affecting central Asia and even North America later in winter are more likely. In a nutshell, Arctic warming tends to strengthen northerly winds over Siberia, which brings snowfall and chilly temperatures earlier than usual to the region. This couplet of warmer sea and colder conditions in northern Siberia tends to amplify a Rossby wave over the area, which can then disrupt the pool of extremely cold air that usually sits high over the North Pole (known as the stratospheric polar vortex). If the jet-stream waviness is sufficiently strong and persistent, the pool of cold air can be unleashed, bringing severe winter conditions to northern hemisphere continents. The extreme cold spell that caused widespread disruption to the US south-central states in February 2021, for example, was intensified and prolonged by a distortion of the stratospheric polar vortex. The severe freeze penetrated unusually far southwards into regions unaccustomed to and unprepared for the prolonged cold and icy conditions which crippled power supplies to nearly 10 million customers and froze water pipes serving 12 million people. Dallas, Texas, set a record low temperature of −19°C, 24°C colder than the February average low.

A new line of research has also uncovered a summer connection that contributes to recent severe heatwaves, wildfires, droughts and deluges. These extreme events are more likely when the jet stream splits, with one branch flowing across the middle of the continent and another along the Arctic coast. These splits tend to occur when the spring snowcover on high-latitude land areas melts earlier than normal, a strong trend observed in recent decades. When the snow disappears early, underlying soils dry out and warm sooner, creating a belt of abnormally high temperatures over high-latitude land areas. This warm belt favours jet-stream splits. Rossby waves can then become trapped between the two jet-stream branches, causing stagnant weather conditions that can cause persistent heatwaves, dry periods and rainy spells, often leading to extreme summer events. Split jets likely contributed to a variety of recent summer extremes, including devastating heatwaves that killed thousands of people across Europe in 2003 and 2018, Russia in 2010, the south-central US in 2011 and East Asia in 2018. Extreme flooding events, such as prolonged deluges across Pakistan in 2010 and Japan in 2018, have also been linked to split jet-stream patterns, and evidence suggests the occurrence of these conditions will increase as the Earth continues to warm.

Because change is rampant in many aspects of the climate system, these and other Arctic/mid-latitude connections do not occur every year, nor do they occur in the same locations or seasons. But there is no doubt that these disruptions will become more frequent and more intense, stressing infrastructure, ecosystems and our sense of normalcy beyond their limits. The escape hatch out of this predicament is in plain sight. Our efforts today to reduce emissions and concentrations of heat-trapping gases – if effected swiftly and broadly – can avert the worst escalations of weather extremes. And in the near term we must prepare for the impacts of worsening extremes until (and if) the climate is stabilized. We have no time to lose. /

When the Arctic is abnormally warm, weather conditions become more persistent – be they hot, dry, wet, cold, or even drizzly.

Dangerous Weather

Friederike Otto

Today, those of us who are not completely delusional have realized that climate change is not something happening somewhere else, at some point in the future, not merely some abstract concept related to jargon like 'global mean temperatures', but a phenomenon that is killing people here and now. There is no way to ignore the evidence any longer. Every region in the world is seeing its impacts – whether it's changing seasons, melting glaciers or rising sea levels – but most people are experiencing it through extreme weather events.

Long before we began observing this impact on weather on a daily basis, climate scientists, and everyone with a grasp of basic physics, knew that a warmer climate would mean a higher likelihood of heatwaves and fewer cold events. Given that a warmer atmosphere can hold more water vapour, we could also expect more heavy rainfall. From the same basic relationship we also knew that in a warmer climate more extreme heatwaves would occur. And the faster you warm the climate, the faster you change how intense extreme events are.

By changing the composition of the atmosphere, we have not only warmed the world as a whole but also changed atmospheric circulation. In other words, we have altered how and where weather systems develop and how they move. These changes can increase the effects from the warming alone, or act in the opposite direction, decreasing the risk of some extreme weather events in some locations. Both aspects of climate change – warming and atmospheric circulation – can interact in complex ways and for some of the most devastating extremes, like storms and tropical cyclones, it is not straightforward how these aspects interact.

This does not mean that we cannot know how these more complex events are changing. In fact, the emerging science of extreme event attribution is doing exactly that. Attribution science is straightforward in theory: it estimates what weather events are possible in a world with climate change and compares these to possible weather in a world without anthropogenic climate change. In practice, these methods require weather observations and climate models that can reliably simulate the extreme event of interest.

This can be done for most heatwaves and heavy rainfall – and, to some degree, droughts – but it is much more difficult for events where wind must be taken into consideration. Over the past decade the science has progressed meaningfully and an increasingly large number of individual weather events have been attributed to climate change, underpinning the key finding of the most recent IPCC report that 'Human-induced climate change is already affecting many weather and climate extremes in every region across the globe.'

From this science we also know with high confidence that, when a storm happens, the associated level of rainfall is higher than it would be in a world without climate change. In the case of Hurricane Harvey, which catastrophically flooded Houston, Texas, in 2017, this means that it would have rained 15 per cent less without human-caused climate change. Fifteen per cent might not sound a lot, but when translating it to the costs it becomes clear just how catastrophic human-made climate change is even in a single storm. The overall costs of the rainfall associated with the storm were estimated at $90 billion, $67 billion of which can be attributed to the extra rain due to climate change. It is important to highlight that these are only the economic damages. The impacts on individual lives – from fatalities to livelihoods lost – are much harder to quantify, but there is considerable suffering, especially for the most vulnerable in society.

Rising sea levels as a result of climate change are also amplifying the disastrous impacts of storms. Most storms develop over water and make landfall afterwards, accompanied by storm surges, which are increasing since sea levels have risen and will continue to rise for centuries due to our warming climate. A prominent example of this is Hurricane Sandy hitting New York in 2012. In this case the damages were estimated to amount to $60 billion from the storm surge alone, $8 billion of which can be traced to the higher sea levels from human-made climate change. If we hadn't burned fossil fuels, 70,000 fewer people would have been affected by Sandy's storm surge. It is important to highlight that, even if we stop emitting greenhouse gases immediately, sea levels will continue to rise. But the sooner we stop, the slower they will rise and the lower the levels they eventually reach will be.

The warming of the planet has also altered the speed at which storm systems move across the Earth (translation speed). For the parts of the oceans where we have data, translation speeds have slowed down. And slower-moving storms mean that more rain can be dumped at any single location. Thus, taking together everything that we know from physics, statistics and observations, the storms we see today are more damaging than they would have been without climate change.

Pinpointing the role of climate change on individual extreme weather events is an incredibly valuable source of information for decision-makers as they try to rebuild after disasters and plan for the impacts of tomorrow's extreme weather. Unfortunately, access to this information is not equal for all. In many cases – for instance, Cyclone Idai, which devastated Mozambique in 2019, or Cyclone Amphan, which hit Bangladesh and India in 2020 – the models are inadequate or inaccessible to scientists from the Global South. Our knowledge of how weather is changing and what our societies are most vulnerable to is dominated by the research and experiences of the Global North. Given the increasing speed with which our climate is warming, these inequalities need to be addressed. Whether a storm turns into a catastrophe ultimately depends on who and what is in harm's way; and, even if most changes in the climate system are linear, the impacts and damages are absolutely not. Small changes in the climate can have catastrophic consequences. /

Human-made climate change is catastrophic even in a single storm.

Next pages:
A large rainstorm crosses the Irrawaddy Delta in Myanmar in May 2008, four weeks after the river was struck by a storm surge from Cyclone Nargis which killed over 100,000 people.

The snowball has been set in motion

Greta Thunberg

Maybe it is the name that is the problem. *Climate change.* It doesn't sound that bad. The word 'change' resonates quite pleasantly in our restless world. No matter how fortunate we are, there is always room for the appealing possibility of improvement. Then there is the 'climate' part. Again, it does not sound so bad. If you live in many of the high-emitting nations of the Global North, the idea of a 'changing climate' could well be interpreted as the very opposite of scary and dangerous. A changing world. A warming planet. What's not to like?

Perhaps that is partly why so many people still think of climate change as a slow, linear and even rather harmless process. But the climate is not just changing. It is destabilizing. It is breaking down. The delicately balanced natural patterns and cycles that are a vital part of the systems that sustain life on Earth are being disrupted and the consequences could be catastrophic. Because there are negative tipping points, points of no return. And we do not know exactly when we might cross them. What we do know, however, is that they are getting awfully close, even the really big ones. Transformation often starts slowly, but then it begins to accelerate.

Stefan Rahmstorf writes that 'We have enough ice on Earth to raise sea levels by 65 metres – about the height of a twenty-story building – and, at the end of the last ice age, sea levels rose by 120 metres as a result of about 5°C of warming.' Taken together, these figures give us a perspective on the powers we are dealing with. Sea-level rise will not remain a question of milli-, centi- or decimetres for very long. Even if the change takes time, we must realize that this is not something we can *adapt* to.

The Greenland Ice Sheet is melting, as are the 'doomsday glaciers' of West Antarctica. Recent reports have stated that the tipping points for these two events have already been passed. Other reports say they are imminent. That means we might already have inflicted so much built-in warming that the melting process can no longer be stopped, or that we are very close to that point. Either way, we must do everything in our power to stop the process because, once that invisible line has been crossed, there might be no going back. We can slow it down, but once the snowball has been set in motion it will just keep going.

Billions of people all over the world are dependent on the cryosphere, relying on glaciers for drinking water and irrigation. And these are melting too, rapidly. Here we have already passed a number of irreversible tipping points which will bring enormous challenges in the decades to come. The Himalayan glaciers, also known as the Third Pole, are particularly crucial, as 2 billion people across Asia rely on them for their water supply. These glaciers are currently melting at an exceptional rate; one landmark study, requested by the eight nations spanning the area and carried out by 200 scientists, finds that even if we limit warming to 1.5°C, one third of the ice mass will be lost.

Not only are we losing this vital resource, we are doing so at a pace which in itself is a problem – because the faster melting speed is making us used to unnaturally high levels of water flow. When all that water starts to run out, we will be in even more trouble. Our infrastructure and societies were built for the Holocene, which is now becoming a geological epoch of the past. The world we used to safely inhabit no longer exists. /

Sea-level rise will not remain a question of milli-, centi- or decimetres for very long. Even if the change takes time, we must realize that this is not something we can *adapt* to.

2.9

Droughts and Floods

Kate Marvel

The Earth does not, in general, make its own water. It doesn't have to. Plenty arrived from space at the planet's formation and, essentially, the same amount has remained ever since. Billions of years from now, when the sun burns through its own store of fuel and dies, the Earth's moisture will disappear into space, ready to water the surface of some distant planet.

What this means is that the water we drink is the same stuff that quenched the dinosaurs' thirst and nourished the first stirrings of life on the young world. It morphs from ice to liquid to vapour and back again, rises from humid forests and sinks into cold ocean abysses, moves from the tropics to the poles and back. Sometimes, when the planet wobbles a little in its orbit, some of the water finds itself locked in glacial ice for an aeon or two. When the ice age ends, it escapes in a fresh torrent that pours into a growing ocean. On shorter timescales – afternoons, months, human lifetimes – it cycles from the ocean or land to the sky and back, not created, not destroyed, always changing.

Shape-shifting is tiring work. It takes energy to turn liquid into vapour, which is why on hot days your body makes you wet and clammy. Evaporation wicks energy away from the surface and up into the sky. Condensation warms the upper atmosphere, which in turn spits heat outwards to cold space. Water in vapour form is invisible, but the sky is visibly painted with white and grey clouds, collections of tiny liquid drops and ice crystals. The Earth sweats in the heat. The cold upper atmosphere wraps itself in a blanket of cloud. Everything is in balance, until everything is disturbed.

As the temperature increases, the world sweats more. The air demands water from the surface, which yields up its moisture to the thirsty sky. The oceans can easily handle the increased demand. But on land, the water is stored in soil like a sponge. Even in years with average rainfall, the greedy air can suck the lifeblood from the surface, leaving it arid and dead. The North American south-west is experiencing the worst megadrought on record, with more drying to come. Southern Europe, the Levant and south-western Australia are drying, too, as expected when temperatures rise. Drought is the consequence of a planet desperate to cool itself off.

In the process of evaporation, liquid morphs into vapour: colourless, odourless, but far from weightless. There are 10 million billion kilograms of water vapour in the atmosphere, pushing up and down and to the side, exerting pressure everywhere. Eventually the pressure becomes unbearable, and some of that vapour escapes the sky, condensing back into liquid. The threshold where this happens increases rapidly with temperature: hot air can hold more water vapour. There is a bank of water in the sky, receiving credits of vapour, spending debits of rain and saving a little in reserve. As the temperature increases, the reserves pile up. There is more moisture in a warming sky, a 7 per cent increase for every degree Celsius of warming. On a hotter planet, when it rains, it pours. A warmer world will suffer from drought but, by the cruel logic of the water cycle, it will flood too.

Our worsening droughts and our catastrophic floods are characteristic fingerprints of human interference, a record of our post-industrial existence etched in the water flows of the planet. Attribution science has now advanced so much that we can quantify the human contribution to individual downpours and droughts. But our fingerprints are visible on much larger scales than that, etched on sky, sea and land. Satellite observations show long-term shifts in rainfall patterns that are corroborated by the ocean. The waters of the Southern Ocean and the north Atlantic have freshened as the rainfall in those regions has increased, while the Mediterranean and subtropical seas have become saltier under drier skies. On land, old trees give long context to our current moment. Their inner rings tell a story of wet and dry years past, of changing moisture in the soil that feeds them.

Together, the tree rings of the world form a pattern, a record of moistening and drying that stretches back for centuries. These changes are natural. But now, something unnatural is beginning to emerge. As we look at the rings of the last century, we see drying soils in the thin rings of thirsty trees. It's not unusual for the American south-west to be dry, or the Mediterranean, or Australia. Droughts would happen even in a world without us. But it is unusual for all of these places to be dry at once. Nature can't do that. But we can.

We now live in a world largely of our own creation. What will we do with it? We will not sit back to wait patiently for calamity. We will rethink the world we've made. We will draw our energy from the sun and the wind that drive the dance of water from surface to atmosphere and back. We will endure and change, like the water on which we depend. We must. /

2.10

Ice Sheets, Shelves and Glaciers

Ricarda Winkelmann

December, 2010: minus 32°C. Our research vessel has reached 71°07 S, 11°40 W – Antarctica. It's 4 a.m. and as bright as day. I look out at the ice shelf in front of us, which is protruding roughly 30 metres from the ocean water. I am stunned by its beauty, by the complex structures within the ice, and I can barely wrap my mind around its vastness: covering almost 14 million square kilometres, it is, in many places, more than 4,000 metres thick. If all this ice were to melt, sea levels would rise by almost 60 metres worldwide. Looking up, I think to myself: much of this ice was formed hundreds of thousands of years ago. Humans, on the other hand, only set foot on Antarctic ice in the early twentieth century. How is it possible that, in this short time, we have become the dominant force determining the future evolution of this majestic giant?

I will never forget this moment from my first scientific expedition to Antarctica. It was then that I sensed what it truly means to have entered the Anthropocene – that humans have become a geological force.

Increasingly, our actions are affecting all parts of the Earth system, including the planet's two ice sheets, in Greenland and Antarctica. Over the past few decades, both ice sheets and their surrounding ice shelves – which are like floating tongues of ice jutting into the ocean – have been losing mass at a rapidly accelerating rate. In total, 12.8 trillion tonnes of ice were lost between 1994 and 2017. To put this into perspective, 1 trillion tonnes of ice can be envisaged as an ice cube measuring 10 cubic kilometres, taller than Mount Everest.

In the future, the ice sheets are expected to become the largest source of sea-level rise. Because of their massive size, even modest losses from them can significantly increase the risk of flooding in coastal communities, with severe consequences for society, the economy and the environment.

Drastic changes in the polar regions are already beginning to unfold. In 2020, temperatures hit a record high in both polar regions, reaching +18.3°C on the Antarctic Peninsula and +38°C in the Arctic. In 2021, two near-record melt events occurred on the Greenland Ice Sheet – following a series of extreme melt events in the years 2010, 2015 and 2019. On the other side of the planet, the largest iceberg in the world calved from the western side of the Ronne Ice Shelf in the Weddell Sea. Analysis of satellite images revealed additional large icebergs breaking off at the edge of the ice shelf next to Pine Island Glacier, leading to further acceleration of what is already one of the fastest-moving glaciers in Antarctica.

While these are only snapshots in time, they reveal the radical and impactful shifts currently occurring in and around the ice sheets. The polar regions are our planet's most efficient early-warning systems for progressing climate change – and these early-warning systems are now raising the alarm.

And we'd better listen to this alarm: with unmitigated climate change, we will push the ice sheets further and further out of balance, potentially unleashing self-perpetuating processes that cannot effectively be halted.

One of these self-perpetuating processes, or positive feedbacks, is linked to melting at the surface of the Greenland Ice Sheet: with increasing melt, its surface is slowly sinking to lower heights. At lower elevations, the air is generally warmer, which can in turn lead to more melt, making the surface descend into even warmer air masses, causing even more melt, and so on. Once a critical temperature is exceeded, this melt-elevation feedback could lead to sustained ice loss until Greenland would eventually be almost ice-free.

Because it is colder in Antarctica than in Greenland, it is not so much the surface melt that threatens the stability of the Antarctic Ice Sheet but rather what is going on underneath. Much of the observed ice loss in Antarctica happens as a result of the melting of the floating ice shelves that surround the continent. These get thinner when they come into contact with warmer ocean waters, causing the continental ice further inland to acceler-ate towards the ocean, and potentially leading to self-perpetuating ice loss.

These positive feedbacks are why both ice sheets are considered tipping elements in the Earth system. Once they are close to a critical warming threshold, or tipping point, a tiny perturbation can suffice to trigger abrupt, widespread, unstoppable ice loss.

The risk of crossing such a tipping point increases starkly when exceed-ing global warming of 1.5–2°C. Above these temperature levels, large parts of the Greenland and Antarctic ice sheets will be lost, and long-term persistent sea-level rise of several metres will be unavoidable. Even if temperatures eventually were to sink again, cooling well below today's temperature would be required to regrow the ice sheets to their present-day size. In other words, parts of the ice sheets, once lost, might be lost forever. /

Warming Oceans and Rising Seas

Stefan Rahmstorf

In 1987 one of the great pioneers of oceanography sounded a warning in the top scientific journal *Nature*:

> The inhabitants of planet Earth are quietly conducting a gigantic environmental experiment. So vast and so sweeping will be the consequences that, were it brought before any responsible council for approval, it would be firmly rejected. Yet it goes on with little interference from any jurisdiction or nation. The experiment in question is the release of CO_2 and other so-called greenhouse gases to the atmosphere.

Wallace (Wally) Broecker wrote these words, and I was lucky enough to work with him for years on the Panel on Abrupt Climate Change before he sadly passed away in 2019. Here, I'll take a look at the consequences of this 'gigantic environmental experiment' for the physical aspects of the ocean – 'physical' here meaning physics rather than marine biology or chemistry, which are covered elsewhere in this book.

The warming ocean

The oceans have absorbed over 90 per cent of the extra heat on our planet that has been trapped by increasing levels of greenhouse gases. That's not because the oceans heat up more than the air but because more energy is required to heat water than to heat air (in other words, water has a much larger heat capacity). The oceans absorb this heat at their surface, which sees the highest increase in temperature; it penetrates more slowly into the ocean's depths. Ocean heat content is increasing at a rate of 11 zettajoules per year, twenty times the amount of energy used by humans.

Despite the oceans absorbing 90 per cent of the additional heat, sea surface temperatures have risen only about half as much as air temperatures over land: by 0.9°C since the late nineteenth century, compared to 1.9°C for

Changes in global sea surface temperature and air surface temperature over land

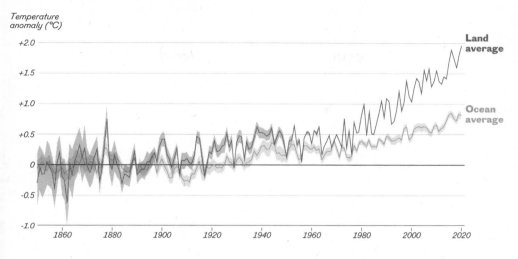

Temperature anomaly (°C)

Land average

Ocean average

Figure 1: Temperature anomalies for sea ice regions are calculated separately and not shown.

temperatures over land (Fig. 1). Given that 71 per cent of the Earth is covered by ocean, that makes for a global average warming of 1.2°C.

By the time global warming has reached 1.5°C, temperatures over land will have warmed by approximately 2.4°C. So when we talk about the 'global average temperature' we make the impact of warming on us land-dwellers appear much smaller than it actually is. However, the relatively large heat capacity of the ocean does mean that our planet takes time to warm up and is lagging behind the equilibrium warming that will eventually be reached.

Many people think that more warming is 'baked in' from our past emissions, making it impossible to limit warming to 1.5°C, which is one of the primary goals of the Paris Agreement. Thankfully, that is not the case. Greenhouse gases start to decline in the atmosphere after we have reached zero emissions, counteracting the thermal inertia effect, so it is still possible to stop warming at 1.5°C – but only if we cut emissions down to zero fast enough.

The warming of the ocean causes a number of alarming problems. First, it provides more energy to tropical cyclones, which are getting stronger and intensifying faster. Second, more water evaporates from warmer oceans, which in turn increases global rainfall. Unfortunately, this adds to heavy rainfall events that can cause flooding rather than alleviating droughts. Third, warming tends to diminish the oceans' ability to act as a sink for carbon dioxide. Currently, the oceans absorb about a quarter of our carbon dioxide emissions, a tremendous contribution, but warmer water doesn't hold carbon dioxide as well (just try heating mineral water). Fourth, ocean warming has a detrimental effect on marine biology, causing calamities such as coral bleaching. Fifth, water expands when you warm it – which brings us on to the next difficulty: rising sea levels.

Rising seas

A warming global climate will inevitably cause sea levels to rise, for two main reasons. First, water in the ocean expands when it is warmed and, given that the oceans are thousands of metres deep, even a tiny percentage of expansion can cause a few metres of sea-level rise. Second, land ice masses are shrinking, adding more water to the oceans. We have enough ice on Earth to raise sea levels by 65 metres – about the height of a twenty-storey building – and, at the end of the last ice age, sea levels rose by 120 metres as a result of approximately 5°C of warming.

Compared to that, modern sea-level rise is still relatively small, at about 20 centimetres globally since the nineteenth century (Fig. 2). That's because it takes a long time for heat to penetrate into the deeper parts of the oceans, and for large ice masses to melt. However, we are seeing only the beginning of a much larger sea-level rise which is already 'baked in' and will unfold over centuries and millennia to come, even without further warming.

So far, the sea-level rise we have observed matches independent data on the different contributing factors. Since 1993, when satellite monitoring of sea levels began, these have been:

- **Ocean thermal expansion**   42%
- **Glaciers**   21%
- **Greenland Ice Sheet**   15%
- **Antarctic Ice Sheet**   8%

(The remaining portion can be attributed to groundwater pumping for agriculture, and the fact that some of the data is not precise.)

The IPCC's Sixth Assessment Report predicts that we will reach between half a metre and a metre of sea-level rise by the year 2100, depending on greenhouse gas emissions. As the small rise we have seen so far is already causing significant flooding, a metre would have catastrophic consequences in many coastal areas. And on top of that, there is a large, one-sided uncertainty: the IPCC cannot rule out more than 2 metres of sea-level rise by 2100, or even 5 metres by 2150. This could happen if large ice masses become unstable and start to slide rapidly into the ocean, a process that cannot be

Observed global sea level changes

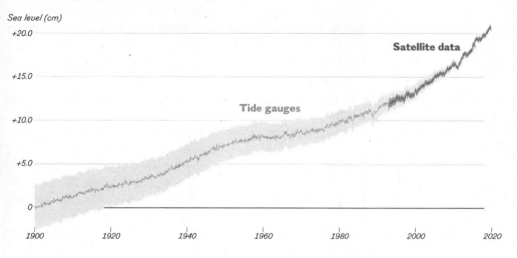

Sea level (cm)

Satellite data

Tide gauges

Figure 2

reliably simulated by current scientific models. Earth's history provides us with a stern warning: this kind of ice-sheet instability occurred repeatedly during the ice-age cycles of the past.

Even though the oceans are connected to form one world ocean, the surface is not flat and sea levels will not rise uniformly everywhere. In places such as Venice or New Orleans, coastal land is subsiding; meanwhile, it is rising in Scandinavia, where the ice sheets during the last ice age had been weighing the landmass down. But even the ocean surface itself can differ in various regions, for example due to the reduced gravitational pull of shrinking land ice masses or changes in prevailing winds or ocean currents.

Changing ocean currents

Global ocean circulation plays a key role in influencing climate by transporting heat. It is driven by wind and by density differences in the water (thermohaline circulation), as water density is determined by its temperature and how salty it is.

Wind patterns may change in response to global warming, subtly altering these wind-driven currents. However, a much more alarming disruption of ocean circulation looms over the thermohaline circulation, particularly in the Atlantic, where the ocean-spanning system of currents called the Atlantic Meridional Overturning Circulation (AMOC) – sometimes referred to as 'the conveyor belt of the ocean' – acts as a major heat-transporting system, bringing warm water from the tropics to the north Atlantic and cold water back to the southern hemisphere towards Antarctica (Fig. 3).

The AMOC is the main reason why the northern hemisphere is warmer than the southern hemisphere. The massive movement and release of heat makes the northern Atlantic and surrounding land areas – like much of Europe – several degrees warmer than they would otherwise be.

Climate models have long predicted that in the course of global warming the part of the northern Atlantic just south of Greenland will warm only a little, or perhaps even cool, because the AMOC is expected to weaken. Warming, combined with increases in rainfall and meltwater from Greenland, makes the surface water less dense, and, in turn, this water will not sink as deep as it did before. Disturbingly, this is happening now: the northern Atlantic is the only region of the planet that has cooled since the late nineteenth century (note the cold blob south of Greenland in Fig. 3).

This is especially troubling because the AMOC is known to have a tipping point beyond which it cannot be sustained and will collapse. The AMOC has collapsed a number of times during Earth's history, disrupting weather patterns around the globe.

Sea surface temperature change of the Atlantic Meridional Overturning Circulation

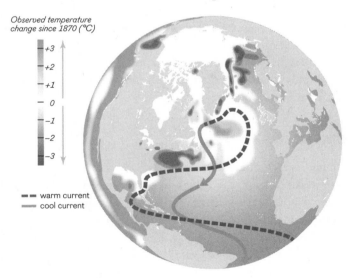

Observed temperature change since 1870 (°C)

- +3
- +2
- +1
- 0
- −1
- −2
- −3

▬ ▬ warm current
▬ cool current

Figure 3: The Atlantic Meridional Overturning Circulation has warm surface currents that flow north, and release heat into the atmosphere, before sinking down to 2,000–4,000 metres deep and returning south as a cold deep-water current. It moves nearly 20 million cubic metres of water per second, almost a hundred times greater than the Amazon River's flow.

In practice, it works like this. The AMOC brings salty waters from the subtropics to the northern Atlantic, helping to make the waters there dense enough to sink. When it weakens, less salt is transported north, which slows the AMOC even more by making the waters even less dense. At a certain point this becomes a vicious circle and the AMOC grinds to a halt.

Ever since Wally Broecker warned in 1987 of 'unpleasant surprises in the greenhouse' it has been feared that greenhouse gas emissions could push the AMOC past its tipping point. This is indeed one of the mega-risks of global warming. It is still unknown how far from this tipping point we are. On the one hand, climate models suggest the risk is small within this century. On the other hand, these models struggle to accurately represent AMOC stability and there are credible warning signs from observational data that we may be dangerously close.

Crossing this tipping point would not only cool north-western Europe, it would dramatically raise sea levels on the American East Coast, cause the collapse of marine ecosystems, reduce the carbon dioxide uptake of the ocean and warm the southern hemisphere even more, and it could also shift the tropical rainfall belts and disrupt the Asian monsoon. And we know from Earth's history that it takes about a thousand years for the AMOC to recover. /

We have enough ice
on Earth to raise sea levels
by 65 metres – about the height
of a twenty-story building.

2.12

Acidification and Marine Ecosystems

Hans-Otto Pörtner

Currently, the amount of carbon dioxide in the atmosphere is increasing about a hundred times faster than it was at the end of the last ice age when atmospheric CO_2 content increased by about 80 ppm (parts per million) in 6,000 years. And already, at approximately 416 ppm, it has reached its highest level in the last 2 million years.

Carbon dioxide produced by human activities not only penetrates the surface layers of the ocean but also, with the help of ocean biology and ocean currents, reaches its deeper layers. As much as on land, photosynthesis is the primary biological process involved in storing CO_2 in the ocean. The ocean has absorbed 20 to 30 per cent of human-made CO_2 emissions, dissolving, buffering and exporting the absorbed carbon dioxide to its depths. However, the capacity of the ocean (and land) to absorb CO_2 decreases as atmospheric CO_2 levels rise, partly due to global warming. At the same time, CO_2 not only enters the water but also the body fluids of marine organisms, for example the blood of fish, where it forms a weak acid. The enrichment of CO_2 in ocean water and the subsequent fall in pH is called ocean acidification.

Rising CO_2 levels and the resulting ocean acidification place marine organisms and ecosystems in peril, adding to the dangers of warming and oxygen loss. Acidity has already risen by approximately 30 per cent. Even if current efforts to reduce and finally stop CO_2 emissions are fully successful, some ocean acidification and the hazards it presents to marine organisms and ecosystems will remain long term.

Thus far, we have observed that ocean acidification often causes reduced calcification – leading to, for instance, the thinning or fracture of shells on organisms – or the destabilization of carbonate-based ecosystems such as coral reefs. Calcification processes are negatively affected in marine phytoplankton and foraminifera (single-celled organisms with shells), and also in corals and shell-forming animals such as mussels and sea urchins. There is a related decrease in growth and survival, as seen in echinoderms (i.e. starfish and urchins) and gastropods (i.e. snails and whelks). Corals, molluscs and echinoderms are especially vulnerable. Some fish show strong behavioural disturbances in response to elevated CO_2, but it is unclear to what extent these are long-lasting and have long-term consequences for ecosystems. So far there is limited evidence of whether organisms are able to evade functional impairments through adaptation. We do, however, know that all marine organisms are directly affected by changes in the chemistry of the ocean, while animals which feed on other organisms are also indirectly affected through changes in the food chain.

So the oceans are simultaneously warming and acidifying – and it is as yet unclear to what extent the effects of oxygen deficiency and ocean acidification are already influencing or exacerbating the impacts created by warming oceans. Complex organisms such as animals and plants thrive in a relatively narrow temperature range and so respond strongly to this warming. Warming is a key driver of ongoing shifts in biogeography, and mortalities are caused when extreme temperatures exceed the tolerance limits of species. Cold-blooded, water-breathing animals from the Antarctic (e.g. icefish) or the High Arctic (e.g. polar cod) live in particularly narrow temperature ranges and are especially vulnerable to the strong warming trends in polar areas as they have nowhere to go. In the warmest ocean regions, individual species and – in the case of coral reefs – even ecosystems are progressively wiped out as temperatures rise. It is increasingly thought that elevated concentrations of CO_2 and declining oxygen levels in seawater influence the temperatures that species can tolerate, which in turn influences biogeography, as well as the survival of species and populations. The state of ecosystems and their species composition is changing, and their future is unclear. Extinctions exclusively triggered by climate change are few, but projections indicate that species losses elicited by human-induced habitat destruction and environmental degradation will be exacerbated by climate change.

We must take action to strengthen the marine biosphere and enhance its capacity to absorb, convert and store CO_2. It is vital to restore healthy ecosystems and establish networks of protected areas that cover 30 to 50 per cent of the ocean. Doing this would support the effective conservation of biodiversity and increase populations of mangroves, seagrass meadows, salt meadows, seaweeds, whales and fishes, which play a valuable role in carbon cycling and in reducing acidification. Above all, we must prevent our world from overshooting the 1.5°C goal of the Paris Agreement. This has to be done to enable these marine species to thrive and to safeguard their role in climate mitigation and in providing food and coastal protection to humankind. /

Microplastics

Karin Kvale

Microplastics are not unlike manmade carbon dioxide (CO_2) emissions. They are largely derived from the same sources as carbon-based fuels. Like CO_2, they are long-lived pollutants and, like CO_2, they arise from ubiquitous human activities. Both are accumulating in the atmosphere and the ocean, thanks to a combination of individual human contributions (such as exhaust emissions from tailpipes in the case of CO_2, or degradation of tyres and brake pads in the case of microplastics), and collective ones (such as agricultural and industrial activities).

The global ocean, by virtue of covering 70 per cent of the Earth's surface and being downstream of nearly all the world's rivers, will be the ultimate resting place for a still poorly quantified fraction of the plastic that escapes from human control. Fifteen to 40 per cent of mismanaged plastic waste from coastal countries enters the sea each year, according to one estimate. Repeated sampling along beaches and in the open ocean suggests that the amount of plastic in the ocean is increasing (although not uniformly), but simple budget calculations indicate that it does not stay at the surface. Notably absent in surveys of the surface ocean are the smallest plastic fragments, pieces less than half a centimetre in length. In recent years these tiny plastics have been found at astonishing concentrations in the deepest reaches of the ocean, in offshore sediments along the continental shelf, and just below the ocean's surface, where light does not penetrate. Microplastic particles have been found at high concentrations around the Arctic Ocean, in locations far away from human settlements. Beyond these and other known hotspots, such as the mid-ocean gyres, the Mediterranean, the Sea of Japan and the North Sea, microplastic fragments have been found nearly everywhere that has been sampled. It is time to consider them a new component of seawater.

There are a number of significant risks from plastic contamination of seawater. Some of the local risks are familiar: we know big plastics such as shopping bags or fishing nets can result in entanglement, suffocation and starvation for whales, sea turtles, birds and other animals. The same thing happens with microplastics: copepods, tiny swimming predators, have been found with their legs ensnared in microplastic fibres and they fill their stomachs with microbeads. For potentially invasive but normally sedentary species, floating larger plastics are also an easy means to hitch a ride; likewise, surface ocean microplastics have been found that are coated with pathogenic bacteria and toxins. If shelled creatures such as oysters eat the coated micro-plastics (and they do), then the toxins can accumulate in their tissues and be passed on (along with the bacteria) to the person who eats them. Beyond this, eating plastic stresses sea-floor creatures and makes them less productive, which might damage the functionality of the whole ecosystem.

What might plastic in the oceans do on a global scale? It was recently revealed that microplastic particles in the atmosphere both scatter and absorb radiation, but it is still uncertain if the net effect is warming or cooling the planet. The ocean is a significant source of atmospheric microplastic because particles are tossed into the air by sea spray – so our continuing contamination of the ocean with plastics in the coming decades might pose some risk to climate commitments. However, modelling has shown that microplastics are potentially as disruptive to ocean oxygen levels as global warming. This is because tiny predators at the base of the food web some-times eat microplastics instead of phytoplankton (tiny plants – their normal food), which has consequences for the function of the ecosystem as a whole. Thus, while the absolute quantity of plastic in the natural environment is at present only a tiny fraction of the CO_2 problem, it might be having a disproportionately large influence on planetary function.

The unfolding plastics contamination problem looks set to intensify as petrochemical companies are targeting plastics as a major growth sector into the future. The cheap and convenient packaging omnipresent in our daily lives is plastered with claims of recyclability or compostability, encouraging us to buy more and feel good about it. But worldwide, our waste manage-ment systems remain both leaky and incapable of recycling many of the products on the market. The lack of regulation of package labelling encour-ages the phenomenon of 'aspirational recycling', which contaminates the waste stream and results in plastics that might otherwise be recycled ending up in landfill or worse. Meanwhile, a lack of accountability in multinational waste export systems results in a global flow of post-consumer plastics to countries where regulation and law enforcement are insufficient to prevent mismanaged waste entering the environment. Each country may solve its waste management problems or regulate plastic production independently, but there grows a tragedy of the commons in our oceans and atmosphere that, to be solved, requires urgent intergovernmental coordination. /

Fresh Water

Peter H. Gleick

Water connects us to everything on the planet: our food and health, the well-being of the environment around us, the production of goods and services, and our sense of community. And water is central to the climate – the entire hydrologic cycle of evaporation, precipitation, run-off and all the stocks and flows of water around the world lies at the heart of our climate system. In turn, our use of water has implications for the climate crisis. As long as fossil fuels power our energy systems, the use of water will also mean the production of greenhouse gases. For example, as much as 20 per cent of California's electricity and a third of the state's non-power-plant natural gas is used for the water system, including heating the water we use in our homes and businesses. Decarbonizing our electricity sector and getting fossil fuels out of our homes can help break this energy–water–climate link.

Humans are already changing the climate, which means we are already fundamentally changing our water system. As temperatures rise, evaporation of water from soils and plants is increasing, driving more water into the atmosphere and leading to more intense rainfall in some places and worsening droughts in others. Snow in the mountains – a leading source of water for billions of people – is falling as rain or melting sooner than usual, worsening flooding and reducing water availability in warm periods. Rising seas are pushing salt water into coastal fresh-water aquifers, making them unusable for drinking water. Warming and disappearing rivers are harming fisheries and other aquatic ecosystems.

These impacts have long been predicted by climate scientists and now they are appearing as the world has dithered and procrastinated and argued. And these impacts are worsened by the fact that our water problems are already profound, even without the challenge of climate change. Billions of people still lack safe and affordable drinking water and sanitation. Industrial and human waste pollutes our waterways. Human withdrawals of water are harming aquatic ecosystems around the world. Violent conflicts over water are increasing in number and severity, with recent riots in India and Iran over drought and water availability, disputes in sub-Saharan Africa between farmers and pastoralists over access to land and water, and the growing use of water as a weapon or tool of conflict. More and more regions are at or approaching 'peak water' – the point where taking more water from the environment is impossible, physically, economically or environmentally. Some rivers are literally entirely consumed by human use, such as the Colorado River, shared by seven states in the US and Mexico. Many groundwater basins are over-drafted in China, India, the Middle East and the United States, leading to land subsidence, higher and higher costs for pumping and unsustainable agricultural production. These peak water limits, combined with the growing impacts of climate change, mean we must rethink our relationship with water.

The good news is that a new approach is possible, a 'soft path' for water that can move towards both addressing global water problems and reducing our vulnerability to climate change. The soft path requires moving away from sole reliance on hard, centralized infrastructure like dams, aqueducts and large water treatment plants to a more integrated reliance on treat-ment and reuse of water, better capture and use of stormwater, smaller-scale distributed water systems and, when economically and environmentally appropriate, the desalination of brackish or ocean water. It also calls for us to reconsider how we use water and to maximize the benefits water provides while minimizing the amount of water and energy we use. The soft path is a more equitable path, acknowledging the value of healthy ecosystems and healthy communities. We must address gross inequities in our water and energy system and reduce the disproportionate impacts that climate change will have on already marginalized and vulnerable communities. Providing safe water and sanitation for all, protecting and restoring damaged ecosys-tems, and building resilience against now-unavoidable climate impacts will help us address these inequities and move to a more sustainable water future. /

2.15

It is much closer to home than we think

Greta Thunberg

> Environment ministers from almost 200 nations agreed late tonight to adopt a new United Nations strategy that aims to stem the worst loss of life on Earth since the demise of the dinosaurs. With a typhoon looming outside and cheering inside the Nagoya conference hall, the Japanese chair of the UN biodiversity talks gavelled into effect the Aichi targets, set to at least halve the loss of natural habitats and expand nature reserves to 17% of the world's land area by 2020 up from less than 10% today.

These words come from an article by Jonathan Watts published in the *Guardian* in 2010. It ends with a quote from Jane Smart, the then director of conservation policy at the International Union for Conservation of Nature: 'There is a momentum here which we cannot afford to lose – in fact we have to build on it if we stand any chance of success in halting the extinction crisis.'

One of the non-binding commitments that was signed on that late-autumn evening in Japan was to 'halve the rate of forest loss by 2020'. But by the time the deadline for the Aichi goals came round, it was clear that the world had failed to reach every single one of its targets. This may sound like a unique failure, but what happened to the efforts of the UN in 2010 is far from an isolated event. In 1992 the United Nations Environment Programme stated in Agenda 21 its aim to combat deforestation. The New York declaration of 2014 pledged to reverse deforestation by 2030. The UN pledged as one of its Sustainable Development Goals of 2015 to 'protect, restore and promote sustainable use of terrestrial ecosystems, sustainably manage forests, combat desertification, and halt and reverse land degradation and halt biodiversity loss'. These plans are all on a clear path to failure, if they haven't failed already.

There is a pattern here. Every once in a while, our leaders make a few pledges and set a variety of vague, non-binding, often distant targets. Then, as soon as they have failed to reach them, they immediately set some new ones. And on it goes. It may seem absurd, but it most definitely works – if your aim is to maintain business as usual, economic growth and high popularity ratings. Since the level of public interest in and awareness of these failed climate and biodiversity pledges is close to non-existent, and since the media long for positive news as part of their policy of both-side reporting – It can't be all doom and gloom! – the overall message that is conveyed, if any, is that action is being taken. It may not always go all that well, but hey, they are actually trying really hard and there has definitely been a lot of progress made, so stop being so negative all the time!

When the media in wealthy nations cover the problem at all, they don't show us pictures of what is causing it, for example an SUV factory in Germany, a dairy farm in Denmark, a shopping mall in Seattle, a clear-cut forest in Sweden, or a container ship arriving in Rotterdam filled with plastic toys, sneakers and smartphones. Instead we get pictures of polar bears in the Arctic, melting glaciers in Antarctica, the Greenland Ice Sheet collapsing, illegal loggers in the Amazon or thawing permafrost in the remote wilderness of northern Siberia. These are not exactly your average, everyday events. The result is that we forget that the climate and ecological crisis is happening everywhere, all the time. It is much closer to home than we think.

The permafrost, for instance, is not just melting on the shores of the Arctic Ocean. It is also melting in Italy, Austria and other mountainous, alpine nations. In Switzerland, the village of Bondo was levelled to the ground in 2017 by a huge landslide, partly caused by melting permafrost at high altitude.

The same aggressive and irresponsible deforestation that is happening in the Amazon is also happening in the northern boreal forests. And those nations that have not already cut down their woodlands are seeing an unprecedented transformation of their local geography as the last natural forests are clear-cut and replaced by plantations in what can only be described as a biodiversity catastrophe.

The earth and soil across the planet are steadily being degraded, losing resilience and nutrition in a process that is partly driven by a warming climate, deforestation, monocultures and common land-use policies for agriculture and forestry which are not primarily intended to feed us or to care for our needs but to make as much money as possible.

But it is not only money that is driving our ongoing slaughter of nature and biodiversity. The ecological crisis is also – ironically – being fuelled by our pursuit to cut emissions of CO_2. You see, one of the most effective ways of lowering our emissions is to exclude them from official territorial statistics. And burning biomass for energy does exactly that. At least on paper.

Since trees grow back, we have decided that cutting them down and shipping them halfway around the world in order to burn them is considered renewable. A 2018 study estimated that it would take 'between 44 and 104 years' for forests to recapture the carbon released by burning wood — if they ever could, given their increasing exposure to soil erosion, extreme temperatures, fire and disease.

The decision to consider burning biomass as 'renewable' was made long before the timeframe set out by the Paris Agreement began, in what has been called a blind spot of the Kyoto Protocol dating back to 1997. This loophole allows you to create lots of very carbon-intensive energy – burning wood releases even more CO_2 per energy unit than burning coal – while claiming that the emissions are going down and radical action is being taken, just like magic.

Entire nations' climate policies are based on this loophole. In the UK, for instance, the Selby Drax power plant is the biggest single emitter of CO_2, but its biomass emissions are excluded from the UK's national statistics. The EU would not stand a chance of reaching its climate targets without a wide use of clever creative accounting like this. In 2019, 59 per cent of the EU's so-called renewable energy came from biomass. 'To be perfectly blunt with you, biomass will have to be part of our energy mix if we want to remove our dependency on fossil fuels,' the executive vice-president of the European Commission told reporters at the end of 2021.

All this burning of course requires wood – lots and lots of wood. The tree pellets used in the power plants are said to come from forest industry residues, sawdust and leftovers from the manufacturing of long-lasting wood products such as furniture and houses. However, this is often far from true. Evidence from Canada, Finland, Sweden, the US and the Baltic states shows that not only are whole trees being cut down to be burned, in many cases they come from old-growth and primary forests – forests that have never been cut down before. We do not need Sherlock Holmes to tell us why. There is money to be made; there are climate targets to be met. It's all perfectly legal and in accordance with every international body of law and power you can think of. When I visited the Drax power plant I was told that four ships full of pellets came in every week, and seven trains every day. That is an awful lot of sawdust and leftover branches.

So when we say that our leaders have not been taking any climate action during the last thirty years, we could not be more wrong. In fact, they have been very busy. But not in the way that you might think – or hope. They have spent this time actively delaying action, creating frameworks full of loopholes that will benefit their own, national short-term economic policies –

Next pages:
The Batagaika crater in north-eastern Siberia. At half a mile wide (and growing), it is the largest of many lakes and craters formed across the Arctic by the ground collapsing as permafrost laced with buried ice thaws.

and their own popularity. And as long as the level of awareness is as low as it is today, they will continue to get away with it.

At COP26 in Glasgow in 2021, after the complete failure of the 2010 Aichi goals had been revealed, without any media coverage whatsoever, our leaders pledged that they would once again end deforestation, this time by 2030. In the final text – the Glasgow Agreement – the Conference of the Parties also mentioned the f-word (fossil fuels) for the first time ever and, furthermore, they decided that instead of updating the National Contributions every five years, they would now be updated every year. Needless to say, these vague, non-binding announcements generated lots of hopeful media coverage.

In the weeks that followed, however, Brazil reported record levels of deforestation in the Amazon rainforest and the EU voted in favour of a new Common Agricultural Policy which will effectively put its targets for staying in line with the Paris Agreement out of reach. China opened even more coal power plants and the US administration auctioned off a 90-million acre area in the Gulf of Mexico for oil and gas exploration – a sale that could ultimately result in the production of up to 1.1 billion barrels of crude oil and 4.4 trillion cubic feet of fossil gas. To add to this farce, the EU concluded that – in spite of what was agreed upon in Glasgow – the EU will not update its climate targets in time for COP27 in Egypt.

These events were largely followed by a massive media silence. No one was held accountable. No headlines. No front pages. The focus was gone. Again. This is exactly how you create a catastrophe. /

The same aggressive and irresponsible deforestation that is happening in the Amazon is also happening in the northern boreal forests.

Wildfires

Joëlle Gergis

Along with the burning of fossil fuels, humans have spent centuries clearing the land. This has drastically altered the concentration of naturally occurring heat-trapping greenhouse gases like carbon dioxide and methane, imbalancing natural processes that have regulated the Earth's temperature throughout its history. Widespread deforestation has modified the Earth's ability to soak up excess carbon as more and more of the land surface has been converted from natural ecosystems like forests or wetlands into agricultural crops and concreted urban areas. Today, forests cover only around a third of the global land surface, with more than half found in Brazil, Canada, China, Russia and the United States.

Long-term climate trends, local weather conditions and land-use management practices have led to changes in fire activity around the world. When large wildfires occur, burning vegetation releases vast quantities of carbon into the atmosphere. Because their behaviour is a complex interaction of climate, weather, landscape and ecological processes, wildfires are difficult to monitor and predict. This means that they can influence climate change in unexpected, non-linear ways that are currently not well captured by climate models. In addition to their impacts on emissions, wildfires also generate air pollution that can be dangerous to human health, contaminate water quality in burnt catchments, and destroy habitats and wildlife needed to sustain global biodiversity. An example of these complex interactions is the Amazon Basin in South America: a huge carbon sink that is now drying out because of climate change, while being burned and felled to support industrial-scale agriculture. Not only does this pose a threat to the stability of the global carbon cycle, but it also threatens to destroy one of the world's remaining biodiversity hotspots.

While there have always been naturally occurring wildfires, climate change is increasing the Earth's temperature and altering global circulation patterns that influence regional weather and climate conditions. This means that all wildfires are now occurring against the background of hotter temperatures and more erratic rainfall which falls in less defined seasons. Prolonged heatwaves and droughts can lead to warmer temperatures, below average rainfall, low humidity, reduced soil moisture and wind changes that can trigger wildfire outbreaks. Higher temperatures increase the vapour pressure deficit: the evaporative force that regulates the amount of moisture released into the atmosphere from the Earth's surface and vegetation. After sustained hot, dry and windy conditions, vapour pressure deficits intensify, causing soils and vegetation to dry out, turning normally moist landscapes into burnable fuel. Fires can be started by natural ignition sources – such as lightning – or humans can cause them, either accidentally (through fallen power lines, for example) or deliberately.

Increases in the frequency and severity of fire weather have been observed across parts of the globe particularly since the 1970s. Wildfires have intensified across southern Europe, northern Eurasia, the western United States and Australia. The IPCC reports that evidence for dangerous fire weather conditions linked to human-caused climate change is strongest in regions like the western United States and south-eastern Australia, where formal attribution studies have been conducted. Recent research shows that human influence on fire weather conditions has already emerged above natural variability over close to a quarter of the globe, including in places like the Mediterranean and the Amazon. Models show that the area of heightened wildfire risk increases with higher levels of warming, with the area affected at 3°C above pre-industrial levels doubling compared to 2°C of global warming.

Global warming is already resulting in more extreme and longer fire seasons, which are extending into regions not historically considered fire-prone. This is particularly the case during warmer summer months, but pronounced warming in some regions has led to an elongation of the fire season with wildfires now possible throughout the entire year, particularly during severe droughts. For example, in 2019, Australia's hottest and driest year on record, usually wet, subtropical rainforests burned during winter, incinerating over half of the country's ancient Gondwana-era rainforests in one fire season. Although eastern Australia's eucalyptus forests are among the most fire-prone in the world, typically only 2 per cent of them burns during extreme fire seasons. But in 2019–20, 21 per cent of Australia's temperate forests burned in a single event, setting a new global record for the sheer enormity of the blazes. This marked increase in the area burned during extreme fire seasons around the world led to the definition of 'megafires' to describe an individual wildfire or wildfire complexes that engulf more than 1 million hectares. Australia's record-setting megafires burnt a phenomenal 24 million hectares, releasing over 715 million tonnes of carbon dioxide in a single bushfire season; more than all the emissions the country releases in an entire year. A staggering 3 billion animals were killed or displaced by the immense scale of habitat destruction.

Increasingly destructive wildfires have also been observed in the northern hemisphere in recent years. In 2021, the Pacific Northwest region of the United States and south-west Canada experienced extreme heatwaves that shattered historical temperature records. Temperatures in the Canadian town of Lytton in British Columbia soared to 49.6°C on June 29, 2021, before wildfires destroyed close to 90 per cent of its buildings. It was the first time such extreme, desert-like temperatures were experienced so far north anywhere on the planet. California also recorded the single largest wildfire in the state's history, with the Dixie Fire burning more than 400,000 hectares over three months. Further north, record-breaking heat and drought saw the Arctic forests and peatlands of Siberia and eastern Russia ignite, with plumes of smoke reaching the North Pole for the first time in recorded history. The European Union's Copernicus Atmosphere Monitoring Service has estimated that in 2021 wildfires emitted a record 6.45 billion tonnes of CO_2 – equivalent to over double the EU's total carbon dioxide emissions for the year.

The warmer the planet gets, the more frequent and extreme wildfires will become. As the fire season extends into previously cool regions and seasons, more forests will burn, releasing vast amounts of carbon back into the atmosphere, which will further amplify warming. This positive feedback loop is like pressing down on a car's accelerator. Complex, non-linear processes like wildfire dynamics (including lightning strikes) are challenging to monitor and are also difficult to mathematically describe and simulate using state-of-the-art climate models. As a result, carbon cycle feedbacks that amplify warming, like those associated with wildfires, are currently either missing entirely or incompletely represented in the latest generation of climate models. This means scientists don't know exactly how feedback loops will influence the trajectory of future warming. But we do know that the higher the level of warming, the greater the risk of triggering self-reinforcing feedbacks that cause the climate to become unstable. If the world manages to limit warming at well below 2°C, the risk of destructive wildfires will fall, allowing our land ecosystems to rebalance the global carbon cycle and help restore life on our planet. /

The Amazon

Carlos A. Nobre, Julia Arieira and Nathália Nascimento

The Amazon basin encompasses the largest area of rainforest in the world – almost 6 million square kilometres. A critical element of the Earth's climate system, it plays a vital role in global water cycles and regulating climate variability. The Amazon forest is responsible for an estimated 16 per cent of the carbon dioxide removed every year from the atmosphere through photosynthesis, helping store 150–200 billion tonnes of carbon in soil and vegetation. In addition, through evapotranspiration – the forest's capture and release of water into the atmosphere – it works as a giant air-conditioner, lowering land surface temperatures and generating rainfall. This cooling – up to 5°C in forested areas – is essential for minimizing the effects of seasonal droughts and heatwaves in the region.

In recent decades, however, the structure, composition and functioning of the Amazon rainforest have begun to change. The temperature in the region has increased by an average of 1.02°C between 1979 and 2018, and 2019–20 was the second-warmest year since 1960, with an increase of 1.1°C. Over the past twenty years, we have also observed a decreasing amount of moisture in the atmosphere over the south-east Amazon rainforest, especially during drier months (June to October). Both the warming and drying of the air are the result of manmade climate change, aggravated by land-use change – particularly the expansion of farming into forested areas, the burning of agricultural waste, and an increase in forest fires (which, in the Amazon, generally originate from fires escaping managed pastures). Burning biomass emits black carbon aerosols, which have decreased cloud cover over the forest, increased surface warming, and ultimately caused the atmosphere over the Amazon to dry out. Deforestation has played a significant role too, as it has decreased evapotranspiration. Climate variability has also increased the frequency of weather extremes afflicting the Amazon, particularly droughts and heatwaves. Even hotter temperatures and more droughts are expected by the end of the century for Amazonia if greenhouse gas emissions reach very high levels (more than 1,000 parts per million CO_2 equivalent; the current concentration is 414 ppm), which would result in more than 150 days with temperatures above 35°C per year – more than double the annual average for the past two decades of 70 days per year.

Put simply, the situation in the Amazon today appears dire. Approximately 17 per cent of Amazonian forests have been deforested for human use. That is strongly linked to the construction of roads – 95 per cent of deforestation takes place within 5.5 kilometres on either side of roads. And at least an additional 17 per cent of forests have been degraded by selective logging, collection of wood for fuel, fire, and wind damage. In Brazil, this deforestation and degradation is mostly driven by the expansion of pasture and croplands, although in other Amazonian countries the forest has been razed for mineral and oil extraction. Deforestation amplifies climate impacts to the extent that the area may warm by more than 3°C and see a 40 per cent reduction in precipitation between July and November in eastern Amazonia. This warmer and drier climate combines disastrously with the accelerating fragmentation of the Amazon, which has left stretches of forest exposed to increased direct sunlight, higher soil temperatures and increased wind exposure, all of which vastly increases the forest's vulnerability to fires. The result of these fires is a further rise in tree mortality and carbon emissions, setting in motion a feedback loop which is accelerated by extreme weather events such as the intense drought that followed the powerful 2015–16 El Niño episode, whereby 2.5 billion trees perished, emitting approximately 495 million tonnes of carbon dioxide. (For perspective, this is close to the annual carbon dioxide emissions of an industrialized nation like Australia, France or the UK.)

A large portion of the forest is now on a knife edge. The Amazon could be heading for an inflection point, starting a savannization process whereby its vegetation will assume the characteristics of a degraded savannah, seeing a proliferation of grasses and woody plants as it adapts to a longer dry season (with altered seasonal leaf flushing, or sprouting) and increasingly frequent fire events (with new post-fire resprouting strategies). We believe that this transition in central, southern and eastern Amazonian forests will likely occur when the temperature increase in the region approaches 4°C, or as a result of reduced rainfall and longer, harsher, dry seasons, or even once deforestation affects 40 per cent of the total forest area in the Amazon basin. When we consider all of the major ways humans are changing the Amazon – through deforestation, increased fires, global warming and increasingly high concentrations of CO_2 – it seems possible that up to 60 per cent of the Amazon forest could disappear by 2050. The consequences of this massive loss of forest will be irreversible and far-reaching, impacting human well-being in many ways: for instance, our food provision will be threatened by the reduced functioning of its critical ecosystem services, and it will no longer provide us with a 'green barrier' against the spread of infectious disease. We will also see devastating impacts on biodiversity through habitat loss, and the disruption of reciprocal interspecies interactions such as pollination and seed dispersal.

There are an increasing number of signs that the Amazon forest is perilously close to this tipping point. The dry season in the southern Amazon is already three to four weeks longer than it was in the 1980s – mostly in deforested areas – while rainfall has decreased by 20 to 30 per cent, and it is 2–3°C warmer. There has been a marked reduction of evapotranspiration and water recycling by the forest, and some areas of the forest have begun to emit more carbon than they were storing. The Amazon basin as a whole is approaching a point where it will become a carbon source instead of a carbon sink.

We project that the Amazon could be transformed into a degraded savannah – or a degraded secondary forest, with fewer species and with more open canopy – between 2050 and 2070, and that this transformation would affect 60–70 per cent of the rainforest. If the forest reaches this tipping point, more than 200 billion tonnes of carbon dioxide could be released into the atmosphere, making it essentially impossible to reach the Paris Agreement target of limiting global warming to 1.5°C. And the loss of biodiversity would be profound, pushing several thousand endemic plant and animal species to extinction, including the black-shouldered opossum, the pied tamarin and the Kaapori capuchin. And, for the people who call this region home, the Amazon's transformation into a savannah, coupled with soaring greenhouse gas emissions, would mean a future where, for nearly 50 per cent of the year, daily maximum temperatures would combine with high humidity and exceed the physiological heat-stress threshold of the human body, posing a direct mortal threat. /

2.18

Boreal and Temperate Forests

Beverly E. Law

The world's forests are broadly classified as boreal, temperate or tropical, depending on their latitude and climate features (Fig. 1). Boreal and temperate forests cover about 43 per cent of the global forest area, almost as much as the area covered by tropical forests. Although tropical forests harbour more species of animals and birds, the number of subspecies increases in the harsher environments of higher latitudes. Boreal forests, which exist in a circumpolar band across Russia (73 per cent), Canada and Alaska (22 per cent) and the Nordic countries (5 per cent), have evolved under very cold climate conditions with a short growing season. Their evergreen needleleaf tree species – firs, pines, spruce – dominate, along with the hardy deciduous larch. Temperate forests are found between latitudes of 25 to 50 degrees in the northern and southern hemispheres and range from rainforests that grow in a mild, wet climate, like the evergreen needleleaf forests of coastal Canadian British Columbia, to deciduous broadleaf forests in areas with winter temperatures that reach below freezing.

The effect of climate change on forests varies across landscapes and regions, depending on the relative changes in temperature and precipitation, the resilience of forest ecosystems and the vulnerability of individual species. Boreal forests play an important role in climate mitigation and biodiversity protection because they are so vast. They provide habitat for long-distance mammal and fish migrations, abundant populations of large predators, and 1–3 billion breeding migratory birds. They store 367–1,716 gigatonnes of carbon (GtC), most of it in soils. Only about 8–13 per cent of boreal forest area is truly protected, and about half of the boreal forest area is managed for wood production, mostly in Russia. Harvesting has significantly reduced the extent of old forests, destroying habitats, biodiversity and resilience. Combined with increasing disturbance from wildfires over the past thirty years, logging has also reduced the trees' accumulation of carbon. As the climate continues to warm and the area

The global distribution of forests, by climatic domain

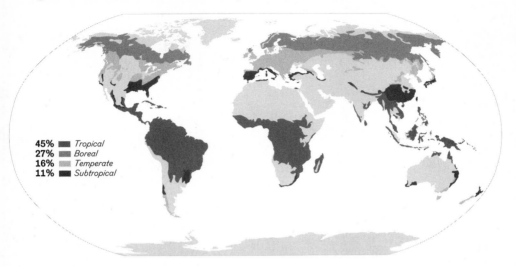

45%	*Tropical*
27%	*Boreal*
16%	*Temperate*
11%	*Subtropical*

Figure 1:
The global forest area in 2020 is estimated at 4.06 billion hectares, which is 31 per cent of the total land area.

burned in wildfires increases, boreal forests' ability to store and accumulate carbon may be further reduced. However, the boreal forest zone is shifting northwards, and the greening area is three times that of the browning caused by mortality in the warmest margins of the biome, which could compensate for the carbon losses from wildfire. The interconnected impacts of climate change, harvesting, homogenization and land-use conversion (such as stripping forests for oil sands mining) have also accelerated biodiversity losses throughout the boreal region. For example, North America's boreal forests are home to herds of migratory caribou that can travel 500–1,500 kilometres in an annual migration, and to migratory and non-migratory wolves. Loss of migratory corridors that allow these animals to move to a better climate and habitat are a major threat to their survival. Sadly, all populations of caribou in Canada are now listed as endangered or threatened.

By contrast with boreal forests, temperate forests hold a large variety of ecotypes. The humid rainforests can be found along the western coast of North America, where conifers dominate, and the wet southern tip of South America, where deciduous broadleaf forests are populated largely by beech species. Temperate forests in the Appalachian Mountains and north-eastern US are like forests in south-eastern and central Europe, composed of broadleaf tree species (oak, ash, beech, elm and maple) and evergreen needleleaf species (pine, spruce, fir). Temperate forests have some of the highest carbon densities in the world. Old temperate forests more than eighty years old with high carbon densities and multiple canopy layers provide critical habitats

British Columbia's managed forests transformed from carbon sink to carbon source in 2002

Figure 2:
Computed from forest growth minus decay, slash pile burning, wildfires, and decomposition of harvested wood products.

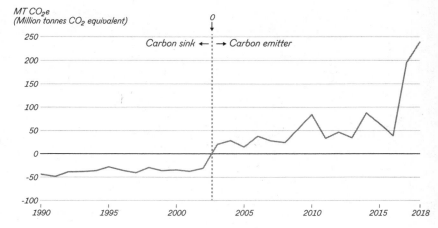

for many threatened and endangered species and have high biodiversity. However, like those in the boreal region, temperate forests are heavily logged – so much so that harvest-related emissions are more than seven times the emissions from all natural causes combined (fire, insects and wind damage).

Globally, the forests of the northern hemisphere regions tend to have larger carbon sinks, with a net ecosystem exchange of forest carbon amounting to about 1.44 GtC per year. The global mitigation potential of natural forest management solutions in boreal and temperate forests combined has been estimated to be about 8.3 GtC by 2100 (0.11 GtC per year). These are rough estimates because of limited data in the remote boreal regions. A majority of this is in temperate forests. In the western US, temperate forests identified as having moderate to high carbon densities and low to moderate vulnerability to drought or fire under future climate would account for about eight years of the region's fossil fuel emissions, or 18–20 per cent of the global mitigation potential of natural forest management solutions for boreal and temperate forests combined by 2100.

Worryingly, tipping points have already occurred in carbon-rich forests around the world, turning them from sinks to sources. Forests in British Columbia switched in 2002, thanks to the combined effects of wildfire, harvesting and the proliferation of insects, particularly the mountain pine beetle and spruce budworm (Fig. 2). The beetles bore through bark to lay eggs, and their larvae kill the trees by consuming and blocking the flow of nutrients. They have benefitted from the changing temperatures in British Columbia's interior mountain ranges, which have risen faster than the global average, especially in winter. The warmer climate has allowed more beetles to survive the cold months and multiply, leading to higher rates of tree mortality. The warmer winters have also allowed the beetles to cross the continental divide, threatening eastern forests in Canada and the US. The warmer, drier conditions, the declining level of snowpack (slow-melting packed snow that provides trees with water in dry summers), as well as the additional dead wood produced by the beetles' spread, have led to more wildfires throughout the region. Astonishingly, British Columbia's forests are now a larger source of carbon in the region than reported emissions from the energy sector, according to the BC Provincial Inventory Report 2021.

Natural forests in boreal and temperate zones can do much to help mitigate climate change and biodiversity loss – but they can only do so when they are allowed to grow for longer. By contrast, the 'sustainable forestry' that is rife within these regions is much less effective, as its emphasis is to provide a sustainable supply of wood, rather than to support sustainable ecosystems. Industrial forestry harvests young trees before they reach their potential biomass carbon. Over time, these trees store less and emit more carbon than do old forests. Limiting forest carbon in this way will not achieve a sustainable climate.

Instead, we need to let mature and old forests grow, and to substantially increase the time between harvests on existing forestry land. This will do the most to increase carbon storage and accumulation. Reforestation and afforestation do also help, but not as much (Fig. 3). Forest protection keeps carbon in the forests and out of the atmosphere, and protects biodiversity and water sources, as shown in temperate wet forests. If we are to mitigate climate change and protect biodiversity, avoiding further loss and restoring carbon- and species-rich forest ecosystems is vital. /

Figure 3: Allowing mature and old forests to accumulate carbon by halving harvests on public land, and doubling the time between harvests on private land, contributed the most significant NECB (net ecosystem carbon balance) or carbon accumulation by 2100.

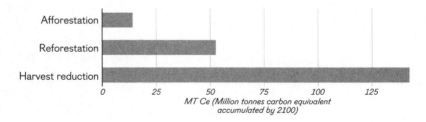

Climate mitigation strategies in temperate forests in the US Pacific Northwest

Terrestrial Biodiversity

Adriana De Palma and Andy Purvis

Biodiversity is the variety of life on Earth, and it is essential for our survival. It provides us with clean air and fresh water, natural control of pests and diseases, healthy soils, food, fuel and medicines; it even promotes mental health. Biodiversity helps ecosystems to slow climate change (by taking CO_2 out of the atmosphere) and to cope with it (giving ecosystems more ways to adapt). It also makes warming easier for us to bear – for example, trees and other plants in cities lower temperatures, helping to protect us against heatwaves.

At the local scale, biodiversity is naturally highest in places that reliably have enough sun, rainfall and soil for structurally complex forests to grow, with enough different niches and enough biomass to support a profusion of different species. At the landscape scale, biodiversity is naturally highest in mountainous parts of the wet tropics, where climates have changed little over millions of years. In these parts of the world, many different climate regimes can be found close together, each one boasting its own set of species that have adapted precisely to the stable conditions and to one another. Many of the species in these ecosystems can be found nowhere else on Earth. The same is true of remote tropical islands; the few species that managed to reach them during Earth's deep history generally found themselves without many rivals, meaning they had sufficient space and time to evolve into many distinct lifeforms. By contrast, in flatter landscapes, the same kind of climate – and hence the same kind of natural ecosystem – can extend for hundreds or even thousands of kilometres, often leading to landscapes with far fewer species overall. Cooler places also tend to have fewer species, because less plant growth is available to power food webs, as do harsh environments, because most kinds of organism lack the adaptations to survive extreme cold, heat, aridity or natural fires.

This natural global pattern of biodiversity on land reflects processes that have operated over many millions of years – but most of us now live in places where biodiversity loss has been driven by three waves of human-caused change.

The first wave of human-caused change occurred far back in prehistory – at the time of our first contact with many species around the world. Our hunting helped to wipe out many large species of mammals and birds (the megafaunal extinction), while the rats and cats we spread to countless islands killed off many bird species which, evolving in a predator-free environment, had become flightless.

Settled agriculture began to replace nomadic lifestyles around 10,000 years ago, beginning the second wave of change. We started deliberately reshaping ecosystems to better meet our needs for food and materials, transforming the world into an easier place for us to live. The resulting agricultural landscapes were typically a complex patchwork of different crops (often changing each year), fallow, grazing land and more natural areas. This heterogeneity, and the fact that only a small fraction of the landscape's biomass was harvested, enabled many species to persist alongside humans. Many Indigenous peoples around the world still manage their land like this today, and moves towards more nature-friendly agriculture typically adopt many of the same features.

From the mid-eighteenth century, linked revolutions in farming and manufacturing ushered in the third wave of human-caused change: ecosystem management turned into ecosystem domination. The resulting population boom required more land for farming and more wood for construction and fuel, driving more deforestation. We now use fossil fuels to power almost every sector of our economies, producing CO_2 far more quickly than ecosystems can absorb it. Our imprint on around 75 per cent of land is visible even from space, and many regions face multiple intense threats (Fig. 1). Most obviously, we farm over 30 per cent of land, increasingly intensively; an area equal to the whole of North America and South America combined is used just to produce livestock.

The impacts of these threats to nature depend very much on where you look. In the few regions with no history of settled agriculture, hunting is often still the main driver of biodiversity loss, so the impacts echo the first wave's megafaunal extinction. For example, in remote parts of many tropical rainforests, wild-meat hunting has largely or totally removed the large mammals; poaching likewise threatens the large mammals in many legally protected areas. Where subsistence farming is widespread, the impacts are more like they were during the second wave: there is a local pulse of biodiversity loss when natural ecosystems are converted to simpler agricultural ones, but the resulting landscapes – a complex,

The number of severe threats to biodiversity around the world

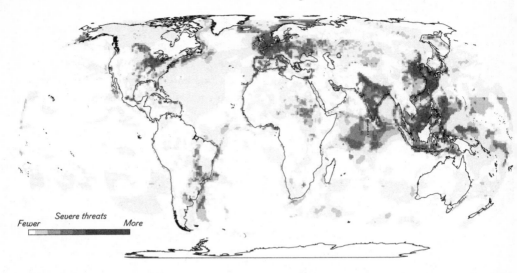

Severe threats
Fewer   More

Figure 1:
Throughout the world's lands and oceans 16 driver variables of biodiversity change – including climate change, human use, human population, and pollution – are mapped by number and intensity in 2020.

changing jumble, kept free of agrochemicals – can maintain moderate levels of biodiversity.

In places where the third wave is well underway – the darker regions in Fig. 1 – the fabric of life gets worn so thin that it can fall to pieces. Intensively farmed land is so structurally simple that there are very few niches for wild species. Harvesting so much of the ecosystem's biomass leaves too little behind to support complex food webs: global vegetation biomass and tree cover are both now only around half of what they would naturally be, and the world's cattle easily outweigh the more than 5,000 species of wild mammals put together. Meanwhile, agrochemicals make most farmland (and many of the waterways into which they drain) a harsh environment which most species cannot survive: ironically, the species best adapted to pesticides are the pests themselves, whereas thousands of species that could have contributed to natural pest control, pollination and soil fertility are often devastated. These include many wasp species whose larvae literally eat pests alive; the bees, flies, beetles, moths and butterflies that most crop species need for pollination; and the earthworms and many insects such as springtails that recycle nutrients from dead plants to fertilize the soil. Although intensive farming has increased agricultural production enormously, almost all of nature's other benefits to people have declined globally over the last fifty years.

The newest threat to nature is human-caused climate change. Its impacts have, so far, been small by comparison, but we can already see species trying to escape warmer temperatures. High-latitude species are being found further and further towards the poles, boreal forests are starting to expand into what used to be tundra, and mountain species are being found at ever higher altitudes. In the last fifteen years, human-caused climate change claimed its first known victim: the Bramble Cay mosaic-tailed rat (*Melomys rubicola*). Found only on a tiny low-lying island at the northern tip of Australia's Great Barrier Reef, and last seen in 2009, this rodent probably succumbed to repeated flooding as the sea level rose and storms became more frequent.

Although climate change is not yet driving anything like as much biodiversity loss as humanity's land use, the alarm bells are ringing. High regional biodiversity only arose where the climate was stable; unless global warming can be slowed very soon, it is sure to cause many more casualties. Mountaintop species will see their niches simply disappear. In flatter areas, rapid warming will mean that species have to move across the landscape in order to track their preferred climates: not all will be able to keep up. Crops will also need to move into more temperate areas that were previously wild, driving additional waves of habitat loss, and many currently productive regions will become too arid for reliable farming. This means it will not only be nature that has to move rapidly, but many millions of people too. Biodiversity loss may even form a vicious circle with climate change: ecosystems that have lost biodiversity store less carbon and are less able to cope with extreme weather events and other climate change.

But a sustainable future is still possible, if we are prepared to make more room for nature and to demand less from it. If we are to minimize the number of extinctions over the coming decades (we cannot stop them all) and avoid the worst impacts of warming, the regions with rich concentrations of unique species need to be cherished, their ecosystems restored and protected. Restoring high-carbon, high-biodiversity ecosystems is a true – and urgent – nature-based solution. /

Insects

Dave Goulson

I have been fascinated by insects all my life. When I was just five or six years old, I gathered yellow-and-black-striped caterpillars from weeds along the edge of my primary school playground, carried them home in my lunchbox and fed them until they eventually transformed into glorious red-and-black moths (you might recognize them as cinnabar moths). I was hooked and I have been fortunate enough to make a living out of my childhood passion for insects. For the last thirty years I have specialized in researching the ecology of bumblebees, the large, furry, stripy bees that drone clumsily among the flowers in our meadows and gardens throughout spring and summer. Their bumbling appearance is deceptive, for they are intellectual giants of the insect world, capable of astonishing feats of navigation and learning, and have complex and sometimes bloodthirsty social lives.

I first became interested in insects simply because I found them to be fascinating and beautiful, but I have long since learned that they are enormously important. Insects comprise the bulk of life on Earth; more than two thirds of the 1.5 million known species are insects. They are food for a great many larger animals, including most birds, bats, lizards, amphibians and fresh-water fish. Insects are also important biological control agents of crop pests, recyclers of all manner of organic matter from corpses to dung, leaves and tree trunks, and they help to keep the soil healthy. The majority of the world's wild plant species depend on insects to pollinate them, as do three quarters of the crops we grow. Without insects, our world would grind to a halt; it cannot function without them.

Given the numerous vital roles that insects perform, we should be worried that many species are in rapid decline. For example, in the UK, butterfly populations have fallen by about 50 per cent since 1976. The biomass of flying insects on German nature reserves fell by an alarming 76 per cent between 1989 and 2016. In the Netherlands, caddis flies declined by 60 per cent between 2006 and 2017, and the biomass of moths by 61 per cent between 1997 and 2017. In North America, numbers of the monarch butterfly, famed for its annual migration between Mexico and Canada each year, are down about 80 per cent since the 1990s. A few insect species are bucking the trend, but most seem to be in trouble. Attempts to calculate an average rate of decline suggest that it might be in the range of 1–2 per cent per year – which may not sound like much, but it amounts to an insect apocalypse on the scale of a human lifetime. Worryingly, we do not know when these declines began as we have no data from before the 1970s – it is likely that we are now monitoring the tail of a much longer fall. We also have no idea what is happening to insects in the tropics, the great hotspots of insect biodiversity. Troublingly, the evidence for these population collapses is still too patchy – almost all the long-term studies of insect populations are from Europe and North America.

So what is driving these declines? In 1962, three years before I was born, Rachel Carson warned us in her book *Silent Spring* that we were doing terrible damage to our planet. She would weep to see how much worse it has become. Insect-rich wildlife habitats such as hay meadows, marshes, heathland and tropical rainforests have been bulldozed, burned or ploughed to destruction on a vast scale. Soils have been degraded and rivers choked with silt and polluted with industrial and agricultural chemicals or drained dry from over-use. The problems with pesticides and fertilizers that Carson highlighted have become far more acute, with an estimated 3 million tonnes of pesticides now going into the global environment every year. In the US, the weight of pesticides applied has increased by 150 per cent since *Silent Spring* was published, while at the same time new pesticides have been introduced that are much more toxic to insects than any that existed in Carson's day. For example, the neonicotinoid insecticide imidacloprid is now the most widely used insecticide in the world, despite an EU-wide ban since 2018 brought on because of the harm it does to bees. Imidacloprid is about 7,000 times more toxic to bees than the insecticide DDT which was widely used in the 1960s and '70s.

On top of all these pressures, wild insects now must cope with climate change, a phenomenon unrecognized in Carson's time. Some insects, such as mosquitoes, will benefit from warmer temperatures and more rain, but most will not. My bumblebees are disappearing from the southern edges of their range, overheating in their furry coats as the climate warms. When climates changed in the past, they usually did so slowly, and wildlife existed in much larger populations, inhabiting extensive areas of intact habitat. Populations could easily shift towards the poles as it warmed, and back again when it cooled. Today, most insects persist in massively reduced populations inhabiting small fragments of surviving habitat. To move pole-wards they have to somehow cross tracts of hostile farmland and urban areas and hope to chance upon a patch of suitable habitat somewhere on the other side. Climate change is also bringing with it increased frequency of storms, droughts, floods and fires, all of them bound to severely impact already depleted populations. This may be the final straw for some.

The American biologist Paul Ehrlich likened the loss of species from an ecological community to randomly popping out rivets from the wing of an aeroplane. Remove one or two and the plane will probably be fine. Remove ten, or twenty, or fifty, and at some point there will be a catastrophic failure and the plane will fall from the sky. Insects are the rivets that keep ecosystems functioning.

If we are to reverse insect declines, we need to act, and act now. We need to engender a society that values insects, both for what they do for us and for their own sake. The obvious place to start is with our children, encouraging environmental awareness from an early age. We need to green our urban areas. Imagine green cities filled with trees, vegetable gardens, ponds and wild flowers squeezed into every available space – in our gardens, parks, allotments, cemeteries, on road verges, railway cuttings and roundabouts – all free from pesticides and buzzing with life. We also need to transform our food system. The way we grow and transport our food has profound impacts on our own welfare, and on the environment, so it is surely worth investing in getting it right. There is an urgent need to overhaul the current system, which is failing us in multiple ways, being a major contributor to greenhouse gas emissions, poisoning and eroding vital soils, and wiping out the biodiversity on which food production depends. We need to work with nature, encouraging predatory insects and pollinators, and stop trying to control and kill. Alternative farming systems such as organic and biodynamic farming, permaculture and agroforestry all have much to offer. There is an appetite for change. We could have a vibrant nature-friendly farming sector, with more small farms employing many more people, focused on the sustainable production of healthy food, looking after soils and supporting biodiversity, and producing mainly fruits and vegetables rather than meat, but this needs support from policymakers and consumers.

It is not quite too late. Most insect species have not yet gone extinct, but many now exist in numbers that are a fraction of their former abundance, teetering on the edge of oblivion. The stripy cinnabar moths that I collected as a boy have since declined in number by 83 per cent, but there are still some left, and they could easily recover if we act now. We do not understand anywhere near enough to be able to predict how much resilience is left in our depleted ecosystems, or how close we are to tipping points beyond which collapse becomes inevitable. In Paul Ehrlich's 'rivets on a plane' analogy, we may be close to the point where the wing falls off. /

Nature's Calendar

Keith W. Larson

For many species, their geographic range is the same, year in, year out. However, for certain migratory species of birds, butterflies, whales, and many others, their so-called species range shifts with the seasons. These seasonal patterns of movement are usually driven by changes in weather, habitat conditions and food availability. Similarly, many plant and animal species undergo profound changes across the course of the year – a phenomenon known as phenology. Just like shifts in species range, these significant reoccurring events in the lives of plants and animals take place according to environmental cues such as changes in temperature, precipitation and day length.

A familiar example of phenology occurs in many plants: in the spring, new leaves grow, often followed by flowering; in the late summer, they produce fruit; finally, in the autumn, the leaves change colour and fall to the ground. In mammals, phenological changes can vary: for example, some species hibernate during cold months, while others change their coats of fur to match their surroundings. Because of the regularity of these seasonal events, phenology is sometimes described as 'nature's calendar'. The timing is important, because it allows individuals to synchronize breeding and avoid having unfavourable weather extremes coincide with key stages in their life cycle (for example, raising your young when there is little food due to winter conditions).

Even in tropical environments that appear to have relatively stable climates, pronounced rainy seasons lead to the predictable timing of flowering and fruiting in plants, influencing breeding patterns across a diversity of insects, mammals and birds. But seasonal changes do become more pronounced as you move higher in latitude from the tropics. In Sweden, the spring is when these changes are most spectacular. Birdwatchers gather to record the arrival of migratory birds such as the willow warbler and the pied flycatcher from their distant tropical wintering grounds; those who live in urban areas note the first flowering of the violet queen crocus in their gardens or the wood anemone blanketing the floors of beech woodlands. Squirrels and bears wake from hibernation to take advantage of spring warmth and the impending abundance of food. The mountain hare and the willow ptarmigan shed their snow-white coats to match their newly leafy surroundings.

Both species range and phenology are incredibly sensitive indicators of climate change, and researchers have turned to study them in a bid to detect the early fingerprints of change across ecosystems globally. As our planet warms, plant and animal species have few choices. They can either track the environmental conditions necessary for life, which generally means moving to higher latitudes and altitudes. Or they can change the timing of their phenological events, such as when plants develop leaves and flower earlier in the spring. If they lack the ability to move or to adjust their phenological clocks in the face of rapid climate or environmental change, they face local, regional or global extinction. The rate of change is crucial: if warming is too rapid, then species may fail to respond in time.

We have already found many cases of species shifting their geographic ranges and altering the timing of phenological events – tracking cool conditions as our planet warms. Across Europe, great tits are breeding up to two weeks earlier each summer. In temperate North America, over half of all animal and plant species have shifted their ranges upwards in elevation or northwards. Most dramatically, the cryosphere (the regions where winter dominates the year), which is home to many Arctic and Antarctic specialists such as polar bears and penguins, is shrinking by 87,000 square kilometres a year.

Fascinatingly, some species may be adapting to a warming world not by moving but by reducing their body size. All organisms face thermo-regulatory constraints – that is, the energy they require to maintain their heat balance. Bergmann's rule predicts that populations and species in higher altitudes and latitudes (colder climes) have larger bodies (a smaller surface area to body size) to help maintain their body temperature in colder climates. Recent research has demonstrated that body size is getting smaller in bird species in North America as the planet warms. Again, the rate of human-caused warming matters, as the more quickly we warm the planet, the less likely it is that species will have time to adapt or move.

Crucially, we also need to understand how different species' responses to warming might affect – or be affected by – their complex interactions with other species. For example, plants rely on pollinators, and migratory birds on insects and fruit. How might the change in the timing of flowering or insect emergence cause a mismatch with their pollinators or prey? Many species, for example migratory birds, reduce competition for food and habitat by partitioning the year for annual events, such as breeding. In Europe, pied flycatchers have been arriving earlier from their wintering grounds in the tropics and failing to compete with resident great tits. In the subarctic regions of the Scandinavian mountains, warming winters have seen birch forest shift higher into the mountains – yet while this expansion of the treeline into the alpine zone is doubtless dependent on warming, changes in the grazing by mammals, for example the Indigenous Sámi's reindeer, will also have played an important role.

It is these complexities that make understanding the full impacts of rapid climate change so challenging. In temperate to boreal and Arctic climes, false springs created by extreme warming events in winter can be devastating for plants and their pollinators that take cues from warm spring temperatures. Frost can be an important trigger for leaf-out in trees, so an earlier spring may not mean that leaves appear earlier. If hibernating animals emerge too early due to warm weather, they may find their food and water sources covered with snow and ice. Migratory birds such as barn swallows may arrive too late to fully exploit the seasonal abundance of insects that take their cues from local environmental conditions, whereas the swallows' migratory cues come from long periods of natural selection. These phenological mismatches have the potential to disrupt agricultural systems dependent on pollinators and further challenge the survival of countless species that have already been impacted by changes inflicted by humanity.

Today, our ability to predict the resilience of species and their communities is challenged by the rapid rate at which climate and environmental change is unfolding. Not only are species shifting their ranges, but entire biomes are shifting. In the case of the tundra or Arctic biome, however, moving north is simply not an option. We are fast heading into uncharted territory, where species range shifts and changes in their phenology will transform local ecosystems. At global scales, these changes can lead to feedbacks that alter the carbon and nutrient cycles, which in turn affect our climate system and risk driving even more warming and further deteriorating the conditions for life on our planet. /

Soil

Jennifer L. Soong

Globally, the soil contains over 3,000 gigatonnes of carbon, about four times the amount of carbon in the atmosphere and all the plants in the world combined. This vast underground store regulates the global carbon cycle, while contributing to food production, biodiversity, drought and flood resilience, and ecosystem functioning. Today we know that our dependence on this crucial carbon pool as a reliable net sink of atmospheric CO_2 – lessening the impacts of anthropogenic CO_2 emissions – is threatened by climate change.

Most of the carbon in the soil today originated from the atmosphere. Soil organic carbon forms as plants use photosynthesis to suck in CO_2 to build their tissues while drawing in nutrients from the soil as fuel. As they grow, and after they die, plant tissues are decomposed by micro-organisms in the soil such as bacteria and fungi, which feed on and recycle both the carbon and the nutrients. During decomposition, nutrients are released back into the soil to feed more plant growth, while much of the carbon is completely decomposed by micro-organisms and gets respired back into the atmosphere as CO_2. But not all soil carbon is the same. Some of it remains underground, protected from decomposition by sticky mineral surfaces or held within clumps of soil called aggregates. Being protected by mineral surfaces, by aggregates, or simply by virtue of being buried away very deep underground means that some of the carbon that plants bring into the soil remains sequestered in the earth for decades, centuries or even millennia.

Over time, the amount of carbon deposited in the soil by plants has outweighed the amount of carbon lost through decomposition. This has created the massive bank of soil carbon that we depend on to maintain the global greenhouse gas balance. The breathing of the land surface, whereby plants suck in carbon and plants and soil micro-organisms respire, while leaving a little stored away underground, naturally cycles ten times more CO_2 between the land and the atmosphere than all anthropogenic emissions combined. The natural cycling of carbon between the atmosphere and land is critical to regulate Earth's climate – even a small change could have an enormous impact on the climate, tipping the balance of the global carbon cycle.

As temperatures rise, the activity of micro-organisms speeds up and soils begin to emit more CO_2 to the atmosphere. Increased emissions of carbon from the soil could tip the natural carbon cycle into positive feedback, where warming increases soil CO_2 emissions, which further increases global warming, which further increases soil CO_2 emissions, and so on. This positive feedback could be especially detrimental in northern ecosystems, where warming is occurring fastest, and where cold conditions have allowed a vast pool of permanently frozen soil carbon, called permafrost, to accumulate. Although permafrost is typically too cold to be decomposed, warming temperatures are causing it to thaw and become vulnerable to microbial decomposition and carbon emission into the atmosphere.

To avoid a possible tipping point where soil carbon and warming shift into a positive feedback loop, potentially leading to runaway warming, we need to take immediate action. First and foremost, we must immediately and drastically reduce greenhouse gas emissions. We should also grow more trees and other deep-rooted plants, and protect them. We should conserve natural ecosystems and adopt sustainable agricultural practices. We must do everything in our power to increase the soil carbon bank and draw down atmospheric CO_2. Our world depends upon it. /

Even a small change in soil carbon could have an enormous impact on the climate, tipping the balance of the global carbon cycle.

Permafrost

Örjan Gustafsson

There are only a small number of processes in nature that could, within decades, cause a vast enough net transfer of carbon from the land or ocean to the atmosphere to significantly accelerate the climate crisis. The top candidates are thawing permafrost and collapsing undersea hydrates – the destabilization of frozen methane – in the Arctic.

Permafrost is a mixture of soil, sediment, old peat, rocks, ice and organic matter that stays frozen throughout the year and is found on both land and underwater. Permafrost in the top few metres of the Arctic landmass holds half of all carbon in all soils globally, containing approximately twice as much carbon as is held in the atmosphere as CO_2 and 200 times as much methane. An astonishing 60 per cent of the enormous land area of the Russian Federation is permafrost. Until recently, permafrost was believed to be a dormant carbon pool, isolated, 'asleep' and not participating in exchange with other carbon pools in the global carbon cycle. However, with temperatures in the Arctic rising two to three times faster than the global average, permafrost carbon pools are now being reactivated.

Hydrates (or clathrates) are frozen methane that have formed over geological time under low temperatures and high pressure in the seabed or deep underground. Over millions of years, they have formed in thick sediment layers on the Arctic ocean floor, generally reaching no higher than 300–400 metres below the sea surface. Some hydrates are also present under shallower waters along the Eurasian Arctic. These would have required much colder conditions in order to form. They first emerged on the freezing ice-age tundra of north-east Siberia, which was then submerged by sea-level rise as the glaciers melted, becoming today's vast East Siberian Arctic Seas. This inaccessible and severely understudied coastal region, the size of Germany, Poland, the UK, France and Spain combined, is estimated to host about 80 per cent of the world's subsea permafrost and about 75 per cent of the shallow hydrates on Earth.

This vast reservoir of ancient carbon and methane which spans the Arctic landscape and seabed represents a 'sleeping giant' – and increasingly there are signs that it is being awakened. During our research expeditions over the past two decades across the entire northern edge of the Eurasian continent, half the Arctic Circle and the largest shallow coastal seas of the world ocean, we have increasingly witnessed carbon that is tens of thousands of years old being released from thawing permafrost, and seen methane vigorously bubbling up from the shallow seabed, likely from thawing underwater permafrost and collapsing methane hydrates.

Across the Eurasian and North American Arctic landmass, some permafrost thaws and refreezes each year. But as temperatures rise, the layer that thaws gets deeper and the permafrost zone shifts northwards. Even if climate warming is globally held within 1.5°C, scientists expect that between a third and a half of the permafrost area will be lost by the end of this century. In addition, rising temperatures and increasing rainfall could cause further collapse of the landscape and degradation of deeper organic carbon deposits.

Along the many thousands of kilometres of coastline of the remote Siberian Arctic lie great ice-rich permafrost deposits ('yedoma', or 'ice-complex deposits') formed during the last ice age. These are particularly vulnerable to collapse, given the growing pressures of warming, sea-level rise and increased frequency of storm-induced erosion.

Beyond the Arctic, the permafrost on the Himalaya–Tibetan Plateau, also known as the Third Pole, is also a concern. This permafrost holds about one tenth of the Arctic land permafrost, yet scientists show that permafrost on the Tibetan Plateau could be even more vulnerable to collapse, given its steeper topography, lower latitude, and proximity to population centres and human activities that directly disturb it through grazing, construction and emissions of climate-warming black carbon aerosols such as soot. While permafrost collapse has been reported to have doubled over recent decades in the Arctic, scientists are now reporting that permafrost collapse with associated greenhouse gas releases is increasing ten times faster on the Tibetan Plateau.

While most research into Arctic methane and CO_2 releases has studied inland permafrost, there is now a growing focus on underwater permafrost and methane hydrates. Subsea permafrost is likely to be even more vulnerable than its land-based sibling. While they both have the same origin, the area that was submerged by sea-level rise at the end of the ice age has not only been warmed during that time by the natural climatic change of the past 10,000 years, it has also been warmed by around 10°C by the overlying seawater. Anthropogenic warming may further cause this permafrost to thaw.

The extensive methane hydrates that lie at a depth of around 300–400 metres along the upper continental slope of the Eurasian Arctic are

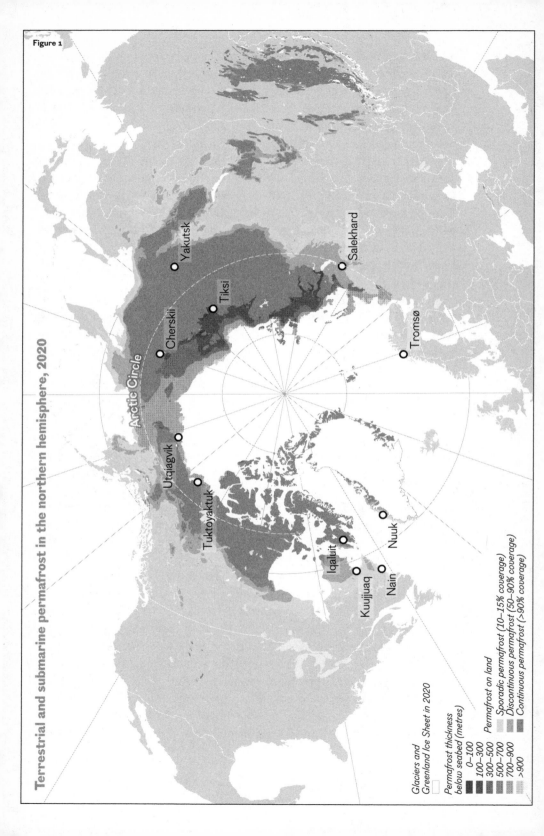

Figure 1

Terrestrial and submarine permafrost in the northern hemisphere, 2020

Arctic Circle

Yakutsk
Salekhard
Tiksi
Cherskii
Tromsø
Utqiaġvik
Tuktoyaktuk
Iqaluit
Nuuk
Kuujjuaq
Nain

Glaciers and Greenland Ice Sheet in 2020

Permafrost thickness below seabed (metres)
0–100
100–300
300–500
500–700
700–900
>900

Permafrost on land
Sporadic permafrost (10–15% coverage)
Discontinuous permafrost (50–90% coverage)
Continuous permafrost (>90% coverage)

also under threat, as they are located at the same ocean depth as the inflow of warm Atlantic water that is increasingly entering the region (a phenomenon known as Atlantification). Indeed, over the past decade we have observed that methane levels in the vast, shallow East Siberian Arctic Seas are ten to a hundred times higher than normal levels in other areas of the ocean; we have also witnessed methane bubbling at hundreds of sites. This indicates that the subsea permafrost system is now being perforated, releasing methane in quantities larger than those from the rest of the entire world ocean combined. At present these emissions correspond to only a few per cent of total natural and anthropogenic methane emissions, but there is no doubt that the release of methane from large subsea chambers of permafrost or hydrates is underway.

This sleeping giant is starting to wake up, but it is not taken into account in our carbon budgets. Scientists suggest that, even with today's climate pledges (which we are currently unlikely to reach), the rate at which Arctic land permafrost will thaw will contribute methane and CO_2 emissions this century equalling the emissions from all the countries in the European Union combined. The release of these emissions – not to mention the role of underwater permafrost and methane hydrates – would drastically reduce our ability to stay under 1.5°C or even 2°C of warming.

We must immediately stop extracting fossil fuel from reservoirs in the Arctic and avoid further contaminating the atmosphere with short-lived pollutants such as black carbon aerosols, which are particularly troublesome as they both warm the atmosphere and, when they settle on snow and ice, darken the land surface and increase the ice-albedo feedback. Mitigation of black carbon emissions can readily be achieved by minimizing gas flaring in the Arctic oil and gas industry and by regulating wood fuel burning from open fireplaces in the boreal belt, including Scandinavia, Russia and Canada. There is an urgent need for us to take these mitigating steps and bend the curve of anthropogenic emissions by acting now. It is my personal hope that, as we witness the awakening of the Arctic permafrost/hydrate sleeping giant, we will also awaken our one and only global society. /

This sleeping giant is
starting to wake up.

What Happens at 1.5, 2 and 4°C of Warming?

Tamsin Edwards

We've started to notice. Now we're at just over 1°C of warming. Heatwaves are breaking records. Floods are devastating even the most well-prepared countries. Ferocious fires are burning forests, villages, the frozen north.

It's not just our imagination, or greater media coverage: the climate has changed. The kind of heat that was expected only once a decade before human influence is now three times more likely. Extreme heavy rain is now 30 per cent more likely. Droughts – less rain, and drier soil – are 70 per cent more likely. And scientists can now see a human fingerprint on some of the worst events in living memory, some of which we have made three, or ten, or a hundred times more likely; some of which would have been virtually impossible without our influence.

So how different will our world feel at 1.5°C or 2°C, the lower and upper limits of the Paris Agreement (Fig. 1)? And how will things change if we ignore this global pledge and keep increasing our emissions at the current rate – doubling them by the end of the century – reaching 4°C? Three or four degrees of warming may not sound like much, but the last time global temperatures were more than 2.5°C warmer than pre-industrial levels for a prolonged period was over 3 million years ago. When our ancestors were beginning to craft stone tools.

The changes our planet will experience will increase with each half a degree of heating. Warming will be faster over the land and polar regions. The Earth's water cycle will become amplified: many parts of the world that are already wet will have more heavy rain, and places that are already dry will have more droughts. Monsoons will change.

Many kinds of extreme weather will continue to worsen (Fig. 2). At 1.5°C, the kind of extreme heat previously seen once a decade will be four times more likely, with hundreds of millions more people exposed to deadly

Which future will we choose?

1.5°C   2°C   4°C

Temperature

Rainfall

Figure 1

Extreme weather events that happened once per decade before human influence will be . . .

	1.5°C	**2°C**	**4°C**
Extreme heat	4 times more likely	6 times more likely	9 times **more likely**
Heavy rain	50% more likely	70% more likely	3 times **more likely**
Drought	2 times more likely	2 times more likely	4 times **more likely**

Figure 2

heatwaves by the middle of the century. At 2°C, this extreme heat will be nearly six times more likely, and at 4°C these temperatures we previously considered extreme will be seen nearly every year. Extreme rainfall and drought will also become more frequent and severe.

Our Earth will look progressively different from space. By 2050, even at 1.5°C of warming, sea ice covering the Arctic Ocean will almost disappear in at least one September, revealing a dark ocean. If the sea ice does disappear, it will regrow the following winter as a thinner, more delicate cover. At 3–4°C of warming it will disappear entirely during most summers, or perhaps in all.

Four degrees is not the most warming we could imagine reaching. In the longer term, there are a huge range of possible futures, depending on our choices. The left panel in Fig. 3 shows the stable temperatures of the past 2,000 years, followed by scenarios of future warming until the year 2300.

We could act to limit warming to the lower end of that range, 1.5–2°C. Even at these temperatures, we would still lose many or most of the world's glaciers, heat the oceans and erode the ice sheets. But sea-level rise by 2300 (the lighter bar in the right panel of Fig. 3) would be somewhat limited. If we're lucky, it would rise by half a metre, although it could reach 3 metres – enough to change coastlines around the world.

Or we could choose to keep increasing the greenhouse gases in Earth's atmosphere, decade upon decade, century upon century, and travel to an unrecognizable planet: 10°C hotter by the year 2300. Every glacier in the world would be lost. Each year, we would increase the risk of destabilizing the Antarctic Ice Sheet – if we have not already – which would rapidly increase sea-level rise for centuries. In such a world, the seas could rise by up to 7 metres, shown by the darker bar in Fig. 3. If we are unlucky, and Antarctica is particularly sensitive, the seas could rise by much more.

What if we could stop our emissions instantly? Some parts of the planet would keep on changing as they continue reacting to our past emissions. The world's glaciers would continue to retreat for decades or centuries, and the oceans would continue to warm. This means the seas will rise and there will be more coastal flooding, no matter what we do.

One day in the distant future, we may be able to reverse some of the impacts of climate change, if we can reduce temperatures back to previous levels by removing the excess carbon dioxide from the atmosphere. Our weather could become more normal again and the Arctic sea ice could return each summer. After a very long time, we might even see some re-advance of the glaciers. However, it will not be possible to reverse many of the other changes to the planet in any meaningful human timescale. The oceans will be warmer, the ice sheets will be diminished and sea levels will be higher for hundreds to thousands of years.

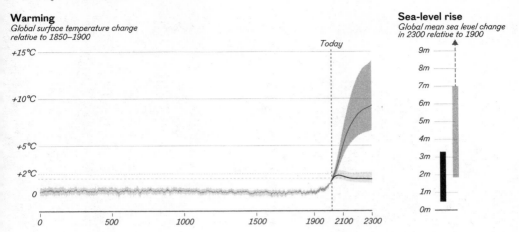

Warming
Global surface temperature change relative to 1850–1900

+15°C

+10°C

+5°C

+2°C

0

Today

0   500   1000   1500   1900 2100 2300

Sea-level rise
Global mean sea level change in 2300 relative to 1900

9m
8m
7m
6m
5m
4m
3m
2m
1m
0m

Figure 3:
Temperature changes of the past 2,000 years, followed by scenarios of future warming until the year 2300. Dashed line indicates 1.5°C. Sea-level rise in metres in 2300 for the same scenarios.

Next pages:
Polar bears living in an abandoned weather station in Kolyuchin, in the Chukotka Autonomous Okrug of the Russian Federation.

So, what do we need to do? It's simple. The IPCC has said that 'unless there are immediate, rapid, and large-scale reductions in greenhouse gas emissions, limiting warming to close to 1.5°C or even 2°C will be beyond reach'.

How are we doing so far? A world of ever-increasing burning of fossil fuels would heat up by 4–5°C by 2100. Thankfully, we have put some policies in place that make it far less likely that we will face that future. We have also made progress in technology and in the way we live. So, if those policies are successfully enforced, warming is now predicted to be less than 3°C this century.

We have also made promises of what we intend to do, including the national pledges under the Paris Agreement for how much each country will cut their emissions by the year 2030, and statements about when they plan to stop adding greenhouse gases to the atmosphere completely. If we stick to those promises, warming will be limited to just over 2°C, and perhaps even less than 2°C. Every time a new policy or pledge is made, these predictions gradually tick down.

Every single new policy – every single tonne of carbon dioxide we avoid emitting – will bring future global warming further down, towards our target of 1.5–2°C. As we have seen, there would still be serious consequences even at these levels of warming, and the last degree or so will be the hardest to avoid. But climate change is not something that is simply won or lost. It is a curve that we can keep bending towards a better world. The future is looking better than we imagined, but it is not yet as good as we hope. What happens next is up to us.

Future generations will know how we did. /

PART THREE /

How It Affects Us

'We fail to connect the dots'

3.1

The world has a fever

Greta Thunberg

The world has a fever. And a fever is usually a symptom of something else, like an infection, a disease or a virus. The climate crisis is also a symptom, or a result if you prefer, of a much deeper sustainability crisis. In other words, it is not the increasing average temperature that is the root cause of the problem. Rather it is the fact that we are living way above our means, exploiting people and the planet. Or, more accurately, a small number of us are doing this. Absurd inequalities divide the world. The richest 10 per cent cause 50 per cent of our CO_2 emissions. The wealthiest 1 per cent are responsible for more than twice as many emissions as the entire poorer half of the world, according to a 2020 report from Oxfam and the Stockholm Environment Institute.

Humankind has not created this crisis – it was created by those in power, and they knew exactly what priceless values they were sacrificing in order to make unimaginable amounts of money and to maintain a system that benefitted them. It is – among other things – the social and economic structures which generate such perverse inequalities that are driving us towards the ecological precipice. It is the idea of infinite growth on a finite planet.

When you heat up a pot of water, you know that the water will boil at 100°C. But you cannot predict exactly where the first bubble will appear, or the second one, and so on. We just know that the water will eventually start to boil. This is how I have heard some scientists describe the process of the climate crisis. And that same allegory can be applied to the larger sustainability crisis.

Many have wondered what the first disaster to bring the modern, globalized world to a temporary standstill would be. Some kind of resource-related conflict, an energy crisis or financial collapse, perhaps. Instead, it was a pandemic that appeared and changed our lives overnight.

In the winter of 2022, when we were finishing this book, it was not possible to say for sure that Covid-19 had been transmitted to humans from other animals, in this case bats. There are still uncertainties. What we do know, however, is that most pandemics do come from animals; they are zoonotic diseases. In fact, 75 per cent of all new infectious diseases originate from wildlife. Natural habitats should work as a protecting shield, but once you strip back too much of that natural barrier we are exposed to increasing levels of risk. So maybe the coronavirus did spread from animals to humans, or maybe it did not. Either way, our destruction of nature is laying the perfect groundwork for the creation of new – and potentially much deadlier – pandemics. Since the global outbreak in February 2020, the scientific community has been clearly spelling this out. Yet hardly anyone seems to be making the connection. So, for the record, let it be known that we were given all the facts. As the executive director of the World Health Organization Health Emergencies Programme put it in a speech in February 2021:

> We are creating the conditions in which epidemics flourish, we're forcing and pushing people to migrate away from their homes because of climate stress. We're doing so much and we're doing it in the name of globalisation and some sense of chasing that wonderful thing that people call economic growth. Well, in my view that's becoming a malignancy, not growth, because what it's doing is driving unsustainable practises in terms of how we manage communities, how we manage development, how we manage prosperity; we are writing checks that we cannot cash as a civilization for the future, and they're going to bounce. And my fear is that our children are going to pay that price. Some day when we're not here, our children will wake up in a world where there is a pandemic that has a much higher case fatality rate, and that could bring our civilization down to its very knees. We need a world that's more sustainable, where profit is not put before communities. Where that is not the bottom line, where the slavery to economic growth is taken out of the equation.

If you were reading this book a few years into the future, perhaps decades from now, you might think that such words would have made some kind of an impact. You might imagine that there would have been articles written, or segments about it on the radio or television news. But let me assure you, basically no one reacted. When it comes to our health, 'we are throwing stones at ourselves', according to Ana Maria Vicedo-Cabrera. Today, 37 per cent of heat-related deaths are caused by climate change, roughly 10 million die each year from air pollution, and as our planet continues to warm, malaria and dengue could put billions more at risk by the end of the century. All for a crisis that perhaps can best be described as the price of pursuing short-sighted economic growth, or simply the result of a world where greed, selfishness and inequality have shifted everything out of place and out of balance. In other words, the sustainability crisis is what you get once you finally connect the dots. /

Health and Climate

Tedros Adhanom Ghebreyesus

The climate crisis is upon us, powered by our addiction to fossil fuels. The consequences for our health are real and often devastating, and they are starting to play out in front of our eyes.

Climate change impacts health in all countries, but people in low- and middle-income countries, who are already struggling with other health, economic and environmental challenges, are being hit the hardest. The risk of vector-borne diseases and mass hunger crises are increasing daily, as water scarcity looms and sea levels rise.

Climate change does not cause diseases, but it affects the way they spread and undermines our work to combat them. Take malaria, for example. Increases in temperature, rainfall and humidity allow malaria-carrying mosquitoes to multiply and expand their range, resulting in an increase in transmission, including in areas in which malaria was not reported earlier. A study by the World Health Organization (WHO) conservatively estimated that climate change may cause an additional 60,000 malaria deaths between 2030 and 2050, even when accounting for the impact of other measures to reduce the burden of malaria. The study also showed that at least 5 per cent of global malaria cases, or 21 million cases, would be attributable to climate change in 2030.

This is only one example, but there are hundreds of other 'climate-sensitive health risks'. For instance, children born after the year 2014 (which would make them eight years old or younger in 2022) will experience thirty-six times more heatwaves than a person born in 1960 (i.e. a sixty two-year old in 2022). Up to one fifth (19 per cent) of the global land surface was affected by extreme drought in 2020, leading to a stark rise in food and water shortages. The list goes on and on.

The extent to which people are vulnerable to these threats is largely determined by social factors: the impacts are disproportionately felt by the most disadvantaged, including women, children, ethnic minorities, poor communities, migrants or displaced persons, older populations and those with underlying health conditions.

Any delay in preventing these health threats from getting worse will disproportionately affect the most disadvantaged around the world. With the poorest people largely uninsured, health shocks and stresses already push around 100 million people into poverty every year, with the impacts of climate change worsening this trend. To fully address the urgency of this crisis, we need to confront the inequalities that lie at the root of the challenge.

In the longer term, however, nobody will be safe from harm. The risks will increasingly depend on the extent to which transformational action is taken now to reduce emissions and avoid the breaching of dangerous temperature thresholds and irreversible tipping points.

Many governments increasingly understand that they need to act swiftly to protect their citizens from growing climate impacts. In a recent survey WHO conducted with its member states, over three quarters of the countries indicated they have developed or are currently developing national health and climate change plans or strategies.

However, while low- and lower–middle-income countries often lack the resources and technical support to implement these plans, only one third are receiving international assistance. Global solidarity, building capacity and sharing technology and know-how will be key to overcoming these barriers.

What gives me hope is our growing understanding of the many benefits that come from taking rapid and ambitious action to halt and reverse the climate crisis, including benefits for health.

For example, many of the same actions that reduce greenhouse gas emissions also improve air quality and contribute towards many of the UN's Sustainable Development Goals. Some measures – such as encouraging more walking and cycling – improve health through increased physical activity, resulting in reductions in respiratory diseases, cardiovascular diseases, some cancers, diabetes and obesity.

Another example is the promotion of urban green spaces, which facilitate climate mitigation and adaptation while also offering health co-benefits, such as reduced exposure to air pollution, local cooling effects, stress relief and increased recreational space for social interaction and physical activity.

A shift to more nutritious plant-based diets could reduce global emissions significantly, ensuring a more resilient food system, and avoiding up to 5.1 million diet-related deaths a year by 2050.

The public health benefits resulting from ambitious mitigation efforts would far outweigh their cost. They offer strong arguments for transformative change and can be gained across many sectors. Research has shown that climate action aligned with the Paris Agreement targets would save millions of lives due to improvements in air quality, diet, and physical activity, among other benefits.

However, many climate decision-making processes do not yet account for these health co-benefits. It can be easy to forget when crunching the numbers and setting distant climate targets that behind those numbers and targets are people whose health and future depend on ambitious actions now.

Delay comes at great cost: every fraction of a degree of additional warming adds a price tag to our own health and that of our children. The catchphrase '1.5 to stay alive' – used by the most vulnerable countries to call for more ambitious climate action – can be taken quite literally from a health perspective.

Health leaders everywhere have been sounding the alarm on climate change and are increasingly taking steps to protect their communities from worsening climate impacts while reducing their own emissions. In October 2021, weeks ahead of the COP26 UN climate conference, an open letter signed by over two thirds of the global health workforce was sent to national leaders. The letter noted that: 'Wherever we deliver care, in our hospitals, clinics and communities around the world, we are already responding to the health harms caused by climate change.'

At the same time, WHO released a Special Report on Climate Change and Health, spelling out the global health community's prescription for climate action, highlighting the priority actions governments need to take to confront the climate crisis, restore biodiversity and protect health.

Putting those recommendations for a healthy future into practice – from committing to a healthy, green and just recovery from Covid-19, and creating energy systems that protect and improve climate and health, to promoting food systems that are healthy, sustainable, and resilient – means investing in a healthier, fairer and more resilient world. Advanced economies, in particular, have a once-in-a-generation opportunity to demonstrate true global solidarity.

The COP26 climate conference did bring us a step closer to that goal, with ramped-up national climate commitments, additional climate finance for vulnerable countries and dozens of commitments from governments to decarbonize and strengthen the resilience of their health systems.

But protecting health requires action in other sectors, including energy, transport, nature, food systems and finance. Unfortunately, most sectors are still terribly underprepared for the changes that need to take place, and harmful fossil fuel industries are still receiving subsidies of $11 million every minute.

We still have a very long way to go to protect our health and ensure a healthy future for our children, but we know what needs to be done. The health arguments for rapid climate action have never been clearer. Let's get to work. /

3.3

Heat and Illness

Ana M. Vicedo-Cabrera

Heat is one of the biggest environmental threats we face. In recent years, historical extreme heatwaves such as the 2003 heatwave in Europe or the 2010 Russian heatwave have offered staggering demonstrations of just how devastating it can be: it is estimated that several thousand excess deaths occurred in these periods. Today, around 1 per cent of all deaths in the world can be attributed to heat, with approximately 7 heat-related deaths per 100,000 people per year, putting it nearly on a par with deaths from malaria (Fig. 1).

One cannot talk about heat without considering its connections to climate change. Anthropogenic climate change is responsible for one in three deaths due to heat today – accounting for 37 per cent of heat-related deaths between 1991 and 2018. Given that this substantial mortality burden is happening at 0.5–1°C of warming, it is realistic to expect that this burden will grow in the coming decades, as warming progresses to levels above 2, 3, or even 4°C. Recent studies have projected that, under the most pessimistic scenario (that is, if emissions persist and no adaptation occurs), climate change will increase the current number of heat-related deaths by ten times by the end of the century in regions such as southern Europe, Southeast Asia and South America. Importantly, current societal trends such as ageing populations and increasing urbanization would act as amplifying factors, since larger heat-related risks are mostly observed in urbanized areas (due, in part, to the urban heat island effect), and among the elderly, who are especially vulnerable to heat's physiological impacts.

When exposed to high temperatures, humans have a variety of systems to maintain body temperature within a safe (and narrow) range close to 37°C. However, these mechanisms might not work properly in some individuals, or they might not be efficient when subject to extreme heat conditions – usually, heat coupled with high humidity. We need the air around us to be cool enough to draw heat away from our bodies. However, we live in environments where the air may frequently be warmer than our bodies. So, we also require humidity to be low enough that we can cool ourselves down by sweating and so continue to draw heat away our bodies. When relative

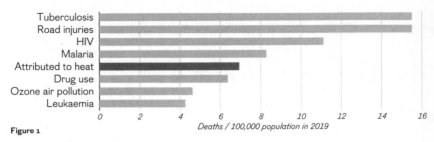

Global annual mortality risk factors

Figure 1

Deaths / 100,000 population in 2019

atmospheric humidity reaches 100 per cent, however, sweat ceases to evaporate efficiently and can no longer cool our skin. Temperatures accompanied by 100 per cent humidity levels are known as wet bulb temperatures. A wet bulb temperature of around 35°C is lethal – but it causes serious problems even before that point.

When extreme heat occurs, the body cannot cope and this eventually triggers a series of mechanisms leading to a variety of adverse health outcomes. Contrary to conventional wisdom, however, deaths from heat stroke are a very small fraction of all deaths due to heat, since heat can act as a triggering factor for a number of acute conditions such as heart attacks, or exacerbate underlying conditions such as chronic obstructive pulmonary disease. Mortality represents just the tip of the iceberg: heat has also been associated with increased risk of hospitalization due to cardiovascular or respiratory diseases and premature births.

Heat affects all populations in the world, but the elderly, pregnant women, children and people with chronic conditions have been identified as particularly physiologically vulnerable subgroups. Impacts also vary considerably across regions, countries and even cities within the same country. For example, in Europe there are larger risks in the Mediterranean region, compared to cities in the north. The magnitude of the impact on a given population depends on how severe the heat is, the proportion of vulnerable people in that region and the resources the population has to protect itself from heat. Recent assessments have found that highly urbanized populations with greater levels of inequality are most impacted.

In today's world, warmer summers and extreme heat events are becoming the norm. There is an urgent need to understand how we can reduce our vulnerability – or, in other words, how we can efficiently adapt to higher temperatures. Although we have partially adapted to heat in past decades, it is still unclear which means of adaptation are most viable for the future. Air-conditioning has traditionally been considered an effective solution, but it is not the only one available to us and we are yet to prove its efficiency in a considerably warmer world – not least because of the implications this approach could have for both energy expenditure and inequality. Air-conditioning is simply not a realistic solution for many. Public health interventions such as heat warning systems have also proved to be useful tools, but even here we should be cautious, because it is likely that interventions that work well today might not be as efficient as we would hope tomorrow.

Rising inequality, accelerating urbanization and the depletion of natural resources are all closely linked to climate change, and are also affecting our health, whether directly or indirectly. It is therefore essential that we consider more holistic, far-reaching and ambitious interventions. This is one of the key messages that scientists have tried to convey since the start of the Covid-19 pandemic, the onset of which revealed clear deficiencies in our public health systems. Despite continual warnings from experts about the risk of harmful emerging infectious diseases, the initial surge of the pandemic took almost everybody by surprise and found governments and public health institutions unprepared. As with climate change, the abundance of misinformation, the lack of trust in research and the absence of community leadership, and the disconnection between policymakers, the scientific community and the general population, all put additional strain on the management of this crisis. This global health emergency has taught us that effective and timely prevention, preparedness and response are key for mitigating health crises in the future. So let's learn from our past mistakes. There is still time for us to build a more resilient, sustainable and equitable world for the next generation. /

Mortality represents just the tip of the iceberg.

3.4

Air Pollution

Drew Shindell

Climate change and air pollution are both primarily invisible killers. We may see a few victims of tropical storms on television, but heat exposure kills hundreds of thousands of people annually. Similarly, not many are aware that roughly 10 million deaths per year from heart or respiratory diseases are due to outdoor air pollution exposure. Exposure to elevated levels of both fine particulate matter and ozone, a component of smog, increases risk for these diseases. Reducing the emissions that drive climate change, which largely come from the same sources as those that cause air pollution, will therefore lead to even greater benefits than most people realize. As damages from both air pollution and climate change fall disproportionately on the poorest and most vulnerable members of society, the health benefits from mitigating climate change also provide a path towards a fairer, more equitable world.

One of the most important actions we can take to combat air pollution and the climate crisis is, simply put, to stop setting things on fire. Stopping the burning of fossil fuels for energy is a critical step that yields enormous, immediate air-quality benefits as up to one in five of all premature deaths stem from fossil-fuel-related pollution. Stopping the burning of biofuels for daily cooking (and sometimes heating) by bringing the world's poorest people access to modern, efficient energy provides outsized health benefits by improving both outdoor and indoor air quality. Indoor air pollution is estimated to kill roughly 3.2 million people prematurely each year, especially women and children who are most often exposed to kitchen smoke. Similarly, stopping the burning of agricultural waste and instead ploughing it under reduces air pollution and returns vital nutrients to the soil. All of these actions will mitigate climate change in the long run.

There are additional important actions we should take so that the air we breathe will be safer and cleaner, such as reducing emissions from landfills and manure. That decreases methane, which is a precursor of ozone, and also reduces noxious local pollution. Additionally, we need to reduce our consumption of cattle-based foods as our enormous population of livestock leads to very large methane emissions (primarily from cow burps and

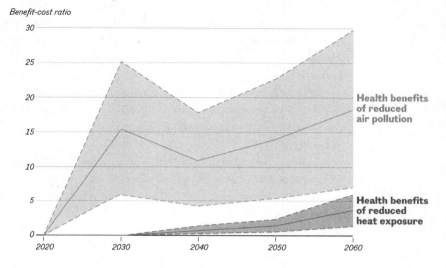

The benefit–cost ratio of the US reducing emissions to meet Paris Agreement goals relative to a high-emissions scenario

Figure 1:
The higher the ratio, the more the financial benefits of mitigating emissions outweigh the costs of limiting warming. Dashed lines show the range; the unbroken line is the mean.

Benefit-cost ratio

Health benefits of reduced air pollution

Health benefits of reduced heat exposure

manure), accounting for roughly 30 per cent of global methane emissions due to human activities. That's important, as methane emissions are responsible for about one third of the warming impact of all greenhouse gas emissions to date and lead to about 500,000 premature deaths a year from ozone.

What's tremendously important about clean air and climate change mitigation benefits is that they complement each other extremely well, and policies should attempt to maximize both simultaneously. Clean-air benefits occur rapidly since air quality responds quickly to emissions changes, as we saw in the blue skies that appeared following Covid-19 lockdowns in normally smoggy cities such as New Delhi, Guangzhou and Cairo. In contrast, climate change mitigation benefits typically take a long time to occur as the climate system responds at a slower pace, but they are crucial in the long term. Similarly, the spatial extent of these two environmental changes is complementary. Air pollution is a problem primarily on a national to regional scale, so countries that reduce their emissions are the ones that receive the bulk of the clean-air benefits. In contrast, climate change is a global problem that requires global cooperation to reduce emissions, which then provides benefits to the entire world. Paying attention to the clean-air benefits highlights the fallacy in arguing that others should be the first to act or that any nation should not act until all others agree. In a 2021 study, for example, we showed that if the United States reduced emissions to meet the goals of the Paris Agreement it would provide clean-air benefits to the country that would outweigh the costs of the societal transition in the very first decade of action. The climate benefits also outweigh the cost of action, but they do so only during the second half of the century and only if the rest of the world is similarly dedicated to pursuing the goals of the Paris Agreement (Fig. 1). In the United States alone, reducing emissions globally over the next fifty years to meet the Paris Agreement target to keep warming under 2°C could prevent about 4.5 million premature deaths, 1.4 million hospitalizations and emergency room visits, and 1.7 million incidences of dementia. If only the United States reduced its emissions in line with the Paris Agreement, 60–65 per cent of these benefits would still be realized.

Hence, taking a broad view that looks at overall societal well-being rather than the effects of climate change alone can motivate action by showing people how they can reap near-term local health benefits in addition to preventing the long-term climate catastrophe that many have a hard time conceptualizing. /

In the United States alone, reducing emissions globally to meet the Paris Agreement could prevent about 4.5 million premature deaths.

Vector-borne Diseases

Felipe J. Colón-González

Vector-borne diseases – those transmitted to and between humans by a range of different organisms such as mosquitoes, sandflies and other arthropods – account for about 17 per cent of all lost life, illness and disability around the world, causing more than 700,000 deaths every year. The major vector-borne diseases that afflict humans are malaria, dengue, chikungunya, Zika virus disease, yellow fever, Japanese encephalitis, lymphatic filariasis, schistosomiasis, Chagas disease and leishmaniasis.

Over 80 per cent of the world's population currently live in an area at risk from at least one of these diseases, and more than 50 per cent of the population are at risk from two or more. These diseases – many of which are chronic, disabling and stigmatizing – are disproportionately linked to poverty and inequality and are a major impediment to socio-economic development.

There are many ways in which climate change might affect the ecology and transmission of vector-borne diseases, and understanding these effects is crucial to anticipating and responding to the potential increases in risk. As temperatures rise around the globe, vector-borne diseases are gradually spreading to regions of the world where they were typically absent, and re-emerging in areas where they had subsided decades ago. For example, malaria is shifting towards higher altitudes in Africa and South America as the climate has become more suitable for transmission. Dengue cases are now being reported in countries such as Italy, Croatia and Afghanistan – countries where the disease had never previously managed to spread.

The success of a vector-borne disease's transmission and spread depends on complex interactions between the climate, the environment and human population characteristics such as immunity levels and mobility. There are growing and valid concerns that climate change will keep triggering the emergence – and re-emergence – of vector-borne diseases. Through laboratory work, empirical modelling and field studies, we have established that changes in temperature can increase the infectiousness of pathogens, since

Effects of temperature and precipitation on the transmission season of malaria

Figure 1:
A warmer climate and increased rainfall can lengthen the transmission season – to year-round, in some cases.

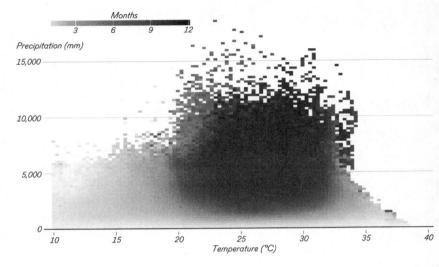

temperature plays a critical role in determining variables such as vector population size, biting rate, survival probability and lifespan.

Generally speaking, warmer temperatures are better for vector-borne disease transmission. Transmission peaks at intermediate temperatures – roughly 25°C – and if it becomes too hot or too cold the risk of transmission is reduced. These effects vary by vector and pathogen but, as climate change takes many regions towards intermediate temperatures (a kind of Goldilocks temperature zone, ideal for disease spread), there are more and more opportunities for pathogens and vectors to thrive.

Precipitation also has important effects, particularly for insects such as mosquitoes that spend phases of their development living in water. Other disease-spreading species, such as ticks or sandflies, which do not have an aquatic stage, are also indirectly influenced by rainfall through changes in humidity. Increased rainfall can create or expand pools of water where such insects can breed; droughts, meanwhile, could also indirectly foster the creation of breeding sites, by prompting people to collect and store water for use in times of scarcity.

A range of studies has focused on the effects of climate change on malaria and dengue, two of the most important global health threats. It has been found that climate change might substantially increase the length of the transmission season of these two diseases and the number of months suitable for their transmission in a given year.

Changes in the length of the transmission season of dengue by 2080

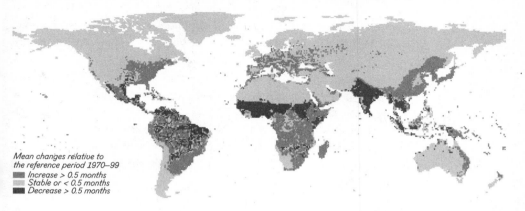

Mean changes relative to
the reference period 1970–99
■ Increase > 0.5 months
▨ Stable or < 0.5 months
■ Decrease > 0.5 months

Figure 2

Malaria's transmission season might increase by up to 1.6 additional months in highland areas in Africa, the eastern Mediterranean and South America by 2080. These are areas where transmission is currently low and so people may be immunologically vulnerable, with public health systems that are unprepared for dealing with the new influx. The transmission season of dengue might increase by up to four additional months in areas in the western Pacific located less than 1,500 metres above sea level.

As temperatures increase, the climate will become more suitable for some disease-carrying vectors and less suitable for others. For example, the species of mosquito that is currently the primary vector of malaria in Africa could be replaced by a more heat-adapted species of mosquito in sub-Saharan Africa. Alternatively, the climate might become less conducive for malaria transmission but more suitable for other diseases, such as dengue, chikungunya or Zika, which are transmitted by mosquitoes that prefer warmer weather.

The range and spread of vector-borne diseases might also change as a consequence of climate change. Malaria and dengue may spread into temperate areas such as France, Bulgaria, Hungary, Germany and the eastern coast of the United States extending from south of Atlanta to north of Boston (Fig. 2). If public health systems identify and suppress infections effectively, these shifts might not translate into an increased number of cases. However, Covid-19 has demonstrated how fragile our public health systems are, even in the wealthiest countries. Epidemiological surveillance, monitoring and early-warning systems will need to be put in place in these potential hotspots to prepare.

As vector-borne diseases reach into new regions, 3.6 billion more people could be at risk of malaria and dengue by 2070 compared with those who were at risk from 1970 to 1999. If emissions are significantly reduced, this number could be as low as 2.4 billion, but with even higher emissions, as many as 4.7 billion more people could be at risk.

The stakes are high for us to limit global warming to 2°C: doing so would significantly reduce the future suffering of communities and economies as a result of vector-borne diseases. While substantial progress has been made against many of these diseases, the impacts of climate change, coupled with drivers of disease such as rising urbanization, migration and international travel will only further complicate our efforts to control and eradicate them in the coming decades. /

Malaria and dengue may spread into temperate areas such as France, Bulgaria, Hungary, Germany and the eastern coast of the United States.

Antibiotic Resistance

John Brownstein, Derek MacFadden, Sarah McGough and Mauricio Santillana

The discovery of antibiotics, one of the greatest medical achievements of the last hundred years, has provided humans with an effective way to combat bacterial infections. Antibiotics have saved countless lives and their use has become essential to the practice of modern medicine. Unfortunately, they appear to be a fading resource. The more we use antibiotics, the sooner they become useless, as bacteria tend to become resistant to antibiotic agents over time. This is 'survival of the fittest' in action: evolutionary selection favours the proliferation of antibiotic resistance genes and bacteria (including highly resistant 'superbugs').

Around the world, infections caused by antibiotic-resistant bacteria are increasing. It is estimated that every year they cause tens to hundreds of thousands of deaths, and billions of dollars in economic losses. Antibiotic resistance has gradually emerged as one of the greatest public health challenges of our time. Precisely how its rise might relate to another critical public health challenge – anthropogenic climate change – is one of the more pressing questions we now face.

Over the last five years, there has been accumulating evidence that the development of antibiotic resistance in bacterial pathogens may be linked with climate, including ambient temperature: studies have shown correlations between higher ambient temperatures and a higher extent of resistance in the bacteria that cause some of the most common infections in humans. Indeed, some of the most resistant bacterial infections of recent decades have been seen in warm central latitudes. The resistant bacteria that cause these infections and, in some instances, their genes, have been known to spread globally from their presumed geographic origins in a prolific manner. Exactly how this happens is still unclear, but it could be occurring through manifold pathways, including in the gut or on the skin of humans or animals, in food products, or through the environment, for example through waterways.

Troublingly, the rate at which antibiotic resistance develops also seems to be correlated with climate, with warmer areas showing more rapid increases in resistance prevalence over time. It is challenging to completely disentangle the role of climate factors from other regional characteristics, but the growing body of evidence suggests that the warming climate may be playing an important role in fostering antibiotic resistance.

These findings seem plausible, especially when we take into account how temperature can impact the life cycle of bacteria and human or animal activities in general. We know that bacterial survival and growth rate are highly dependent upon temperature, so it is not a great leap to expect that human and animal colonization, as well as environmental persistence, are likely influenced by ambient temperatures. This is supported by studies that have shown clear seasonal variations in rates of infections that originate from bacteria commonly found in our own flora, where infections involving the skin, urinary tract and bloodstream have all been shown to increase during warmer months. Temperature can also modulate the propensity for antibiotic resistance gene clusters to be transferred between bacteria. One of the more worrisome resistance genes (NDM-1) has been found in pools of water in city streets and in drinking water from New Delhi. It confers resistance to some of our strongest and most commonly used antibiotics. NDM-1 typically resides in a mobile set of genes that can move between bacteria, and the same study from New Delhi found that this transfer occurred most frequently at common daytime temperatures for the region. It may be that, by aiding the spread of resistant organisms and their genes, warmer temperatures are facilitating the effective selection and propagation of resistance.

There are significant challenges in trying to estimate the current cost of rising antibiotic resistance, let alone the future cost. Some projections suggest that, by the mid-twenty-first century, we could reach millions of attributable deaths per year, and trillions in economic losses. But these numbers are fundamentally limited by imperfect surveillance and economic growth assumptions – and they crucially do not account for factors such as global warming that could accelerate the propagation of antibiotic-resistant bacteria and resistance genes. The imminent patterns of climate change could generate a knock-on effect that would significantly worsen the impact of antibiotic resistance globally over the coming decades, accelerating the loss of our best tool to combat bacterial infections. /

Food and Nutrition

Samuel S. Myers

In February 2020 a Kenyan farmer named Mary Otieno looked out across cornfields blanketed under a voracious swarm of desert locusts that stretched for 2,400 square kilometres. While recent changes in ocean and atmospheric circulations had helped to drive this historic infestation, rising temperatures and more extreme precipitation patterns had already been taking a toll on Mary's harvests, reducing yields and causing crop failure. A month later, following a 'spillover' event that had allowed pathogens to move from wildlife populations to humans, the Covid-19 pandemic arrived in Kenya, threatening the health of Mary's family and creating labour shortages and supply-chain disruptions that impeded their access to the supplies and equipment needed on the farm. And all the while, invisible to Mary, her crops were slowly becoming less nutritious as a result of rising carbon dioxide concentrations in the atmosphere. What each of these disasters – some abrupt and some slow-moving – has in common is that they all arise from human disruption of our planet's natural systems and rhythms. Understanding how these accelerating disruptions are impacting human health and well-being is the focus of the field of planetary health. Planetary health is teaching us that everything is connected – that our transformation and degradation of nature come back to affect us, not always in ways we would expect. And nutrition is one of the most alarming ways in which our actions are coming back to haunt us.

In experiments performed on locations across three continents, my research team has found that staple food crops such as rice, wheat, maize and soy are losing nutrients that play a central role in maintaining human health. We grew crops at a carbon dioxide concentration of 550 parts per million (ppm), the level the world is expected to reach around the middle of this century. We found that crops grown at these elevated CO_2 concentrations had significantly lower amounts of iron, zinc and protein than identical cultivars of those crops grown at today's CO_2 levels. In other words, our steady addition of CO_2 into our planet's atmosphere is making our food less nutritious. Subsequent studies have revealed that several varieties of rice also experience large reductions in important B vitamins such as folate and thiamine in response to higher levels of CO_2.

So how do these nutrient reductions in zinc, protein, B vitamins and iron affect human health? In modelling studies, we found that these nutrient shifts would likely push 150–200 million people into having deficiencies of zinc and protein, in addition to exacerbating existing deficiencies in roughly 1 billion people. Zinc deficiency leads to increased mortality from infectious diseases in children, and protein deficiency leads to increased child mortality as well. When we analysed the impact of reductions in B vitamins in rice, we found that just the effect in rice alone, even assuming no effect in other crops, could lead to an increase of 132 million people suffering from folate deficiency, which causes anaemia as well as neural tube defects in infants. We estimated that 67 million more people would suffer from thiamine deficiency, which can cause nerve, heart and brain damage. For iron, we found that in countries with anaemia rates that are higher than 20 per cent, the most vulnerable populations – 1.4 billion women and children under five – would lose at least 4 per cent of their dietary iron because of this CO_2 effect on crop nutrients. Iron deficiency leads to anaemia, maternal mortality, increased infant and child mortality, and reduced work capacity.

Surging levels of greenhouse gases in the atmosphere are not the only human-caused change threatening our own health and nutrition. We are also driving species extinct at a thousand times the baseline rate and have reduced the populations of birds, fishes, reptiles, amphibians and mammals by two thirds since 1970. Insects have been particularly hard hit. A study across protected areas in Germany, for example, found a greater than 75 per cent decline in flying insects in only twenty-seven years. Some of these insects play a key role in providing humanity with nutritious diets: a large share of the total calories and an even larger share of nutrients in the human diet come from crops that depend on animal pollinators. In our research, we found that a total collapse of insect pollinators would lead to up to 1.4 million excess deaths every year. Most of those deaths are caused by heart disease, strokes and certain cancers that could have been prevented by consuming fruits, vegetables and nuts which require insect pollination. In a study that is currently under review, we estimate that nearly half a million deaths are occurring annually because of insufficient wild pollinators today.

Rapidly changing environmental conditions affect other dimensions of food production beyond agriculture. Fishermen are exploiting roughly 90 per cent of global fisheries at, or well beyond, maximum sustainable limits, and the global fish harvest has been steadily declining as a result since 1996. Ocean warming is likely to worsen these trends by reducing the size and number of fish while driving fisheries away from the tropics and towards the poles. From a human nutritional standpoint, these trends are concerning because over 1 billion people depend on wild-harvested fish for key nutrients such as omega-3 fatty acids, vitamin B12, iron and zinc.

For families like Mary Otieno's, nutrition is being squeezed in a vice of interacting, human-caused global environmental changes. Every other dimension of human health – infectious disease exposure, non-communicable diseases and mental health – is also endangered by these and other disruptions of Earth's natural systems. Protecting our planet is no longer only an environmental priority but has also become central to securing a liveable future for humanity. /

Nutrition is one of the most alarming ways in which our actions are coming back to haunt us.

Next pages:
A fishing village in Andhra Pradesh, south-east India, encroaches on a mangrove forest, revealing the deterioration of this critical coastal ecosystem.

3.8

We are not all in the same boat

Greta Thunberg

Our rapidly disappearing carbon budgets should be seen for exactly what they are: a limited natural resource that belongs equally to all living things and beings. The fact that 90 per cent of the remaining budget for us to have a 67 per cent chance of staying below 1.5°C has already been spent – predominantly by the Global North – cannot in any way be overlooked. Nor can the fact that rich countries – like my own – are currently consuming what is left of that budget at an infinitely higher pace than those who have historically been exploited by those same nations.

If everyone lived like we do in Sweden, we would need the resources of 4.2 planet Earths to sustain us. And the climate targets set in the Paris Agreement would be but a very distant memory – a threshold that we would have crossed many, many years ago. The fact that 3 billion people use less energy, on an annual per capita basis, than a standard American refrigerator gives you an idea of how far away from global equity and climate justice we currently are.

The climate crisis is not something that 'we' have created. The worldview that largely dominates the perspective from Stockholm, Berlin, London, Madrid, New York, Toronto, Los Angeles, Sydney or Auckland is not so prevalent in Mumbai, Ngerulmud, Manila, Nairobi, Lagos, Lima or Santiago. People from the parts of the world that are most responsible for this crisis must realize that other perspectives do exist and that they have to start listening to them. Because when it comes to the climate and ecological crisis – just like most other issues – many people living in rich economies still act as if they rule the world. They may have left many colonies to govern themselves, but instead they are colonizing the atmosphere and tightening their grip on those who are most affected and least responsible.

By using up the remains of our carbon budgets, the Global North is stealing the future as well as the present – not only from their own children but above all from those people who live in the most affected parts of the

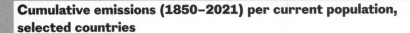

Cumulative emissions (1850–2021) per current population, selected countries

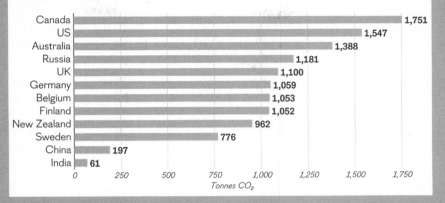

Country	Tonnes CO$_2$
Canada	1,751
US	1,547
Australia	1,388
Russia	1,181
UK	1,100
Germany	1,059
Belgium	1,053
Finland	1,052
New Zealand	962
Sweden	776
China	197
India	61

world, many of whom are yet to build much of the most basic modern infrastructure that others take for granted – roads, hospitals, electricity, schools, clean drinking water and waste management. And still, this deeply immoral theft does not even exist in the discourse of the so-called 'developed' world.

There are numerous things that we can and should be celebrating, like the incredible developments in renewable energy; the fact that more and more people are becoming aware of our situation; that journalism is starting to take its first baby steps towards covering this crisis and holding those responsible to account; the fact that we can spread information, facts, solidarity and ideas around the entire globe within just minutes and hours. And the people who have been most affected by this crisis, which they did so little to create – the communities that Saleemul Huq, Jacqueline Patterson, Hindou Oumarou Ibrahim, Elin Anna Labba, and Sônia Guajajara discuss in their chapters – have shown remarkable leadership and a willingness to teach the rest of us what they have learned. And, of course, we do still have time to avoid the worst consequences of the crisis.

But this is not the main source of optimism in most societies. Instead, when we are communicating the current best available science, we are being told to focus on the possibilities and opportunities – on the 'green industrial revolution' (whatever that means), on positive stories. We want solutions-based reporting, and hope. But hope for whom? The relatively few of us who might initially be able to adapt to a rapidly warming world? Or for the overwhelming majority who will not be so fortunate? What does hope even mean in this context? Is it the notion that we can maintain a system that is already doomed? That we do not have to change? That we can go on living our lives more or less the same way we do today – in a system which most people do not benefit from? That we can 'solve' this crisis with the same methods and mindset that got us into it in the first place?

Progress is being made, they say, and those positive developments are what we should be celebrating. So, if that is the case, what exactly are these improvements? Maybe the fact that we have significantly lowered our emissions while at the same time maintaining economic growth? Yes, of course. But is that really true? Let's look at two examples.

First, the UK. Its governments constantly repeat the fact that the country lowered its territorial emissions by 43 per cent between 1990 and 2018. But, once you include emissions from the consumption of imported goods and international aviation and shipping, that figure is more like 23 per cent. The only reason their consumption based emissions have gone down is because of domestic action in the power sector, not because they have reduced the carbon intensity of imports. In fact, the embodied emissions associated with imported goods consumed by UK residents has only been reduced by 19 per cent. Then there are the missing 13.2 million tonnes of CO_2 released each year from wood burning at Drax – as just one example – to be added into the equation, as well as emissions associated with the military which are very significantly under-reported. The fact that the UK is also currently producing some 570 million barrels of oil and gas each year, and holds a further 4.4 billion barrels of oil and gas reserves waiting to be extracted from the continental shelf, complicates its claim of climate leadership further still. We must also keep in mind that – as in many of other national case studies – the remaining reductions include the easy option of swapping coal for slightly less disastrous fossil gas, locking in many decades of further greenhouse gas pollution.

The second example is the country in which I live, Sweden, which keeps proudly reminding its citizens that we have lowered our emissions by around 30 per cent since 1990. But again, once you include emissions from international aviation and shipping as well as the biogenic emissions that were lost in the loopholes of the Kyoto Protocol, our emissions have not decreased at all. On the contrary, once you add up all available data from this time period, they have actually increased.

So what we are actually being told to celebrate – over and over again – is outsourcing, excluding emissions, clever accounting and the negotiation of global frameworks that make it all perfectly legit. Or, more to the point, we are being told to celebrate the fact that we are cheating. Meanwhile, people around the world are being hit by droughts, crop failures, plagues of locusts and widespread starvation. Other entire nations are drowning in the sea. And this is happening at around 1.2°C of global average temperature rise.

Since the founding of the IPCC in 1988 our emissions of CO_2 have more than doubled. One third of all anthropogenic CO_2 emissions have been emitted since 2005. And, according to a recent investigation by the *Washington Post*, the data on which our climate policies rely are built on flawed, under-reported numbers. There is a clear gap where up to 23 per cent of our total CO_2 emissions are missing. This is the progress that the people in power have created over the past three decades. The progress that they say should not be dismissed as *blah blah blah*.

'We cannot put a price on human lives,' our leaders told us as they shut down our societies in order to gain control over the Covid-19 pandemic. At the time of writing, 5,467,835 lives have been lost in a tragedy that will be remembered for centuries. But every year, 10 million people die from air pollution, as Drew Shindell explains in his chapter in this part. I guess that some human lives are valued less. And if you live in the wrong region, have the wrong nationality or simply die in the wrong part of the world, then the risk is that your death will not matter. Or, at least, it will matter significantly less. There will be no lockdowns to keep you safe, no daily press conferences.

When it comes to the climate crisis, things are no different. With the policies we currently have in place, we are heading for 3.2°C of global warming by the end of this century. This equals disaster. But we are still not reacting. In fact, we are still speeding in the wrong direction. Maybe the reason for this is that the people in power somehow still believe that we can adapt. Inhabitants of financially fortunate parts of the world might feel the same way. And that could well be the reason why referring to the scientific facts is being so broadly dismissed as 'doom and gloom'. *There is no reason to panic or worry – if you live in Germany, Australia or the US you will be fine. Just turn up your air-conditioning or your sprinkler and relax.*

Fridays For Future and the school strike movement exist in all corners of the world. In countries like mine we are being told, *Don't worry, because even if your friends and colleagues can't adapt, you'll be fine.* And if that is not ecofascism – or racism – I do not know what is. We are all in the same storm, but we are definitely not all in the same boat.

The longer we pretend that we can solve this crisis without treating it like a crisis, the more invaluable time will be wasted. The longer we pretend that we can adapt to an interconnected catastrophe, the more priceless lives will be lost. There will only be hope if we tell the truth. Hope is all the knowledge that science has given us to act upon, and the stories from people who are brave enough to speak up, like those in the pages that follow. /

3.9

Life at 1.1°C

Saleemul Huq

The issue of climate change has evolved over time. It rarely stands still, and it's certainly not the problem we thought it was thirty years ago – it's worse. One of its biggest evolutions took place on the precise date of August 9, 2021. This was the day when climate change officially arrived – the date when the IPPC Working Group I, a committee of international scientists, published their sixth assessment report. These scientists are extremely competent, and they are also extremely conservative – they don't stick their necks out. And they'd never said before what they said this time. For the first time, they stated, 'It is unequivocal that human influence has warmed the atmosphere, ocean, and land,' and that, due to human-induced climate change, the global temperature has been raised by 1.1°C. We are no longer just anticipating or planning for climate change, it is here. And its fingerprints can be seen on every region of the planet.

Every year now, across the globe, extreme weather records are being broken, be it a heatwave, a typhoon or a torrential downpour. Somewhere in the world, records are being broken. And every year that will continue to happen, and every year it's going to get worse than the year before. Our global effort to try to keep the temperature rise below 1.5°C is a long-term strategy – it's for the future. But we have crossed the 1.1°C threshold already, and 1.1°C is causing damage now. And therefore, for me, the question of how we deal with 1.1°C is far more important than the question of how we prevent 1.5°C – but we haven't confronted it yet.

The leaders who came to Glasgow for COP26 in November 2021 simply don't understand this. They are living as if the impacts of climate change are something we can still prevent. But they can no longer be prevented. We are now in an era of 'loss and damages'. 'Loss' refers to something completely lost, in the way that a human life is lost: once it's lost, it does not come back; no matter how much money you have, it's gone. It's the same with species loss or the loss of an ecosystem. Once it's gone, it's never coming back. It has disappeared, like an island that's now underwater because of rising seas. 'Damages' refers, on the other hand, to something that can be repaired, if you have the money or the resources. Money is necessary, but it can be done. Crops that are lost can be recovered by the next harvest. A home destroyed in a storm can be rebuilt.

'Loss and damages' is also a diplomatically negotiated euphemism for something we're not allowed to talk about: 'liability and compensation'. Those are taboo words, particularly for diplomats from the United States and other rich countries. Everyone can understand the idea that polluters are liable for causing pollution, and that those who suffer the impacts want compensation. But rich, polluting countries dictated during the discussions for the Paris Agreement that we could not talk in those terms – yet another consequence of the unequal world we live in, whose legacy continues in global talks today. Governments thus far have not demonstrated an ability to act globally; they act nationalistically. The Covid-19 pandemic and subsequent vaccine distribution are clear examples of countries thinking they can look after themselves and in so doing prevent problems from becoming worse. It is morally wrong and scientifically incorrect. However, this way of thinking is very strongly engrained.

We need to think about global injustice now. The manifest injustice whereby polluters – largely rich people around the world, who are most responsible for carbon emissions and environmental harm – are hurting poor people. The communities impacted by environmental degradation and climate change are overwhelmingly poor people of colour, even in rich countries such as the United States. We all saw the tragedy that befell the Black community in New Orleans after Hurricane Katrina. And this disparity in consequences is a global phenomenon. Bangladesh, my home, is facing a slow-moving disaster as sea-level rise threatens the coasts, which may lead to the displacement of millions of people.

But the story of Bangladesh is not a story of victims. It's a story of heroes, a story of the future of the planet. The rest of the planet is going to face tomorrow what we are facing today, and the rest of the planet is going to have to come and learn from us how to deal with this problem. We don't have all the answers, we don't have all the solutions, but we are moving up the learning curve very fast, and I can share a couple of lessons we have learned. The first lesson is that you can have all the money and technology in the world, but it's not going to help you. It's not going to prevent death and destruction arriving in New York City. Hurricane Ida flooded the subway system and a number of poor people died in their basement apartments because they could not get out in time. You can build a barrier, as London has, to protect a city from flood, but you can't build a barrier around an entire country. The UK is very vulnerable to impacts of climate change. Money and technology have a role, but they are not enough on their own.

What is truly important in times of crisis is social cohesion – people helping each other – and we have that in droves in Bangladesh. Whenever we are hit by extreme weather, we go out and we help each other. Nobody is left behind. Schoolchildren have drills so they know where they need to go to evacuate in an emergency, and whom to help – an elderly widow living alone will have two children from the high school assigned to go and pick her up. The hurricane still comes, and it still does a lot of damage, but it won't kill people on the scale it once did. And the primary reason is that we work with each other, we help each other – everybody is in this together. This is not the case in many developed countries. Rich people can lead isolated lives, perhaps not even knowing their neighbours. But working as a community, as we do in Bangladesh, helps to build resilience and the ability to deal with crises when they occur.

The second lesson we have to offer is that young people make all the difference. Once they organize and are given support and guidance, they can be an extremely powerful force. Tackling climate change requires a change in perspective that older people may find difficult; this is one of the reasons why our leaders are unable to understand the paradigm shift that is required. They are the ones who are not changing fast enough, who are preventing change, resisting it, everywhere. Young people can make them change. This is true in Bangladesh, it's true in the US, it's true in Germany, it's true in Sweden. The paradigm shift that we now need is for these young people to be a global force – and this is where we are ahead of the game in Bangladesh. Our kids are not just protesting every Friday, they're spending the whole week going out and helping people, preparing our communities for the impacts of climate change.

As we learn to live with 1.1°C of warming, we need to find ways of thinking about global climate change that empower us, that give us agency. We must recognize that we are part of the problem – we are all polluters by virtue of what we eat and how we live. And this means we can do something about the problem and that we must reduce our emissions wherever we can. However, there is a limit to how much one person can actually do – you can't cut your own emissions to zero, and we don't expect you to. But you need to go beyond simply doing your little bit. You need to do more than just changing your own lifestyle. You need to act with other people, join forces with them, and that's what young people are now doing. Get together with like-minded people in your workplace, in your school, in your village, in your town, in your apartment building, wherever you are – find allies who will join you and then take action: become political. You have to organize at a scale where you can actually change politics. You can influence your political leaders, and no matter the level of democracy or type of government in the society where you live, there's always some opportunity to make a difference and to exert pressure on political leaders. It's a tough task, but it is something that can be done. *You* can make a difference at a global scale. Start local, but aim global. /

We need to think about global injustice now. The manifest injustice whereby polluters – largely rich people around the world – are hurting poor people. This disparity in consequences is a global phenomenon.

3.10

Environmental Racism

Jacqueline Patterson

Serving as a Peace Corps Volunteer in Jamaica in the early 1990s was when I truly began to wake up to the harsh reality of global injustice. I stayed in Harbour View, a community just outside the capital, where the water supply had been contaminated by major transnational oil companies and virtually no recompense had been offered. For my volunteer assignment, I was working with hearing-impaired toddlers who had been ravaged by a rubella outbreak that could have been prevented with vaccinations. Squatter settlements, a sign of extreme poverty, dotted the area – despite the tremendous wealth from the tourism industry which was going into the hands of a privileged few. These were manifestations of a centuries-old global capitalist system predicated on ruthlessly exploiting human beings and extracting natural resources – that is, a white-supremacist economy where there are winners and losers, with a consistent separation along lines of race, gender and nationhood clearly delineating who dominates and who is oppressed.

Jamaica shares historical underpinnings with the country I live and work in – what's now called the United States. In the founding myths of both countries, sanitized fantasies prevail: European adventurers valiantly sail across the seas to discover new lands and bring back rich cargoes of silks and spices. But this leaves out the reality of murders, thefts, diseases and displacements. Soon after arriving to stolen lands across the Americas, white explorers deemed the original Indigenous inhabitants to be inferior and disposable. As such, they proceeded to murder and enslave the Indigenous communities they encountered, driving them off the land. Meanwhile, in sub-Saharan Africa, people were being stolen from their homelands, placed as cargo on ships and brought to the western hemisphere to be the enslaved labour that built the infrastructure and worked the land, laying the foundations for the Industrial Revolution and the modern capitalist economy. While enforcing dominion over people, these colonists institutionalized a relationship with the land and its bounty that was rooted in reckless extraction.

Fast-forward to today, and the ravages of white supremacy, upheld by the extractive economy, persist. Dehumanization and exploitation based on race continue to be the modus operandi. In the United States, beyond the profound suffering of Black people at the hands of the police forces and the prison industrial complex, BIPOC (Black, Indigenous and People of Colour) communities are considered disposable in service to moneyed interests across the country. Sacrifice zones – areas that border environmental threats or dangerous pollution – are overwhelmingly populated by low-income families and people of colour. These hazardous areas have developed in Crossett, Arkansas; East Chicago, Indiana; Wilmington, Delaware; and beyond. One sacrifice zone, Reserve, Louisiana, an African American community, has been dubbed 'Cancer Town'. With emissions of chloroprene (a known carcinogen) at 755 times the guidance put out by the Environmental Protection Agency, residents have a risk of cancer that is the highest in the country, at fifty times the national average. The chemical plant responsible for some of the most toxic air in the country was built on the land of a former plantation that was worked by enslaved persons. Meanwhile, extreme storms and flooding have ravaged BIPOC communities in Alabama, New York, Louisiana and Florida, time and time again. And urban, primarily BIPOC communities are being baked by the increased heat island effect as temperatures rise and heatwaves become more frequent and intense.

To illustrate the front-line impacts of our extractive economy, the story of a young boy named Chauncey, of Indiantown, Florida, provides a cautionary tale. Chauncey is caught in the crossfire of environmental racism and climate injustice. His home is 2 miles away from a coal-fired power plant. Chauncey, like 71 per cent of African Americans, lives in a county that is in violation of federal air pollution standards. He is dependent on a bag of medicines for his lungs to function well enough for him to draw life-sustaining breath. Chauncey lives with asthma, and black children in the United States are three to five times more likely than white children to enter hospital with an asthma attack and two to three times more likely to die from one. He is unable to go to school on poor-air-quality days, when even the medicine isn't enough to protect him from an attack. On top of the air pollution, his hometown is deemed to be at 'very high' hurricane risk, exacerbated by climate change, with seventy-seven major hurricanes passing through his community since 1930.

Beyond our borders, the United States makes up merely 4 per cent of the global population, yet it is responsible for 25 per cent of the historical emissions that drive global climate change. There is a direct link between what Americans are doing and the havoc wreaked on the Global South in the form of droughts, floods and other disasters. Yet when people in neighbouring nations are driven from their lands due to US excesses, in border areas in Texas such as Laredo and Del Rio, they are greeted by horseback riders using their reins as whips and officials putting their children in cages.

The good news is that some of the most vibrant and inspiring alternatives to this extractive, racist system are being led by BIPOC communities. Incinerators and coal plants are closing, and the Dakota Access and Atlantic Coast Pipelines have lost their permits, while the struggle against Line 3 – which would double the amount of oil going from the Alberta Tar Sands to northern Wisconsin – rages on. From Brooklyn, New York, to Boise, Idaho, to Laredo, Texas, sparks of light can be seen in the actions of people from the communities that are most at risk. These range from local food movements to recycling initiatives, clean energy projects, and more. One illuminating example comes from the Jenesse Center in Los Angeles, which takes an intersectional approach to advancing liberation and sustainability, showing how we can move from away from an extractive economy and towards a living one.

The Jenesse Center, a domestic violence prevention and intervention organization, serves a primarily African American population of survivors of domestic violence. For years, one of the centre's biggest expenses was the cost of providing electricity for transitional housing, so they made the decision to move to solar energy. The seven residents then living in the centre's temporary homes were trained in solar installation and became part of the workforce that installed the new solar electric system. Now, three years later, these former residents are gainfully employed in the solar industry and living independently with their children. Their new career has afforded them jobs and housing security, as well as making them at less risk of returning to their abusers: financial security is one of the reasons that too many remain in harm's way. This one project has resulted in a decrease in the centre's greenhouse gas emissions, provided more financial resources for domestic violence services and given numerous families a more secure life.

Sustainable, living economies like this one are already forming and thriving. If we are to achieve this on a global scale, where an abundantly diverse society lives in harmony with the bounty of the Earth, we need everyone to get on board with regenerative, cooperative systems rooted in deep democracy. And Mother Earth – together with the communities of Reserve, Indiantown, Los Angeles, Laredo and beyond – is telling us that pivoting from a global extractive economy to a global living economy is a critical imperative. /

3.11

Climate Refugees

Abrahm Lustgarten

When El Salvador turned suddenly arid, Carlos Guevara knew to call it something other than drought. It was more like the world had changed.

The first year, the 4 acres of maize he planted along the Río Lempe, near where it spills into the Pacific, grew only to hip height, then it shrivelled in the heat. His usual forty-bag harvest was reduced to five. The following spring – in 2015 – things got even worse. Throughout May, June, July and August, in this lush and jungle-like region, it did not rain.

Guevara – whose parents had migrated to El Salvador from Palestine during the Second World War – had known hardship. His village is named Catorce de Julio, to commemorate the day in 1969 when the Salvadorean military clashed with Honduras, instigating two decades of violence and civil war. Over those years, 80 per cent of the village's inhabitants – some 7,000 people – were killed or fled the country. Guevara survived all of that, and in the 1990s was one of the first to return home to 14 Julio. He came back because he believed in the land's promise, that the water was plentiful and with hard work the earth would provide maize, cucumbers, chillies and more.

Now even that was being taken away.

'When I lost my crops, I felt like the sky had fallen,' Guevara said. He is muscular, with close-cropped hair and a widow's peak that makes him look younger than his forty-two years, and he gestures energetically when he talks, as if juggling an invisible ball. 'You always have that intention to give something better to your children – or at least for them not to miss out on any meals.'

In 2016, the banks that hold Guevara's land as collateral for seed warned him that the next growing season would once more prove futile. His family was spending their savings in order to buy food he once grew. Meanwhile, violent gangs tried to recruit his children and sought 'rent' from the family. Guevara's wife, Maria, started to sell pupusas from a rented window of a shop facing the road so that they 'at least have a little money left for [their] son's milk'.

Across Central America, more than 3.5 million lives have been upended by the dry years that began in 2014. Half a million of those people across

El Salvador, Guatemala and Honduras faced immediate and acute malnutrition – even starvation – as farmers struggled to produce food; rations of rice had to be distributed. Worse still, the La Niña cycles – the weather phenomena likely to blame for the perilous drought – were increasingly frequent, a trend that will continue as long as fossil-fuel-driven emissions and human industrial activity keep heating up our planet. Others in Guevara's village and elsewhere had started to leave their homes. The land – indeed the entire natural system – that Guevara depended on wasn't just failing him, it seemed to be pushing him out.

One sultry spring evening after that last failed growing season, Guevara told his wife that he saw just one way forward: he would have to travel north to find work.

The following dawn, with little more than a change of clothes and $50 tucked into the sole of his shoe, he walked several kilometres to the nearby town of San Marcos Lempa and caught a bus to San Salvador, then another through Guatemala to the Mexican border, near Tapachula. He took taxis in order to avoid checkpoints, until he reached the town of Arriaga, where he clambered aboard La Bestia (The Beast), a lumbering freight train on which migrants stow themselves away for the harrowing ride north.

For two days, Guevara curled inside a small cage at the end of a cylindrical grain car, the only place he could rest without falling out of the train. Later, as the train passed through Veracruz and the weather turned colder, he crawled inside the tank, burrowing into the maize to keep warm and hide from the cartels that prey on migrants. After weeks of travel, Guevara waded across the Rio Grande and walked into the barren American desert, one of roughly half a million migrants to make their way across the US border from Central American countries that year.

Around the world, rising temperatures and climatic calamity are unsettling ever larger numbers of people. As droughts, floods, storms and heat make it more difficult to farm, work, and raise children, populations are moving in search of temperate conditions, safety and economic opportunity. Food insecurity is fast becoming the planet's most significant human threat, leading the world to the precipice of a great climate migration.

For 6,000 years, humans have lived within a relatively narrow band of environmental conditions, seeking out a mild mix of precipitation and heat roughly equivalent to the climates of Jakarta and Singapore on one end of the scale and to those of London and New York on the other. Today, just 1 per cent of the planet is considered too hot and dry for civilization. But, by 2070, researchers have concluded, 19 per cent of the planet – home to some 3 billion people – might be uninhabitable. This suggests that the world is about to see hundreds of millions of people displaced, and billions more suffer, as the fastest and most disruptive change in recorded history furiously unfolds.

Mass migration at this level will be globally destabilizing. While good can come from such change – the US, after all, is a product of immigration – the enormous scale of what's to come is more likely to foster competition and conflict, as ever larger numbers of people fight over ever scarcer resources while, at the same time, geopolitical powers erect walls, fences and boundaries to keep migrants out. The world's leading security and defence institutions are already warning that climate migration could lead entire nations to crumble, while shifting the balance of power and advantage to other countries, namely Russia and China, that will be willing to leverage such power.

The hotspots for change are just where one might expect them to be: across the equatorial regions that are already warmest, and which also have the largest – and fastest-growing – populations. Sub-Saharan Africa is home to roughly a billion people, and its population could double over the coming decades. The Sahel region in particular, with a population on track to reach 240 million by mid-century, faces the most severe water crises in the world and is already seeing the greatest numbers of internally displaced people. The World Bank estimates that Sahel nations could see as many as 86 million more people displaced by climate stress factors by 2050.

South and East Asia are other epicentres where enormous populations and uninhabitable heat and humidity are on a collision course. The World Bank has calculated that roughly 89 million people in those regions will be internally displaced.

Central America is just one more focal point for that change, and an important one. Climate models project that the region will be among the world's fastest-warming places, experiencing longer droughts, shorter growing seasons and bigger and more destructive storms. The World Bank estimates that Central American nations will see as many as 17 million people internally displaced by climate stress factors by 2050, a figure which does not estimate how many will move northwards, like Guevara, to the United States. But the numbers could be even higher.

To try to understand how future migrants might move, I worked with demographer Bryan Jones at the City University of New York to build a computer simulation like the one used by the World Bank. Adding in the complexity of drought risk and moving across borders, the modelling suggested that by the middle of the century some 30 million Central Americans would migrate to the southern US border, influenced at least in part by climate factors.

But the model also suggested that different policy approaches to the challenges of climate change and migration would lead to distinctly different outcomes, so it is clear that the choices leaders make today will determine how the future plays out. In a harsh world which is seeing maximum climate warming along with strident anti-immigration policies and strict border controls, a world in which less and less money is being sent to developing nations as economic aid, more people will ultimately be displaced and their suffering will be greater. However, in a world where the warming of the planet is slowed and governments continue to support needy regions with foreign aid, there may ultimately be less displacement of people and greater stability.

Almost immediately after his arrival in the US, Carlos Guevara was captured and sent back home. He had hitched a ride in the desert, and the driver, speeding, was pulled over by the police. Guevara arrived back in El Salvador to find that his village had changed. Others had also fled the drought, migrating to the US and nearby cities, and the village seemed empty. However, at the same time, a United Nations World Food Programme (WFP) project had come to 14 Julio, offering farm aid and irrigation, and the hope that it might be possible for Guevara and others to dramatically improve their prospects for survival.

Guevara and I met on a bright, hot morning, in one of his fields. Leaves crunched beneath the cracked soles of his boots as he walked along rows of failed crops marked by grape stakes, trailing his fingers along once-supple vines that had become brittle. The dry field was a monotone brown. His son threw a stone into a shallow well. It landed with a thud, not a splash.

But as we continued past that field a new structure came into view: a metal-framed, plastic-walled greenhouse. The project was built as a WFP pilot project to build communal farms across El Salvador, and inside, humid air and orderly drip lines surrounded bountiful rows of broadleafed, healthy pepper plants and juicy tomatoes – more than enough to feed Guevara's family, and plenty to sell for profit. Guevara had already reinvested the revenue from the first crop to expand his farm and buy a cow for milk, and his family was doing better than they had throughout the previous five years.

His future, though, remains precarious. The fate of the WFP project depends on the willingness of foreign donors to send more money. And Guevara knows that in five more years the climate will have changed for the worse. The greenhouse gives him reason not to try migrating north again – for the moment. But he has learned that he can't trust in what the future might bring.

'Hope is the last thing that you lose,' he said. 'As long as climate change continues, we will never have confidence that we can be fed.' /

Sea-level Rise and Small Islands

Michael Taylor

Sea-level rise is one of the foremost challenges that climate change poses to small islands like the one which I call home in the Caribbean. Very often, the associated image is of entire islands on the verge of being swallowed up by the oceans. This is not an unreasonable picture: if emissions continue at current rates, sea-level rise of a metre or more is projected by the end of this century. Even if our efforts to limit global warming are moderately successful, some of that future rise is already locked in, and swathes of low-lying islands will be engulfed. This means that the existential threat from sea-level rise is all too real, and the image of disappearing islands in the future should be sufficient to rally global action around climate change. Yet even before we reach that calamitous stage, sea-level rise entails serious losses for small islands, which we can observe around us today.

Every 'small islander' can identify a point in the ocean which once used to be on land. Rising waters are eroding island beaches and coastlines which directly or indirectly support island livelihoods: much of the tourism in the Caribbean is beach-related, with the sector accounting for 7–90 per cent of GDP and, on average, 30 per cent of employment, directly and indirectly. In recent years, many of the highly prized Caribbean beaches have narrowed as they become trapped between rising seas and coastal development. This diminishes their attractiveness to tourists, with cascading impacts on the many people whose livelihoods depend on the viability of the industry. In a bid to protect the beaches and the jobs they create and sustain, Caribbean countries are resorting to expensive infrastructure, including breakwaters or sea walls – though the value of doing so is not yet fully known.

But the impacts of erosion extend beyond tourism, as many small communities are reliant on coastal resources for their survival. Fishing communities develop around beaches, which serve as housing and landing sites and informal trading hubs. Their options are limited when the beaches begin to recede. Fish shops and vending businesses are shuttered up and residents migrate inland in search of alternative livelihoods. As fishing ceases to be sustainable, entire communities eventually relocate. For small islands in the Caribbean, the image of sea-level rise is not just that of disappearing islands in the future. It is one of disappearing beaches, livelihoods and communities in the present.

Increasingly, we are witnessing even graver consequences of present-day sea-level rise. In some places, it is delaying or even undoing countries' development, as it exacerbates the flooding from storm surges due to more intense storms and hurricanes in our warmer world. In the Bahamas, extreme flooding from superstorm Dorian in 2019 caused more than seventy deaths and substantial damage to the low-lying islands of Abaco and Grand Bahama, damage equivalent to a quarter of the country's GDP. Unfortunately, such extreme events no longer seem rare. Only two years prior, three Category 5 hurricanes tracked through the Caribbean, including Irma – which, at the time, was considered the strongest ever open-Atlantic storm – and Maria, which followed just two weeks later. Among the countries impacted were the small island states of Barbuda, Anguilla and the British Virgin Islands, for whom recovery is measured in years due to their contracted economies, depressed living standards and delayed development. Hurricane Irma destroyed 95 per cent of homes on Barbuda and left a third of the country uninhabitable. Even in the absence of an intense hurricane, damage penetrates further inland now than a few decades ago, posing a direct threat to these islands' populations and infrastructure. In the Caribbean, most urban centres are coastal. More than 50 per cent of the region's population live within 1.5 kilometres of a shoreline. If coastal waters rise by a metre, it is estimated that up to 80 per cent of the lands surrounding the region's ports may be inundated.

To envision the impact of sea-level rise is also to picture an inheritance denied. It is contracting habitats, shifting the geographical location of coastal species, diminishing biodiversity and reducing ecosystem services. It is increasing the salinity of coastal aquifers, which are often relied upon by local people as their only source of fresh water. It is threatening many cultural heritage and ceremonial sites, which are located in coastal areas and cannot be moved in response to flooding. It is limiting the availability of beaches as public spaces for relaxation and enjoyment. Access to clean water, vibrant ecosystems, cultural heritage and recreational spaces are all reasonable expectations for the next generation to hold. It is the least we owe them.

With few exceptions, small islands have contributed least to climate change. But they bear the brunt of its consequences. This is not just a question of disappearing islands in the future: it is about threatened livelihoods, delayed development and a generational inheritance being denied today. /

Rain in the Sahel

Hindou Oumarou Ibrahim

In the Sahel, rain is everything. In my community of nomadic pastoralists who live around Lake Chad, we have many words to describe the rain. There are those that announce the beginning of the rainy season and the start of our migration with our herds and those which tell us the dry season is coming and lead us to settle down around the lake. We have words to describe the gentle rains that irrigate our crops and words for those which come in storms to destroy our fields.

In this harsh environment, we have learned to live in harmony with nature. We cooperate with our ecosystems. Our cows fertilize the land all along the path of our transhumance. Every three or four days, we move from one place to another to allow nature the time to regenerate. And we are also living in harmony with our neighbours. In our region, where most people are either farmers or fishermen, our cattle are the only source of soil fertilization, and when we leave a place the land is good for agriculture.

Thirty years ago, when I was born, Lake Chad was enormous. And sixty years ago, when my mother was a child, the lake was almost a little sea in the middle of a desert. But today it's a small drop of water in the heart of Africa. Ninety per cent of our water is gone. Our average temperature has risen. We are now living with temperature increases greater than 1.5°C, meaning that my people already live over the threshold of the Paris Agreement. And this is just a preview of what's to come. According to the new report by the IPCC, we are approaching the gates of a climate hell. In the Sahel, our average temperature could be 2°C higher by 2030, and 3–4°C higher by the middle of this century. Over my lifetime, the face of the Sahel will not be the same.

Most of the rain is already gone. The land is often dry and infertile. Our cows used to produce 4 litres of milk each day, but now they barely produce 2 litres or even 1, because of the missing grass. And more and more often, rain, which had been our ally, is our enemy. Over the last five years, floods have repeatedly destroyed our lands, our houses, the culture of my people.

We now live on the edge of climate wars. People are fighting for the few resources left. When nature is sick in a region where 70 per cent of people depend on it for farming, people lose their minds. The old alliance between farmers and pastoralists has been broken in the competition for nature's bounty. In Mali, North Burkina Faso and Nigeria we have seen villages burned by people who want to grab the land of their former friends.

But for me, the Sahel is still a land of hope. We have so many climate warriors fighting back. In my community, women have already implemented solutions to the changing climate. These Indigenous people are using their traditional knowledge to identify crops which can resist droughts and heatwaves so that we can have resilient agriculture. And in the memories of our grandmothers and grandfathers, we possess a map of ancient sources, sources that still provide water in the middle of the worst dry seasons.

Indigenous people's traditional knowledge not only gives us words to describe the rain, it offers us the tools to fight back against climate change. From centuries of living in harmony with nature, observing the clouds, the birds migrating, the direction of the wind, the behaviour of insects, the behaviour of our cows, we are armed to resist. We may not have had the chance to go to school, but our elders have masters and doctorates in protecting nature and they are becoming specialists in climate adaptation.

We don't want to be merely the victims of climate change. We will do our part. We are doing it already. Our way of life is climate neutral. We are the living proof that it is possible to maintain forests and savannahs and increase carbon stocks in nature while producing food. In most industrialized countries, agriculture is a major source of emissions. In my community, it is a carbon sink.

For a long time, we have been taking care of nature not only for us but also for the seven generations to come. This is how decision-making is done in my community. Before deciding anything important, one should consider what the past seven generations would have done in the situation and what the impact of a decision for the seven upcoming ones will be. It's a way to put intergenerational equity at the core of every important decision.

Now the time has come for the global community to listen to my people, and to help them. For too long, Indigenous peoples have been considered representatives of our Earth's history. But we don't belong to the past: we represent the future.

This is the case for Indigenous communities all over the world. Biodiversity is our best partner. Because we don't regard nature as a tool, something you can own, use and destroy. Nature is our supermarket, our pharmacy, our hospital, our school. And for many Indigenous communities, it is even more than that: it is the essence of our spiritual life, our culture, the source of our language. It is our identity. /

Winter in Sápmi

Elin Anna Labba

Sápmi is beautiful now. The trees are heavy with frost, so white they merge with the clouds. Reindeer are visible on the mire. A calf lies in the snow. She has put her head down, curled up like a soft stone with her vertebrae up against the sky. If you stroke your hand over her woolly winter fur, you can feel her thin heartbeat. She looks peaceful, like a breastfed baby fast asleep.

But the people who have followed the reindeer since they were little, they see. A calf curled up in this way will not make it. They know it's too late. The calf has come here from the summer and followed her mother all the way down from the mountains, but she will go no further. They've tried to feed her, but she was already too weak. She's been starving for too long.

Sápmi is a country within four countries. It is located in the northern parts of Sweden, Norway, Finland and the Kola Peninsula in Russia. The Sámi, Europe's only Indigenous people, have a long tradition of reindeer herding, and a strong tradition of animal care. For as long as we can remember, people here have adapted to the reindeer. It is a life shaped around the snow, because summer in the north is only a bright and brief memory. If you live with snow for much of the year, you learn to follow the shape of the snow cover. This is required to survive. Even before climate change was a fact of life in the world, the unrest had begun to spread like a whisper in the Arctic Indigenous world. Something is happening to the snow. It comes early, and then comes the rain. Then it freezes again. Why do the winters now eat their way into the marrow? The dying reindeer calf's hooves have trampled in snow that shouldn't be there this early.

The place where my family lives is called Dálvvadis in Sámi – the 'winter settlement'. In Swedish, this small community is called Jokkmokk. It is located in the forestland in Sweden, not far from the mountains. Not many years ago, large reindeer herds grazed here during winter. They moved freely and dug into the thick, yet still porous, snow cover. When the snow packed harder as winter wore on, the reindeer stretched up towards the lichen in the trees. In the spring they turned their noses towards the mountains again. This year, Jokkmokk is a landscape of enclosures. The whole community is surrounded by pastures where reindeer are fed, to the north, east, south and west. Bony reindeer are brought into garages so they can regain their body heat. The smell hits you when you open the doors, it fills the nostrils, occupies the crevices. Wild animals should not be locked up; they respond by getting sick. Pus bursts from infected eyes. Their stomachs break.

And the panic spreads in our bodies as well. We took the animals' freedom so that they would not die in the forest, but we were not able to protect them. What is it that we are forever changing? There have always been years of famine and desperation, and for them the Sámi has its own word: *goavvi*. Goavvi is a year of difficult grazing conditions, but the word also means 'harsh' and 'ruthless'. It is a mythical word which spreads fear, especially among the elders. During a famous *goavvi* almost a hundred years ago, the forest seemed to be full of new bushes. Only when you got closer did you see that the bushes were the antlers of dead reindeer, sticking up through the snow.

When the ruthless winters came, the reindeer herders scooped up food by hand and cut down lichen to help the animals. You can work such a long winter if you know that the year of emergency is almost over, that better times are coming.

The elders are now talking about an emergency year that has lasted for over a decade, the end of which we cannot see. Climate change is not a future fear; it dwells in the bones and the skin. 'The world has changed. We keep thinking that's not true, even though we know it's true,' says an elderly reindeer herder. When she was young, there were still pristine forests. Now the clear-cut forests and the wind farms stand on the mountains where the reindeer used to graze. The last migration routes may end up underneath a mine. The ice covering the hydropower reservoirs is weak and unpredictable. The ground is so faint that sometimes it feels like its heart is pounding as weakly as that of the sick reindeer. Yet Sweden still looks to the north and believes that there is more to harvest. Sápmi is the *terra nullius* of the Nordic countries; it is considered empty enough for both green and grey industries. In countries where you have not yet come to terms with your history, you become blind to how it is repeated, how old colonialism just changes shape, finds fresh arguments, new forms. Nowhere in the world can those who are already most affected by climate change control their own story. Past losses of land, language, family and faith sadly prepare Indigenous peoples for this. History and the present walk side by side.

In the classic Sámi poem about the children of the sun, the sun's daughter is troubled. What will happen to the humans? In the poem, the sun is setting, the wolves are coming, sneaking around in the darkness of the night. The sun is sinking, the flock is decreasing. But she is also hopeful, being the daughter of the sun; she has to be. We cannot guard the land if we do not believe in our power to protect it, and the sun's daughter asks herself hopefully, 'Morning is coming, isn't it?'

I think the daughter of the sun was referring to the young, who are now rising up for the morning to come. Here in the north, there are no longer any doubts. The last decade has taught us to wrap down jackets over Arctic animals and to mix sugar dissolved into water with the lichen. Even young children learn how to heal. But, above all, they learn to fight for the forest and the mountains as if they were the last ones, because that's what life tells them when they squat next to the dying calf. Fight for everything as if it were the last, because that's what it actually is. As children of the sun, people must guard the land, otherwise we wouldn't be here. /

In countries where you have not yet come to terms with your history, you become blind to how it is repeated, how old colonialism just changes shape, finds fresh arguments, new forms.

Fighting for the Forest

Sônia Guajajara

The fight against the climate apocalypse is a global struggle that depends on all of us to defend our territories. In every corner of the world, it is fundamental that we fight for the preservation of our ecosystems and allow them to recover from the damage caused by the excessive greed of those who, instead of forest, can see only profit.

I am an Indigenous woman, born in the Amazon. From an early age I understood the fundamental importance of keeping our territories protected, because our people's lives, bodies and spirits are deeply connected to the relationships we have with our land.

Our path has always been the defence of life. Since the first invaders set foot on these lands, which were not then called Brazil, we have lived in a state of vigilance in the face of repeated and constant attacks. The colonizing project usurped our territories, bringing diseases and death to our bodies, and fire and destruction to our biomes. If we have survived to this day, it is because we are tireless fighters, and because we rely on the strength of our ancestors to defend our Mother Earth.

In Brasilia in September 2021, during the Second Indigenous Women's March, we launched the Reflorestarmentes platform, which was created to connect innovative community-based projects on environmental protection and to share Indigenous women's knowledge and wisdom with the world. This is a time when several global crises are devastating humanity and our Mother Earth. The climate and environmental crises, the crisis of an exclusionary and unequal economic system, the crisis of hunger and unemployment, the crisis of hate and hopelessness. These overlapping crises most acutely affect the original peoples of the world, who depend intimately on their relationship with their biomes.

That is to say, those who care most for our planet, our forests, our sources of fresh water, are those who are most impacted by their destruction. And this is an undeniable fact, reinforced by numerous scientific institutions:

Next pages:
Indigenous woman, geographer and environmental activist Hindou Oumarou Ibrahim leads a group of pastoralists in Chad, advocating for ancestral agroecological practices.

the true guardians of the forests, and of the planet, are the Indigenous peoples. Indigenous peoples represent approximately 5 per cent of the global population and occupy no more than 28 per cent of world territories. However, they are responsible for guarding and preserving 80 per cent of the biodiversity that, alongside us, lives upon Mother Earth.

These statistics reinforce what we have been repeating for centuries: there is no possible future for humanity that does not pass through us, the Indigenous peoples. Now I will take this a step further and claim the place of Indigenous women at the heart of the struggle to guarantee a future for humankind. For in many original communities it falls to us, Indigenous women, to manage and preserve our ecosystems and to preserve our knowledge through memory and custom. We have lived in harmony with the forests for thousands of years, and we have shaped them to ensure better living conditions for us, as well as for the forests themselves – they are therefore not wild, as the outsider usually perceives them to be, but cultivated.

We will organize and share our thousand-year-old ancestral knowledge in order to offer humanity a broad project for the future – one that will allow life to continue in a more balanced and equitable way. We are not the owners of the truth, but in the time we have occupied this planet, we – our ancestors – have developed knowledge and technologies that are needed today more than ever.

We need to foster a way of living that harmonizes human existence with the full and powerful continuity of our biomes. And Indigenous women know how to do this, for we are ancestral scientists of life on this planet. And we are willing to share our knowledge so that all of us have a chance of life today and in the future. /

There is no possible future for humanity that does not pass through us, the Indigenous peoples.

3.16

Enormous challenges are waiting

Greta Thunberg

'With current warming trends, 1.2 billion people could be forced to migrate by 2050,' writes Taikan Oki in his chapter. This is another one of those figures that you come across while reading about the emerging climate and ecological crisis. It is almost impossible to fathom and translate all these numbers – all these enormous challenges that are waiting for us further down this road we are still choosing to travel on. Most of those 1.2 billion people will probably be displaced within their own countries, but, given how the world has treated refugees in recent decades, there is reason to believe that it will cause unspoken suffering, widespread human disasters and jeopardize our entire civilization as we know it.

Very, very few people abandon their home because they want to. Escape and flight are natural human instincts, and it is fair to assume that the vast majority of us would do the same thing if we were in their position. But I don't think many people whom we would define as climate refugees would call themselves that. It might have been a flood, a drought, conflicts or a climate-related famine which finally pushed them away, but there would also have been a combination of other factors such as poverty, disease, violence, terror or oppression. It is all interlinked, as Amitav Ghosh explains in *The Nutmeg's Curse*.

No wall or barbed-wire fence will keep anyone safe in the long run. Shutting down our ports and leaving people to drown in the Mediterranean Sea or the English Channel will not make these problems go away. They will keep coming back to haunt humanity until we start making efforts to heal our divisions and share our resources in a reasonable and sustainable way.

Democracy is the most valuable tool we have and, make no mistake, without it we will not stand a chance of solving the problems we face. Just imagine communicating disturbing scientific results or speaking truth to power in a dictatorship. There is no question that a destabilizing climate will lead to a destabilizing world, and that will eventually put everything in our societies at risk, including democracy. The climate crisis will amplify conflicts and societal problems. As Marshall Burke writes in his chapter on climate conflicts, 'The total number of organized armed conflicts around the world is also trending up and is now at the highest level seen in nearly a century, leading to record numbers of internally displaced people and alarming levels of global hunger.' If we fail to address all the deeper issues that ultimately make up this sustainability crisis which is emerging all around us, it will without a doubt lead to further erosion of democracy. In many ways throughout modern history, our dependency on fossil fuels has also played a key role in military conflicts. And still, instead of taking steps to overcome our fossil fuel dependency, we are deepening it. In doing so, we are funding geopolitical powers that clearly work against human rights. We are making ourselves even more dependent on oil, coal and fossil gas from authoritarian regimes, from Putin's Russia to the nations around the Persian Gulf.

As things get worse – and they will – we are likely to see more and more authoritarian politicians stepping up to offer easy solutions and scapegoats in response to more and more complex issues. This is usually where fascism appears and escalates. We are already seeing the signs of it across the globe. This is the sum of all the inequalities we have allowed to spiral out of control for so many centuries. And unless we start to address the root causes of these problems and begin building strong, democratic, grassroots movements all across our societies – movements like those we have just encountered, movements that leave no one behind – then everything of beauty and meaning that humanity has ever achieved might be lost – quite literally – forever.

Some of those movements already exist today. Others will form as we go along. They all have huge responsibilities to stay clear of all forms of violence and to avoid creating social unrest which might result in vandalism and destruction that risks doing far more harm than good. We need billions of climate activists. Non-violent, peaceful demonstrations and civil disobedience that doesn't risk the safety of others; strikes, boycotts, marches, and so on. Humanity has succeeded in changing our societies many times before, and we most definitely can do it again.

Just as the climate crisis requires everyone to work together, so does this. The sustainability crisis, the inequality crisis, the democracy crisis – they can't be solved by individual people or individual nations. We must all work together and we must do it in solidarity. When we humans come together for a common cause we can create just, sustainable and equal societies. Just as we can establish selfish, unsustainable and unequal ones. /

Warming and Inequality

Solomon Hsiang

Our world is profoundly unequal. There are wealthy communities today that enjoy opportunities and standards of living that would have been unimaginable a few centuries ago, while at the same time there are other, poorer communities whose access to resources, health care and technology has barely changed since that time.

In the future, as our emission of greenhouse gases alters the climate, these changes will likely reshape global inequality. As environmental conditions shift, the opportunities and resources available to different societies will change too – improving for some, and worsening for others. For example, the livelihood of a community that depends on agriculture for its income will be impacted if the climate is transformed, but whether this impact is positive or negative will depend on what kind of farming they practise, and how the climate changes in their specific location. In a hot and dry region, rainfall might increase, helping farmers. Alternatively, temperatures may rise even higher, which could make agriculture more difficult. Global warming's overall impact on any given community will depend on many factors, including how that community lives, what their climate is today and how we expect it to change in the future.

Given all this complexity, it isn't always obvious *how* climate change will affect inequality. If richer societies were made poorer by warming, and poor societies were made richer, then climate change could *reduce* global inequality. But if rich societies tend to benefit from warming and poor societies tend to be harmed, then we would expect climate change to *increase* inequality. To try to understand which outcome is more likely, many researchers, including myself, have turned to data analysis to understand how different societies are impacted by different climatic conditions.

What we see in the data strongly indicates that climate change will increase global inequality. Depending on which measures of well-being one studies (for example, health, education or income), we see that rich

Figure 1:
The non-linear
effect of warming
can be helpful or
harmful, depending
on where you live.

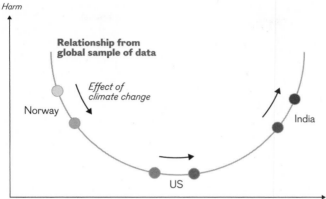

Harm and temperature

Harm

Relationship from global sample of data

Effect of climate change

Norway

India

US

Temperature, from colder to warmer (left to right)

populations are sometimes helped by warming and sometimes harmed. But almost every way we look at the data, we see that poor populations suffer, and usually more than rich populations.

Research suggests that there are two main reasons why poor populations around the world tend to be harmed more by climate change than rich populations. First, poorer communities have fewer resources at their disposal to protect themselves from the effects of climate change. Air-conditioning, sea walls and irrigation systems alleviate the impact of rising temperatures and extreme weather events, but they require a significant investment of money and resources.

The second reason is less widely known but is potentially even more significant: the effect of temperature on many critical human outcomes is *non-linear*. Fig. 1 illustrates this: the effect of warming depends on the current temperature of the location. In general, we see that if a community lives in a cold location (for example, Norway), warming is helpful – heating costs and wintertime respiratory illnesses decline, while labour productivity rises. If a community lives in a temperate location (for example, Iowa, in the US), warming has very little effect on well-being. Many studies find that the 'ideal' average temperature tends to be 13–20°C. If a community lives in a hot location (for example, India), additional warming is very harmful – destroying crop yields, exacerbating vector-borne diseases and slowing economic growth. One additional degree of warming does not have the same effect everywhere, and this has profound implications for global inequality.

The reason that the non-linear effect of temperature matters so much is that today's poor populations tend to live in much hotter locations. The top

GDP per capita in 2019

US$ x thousand

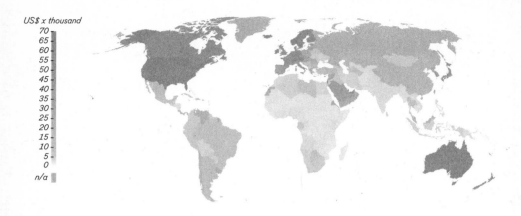

Impact of climate change on mortality rates in 2100

4°C warming scenario

Deaths per 100,000

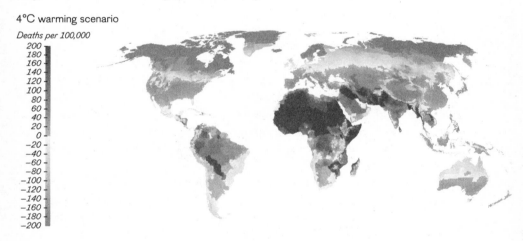

Impact of climate change on GDP per capita in 2100

4°C warming scenario

Percentage change

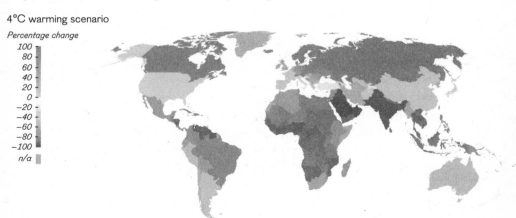

Figure 2

panel of Fig. 2 shows the average GDP per person around the world today. What you see is a familiar pattern in which countries in cold or temperate regions have higher average incomes, while hotter countries closer to the equator in tropical and subtropical regions tend to be much poorer. This puts poor populations in a worse starting position when it comes to climate change, because they live in hot locations, where warming is especially harmful, while rich populations live in cooler places, where warming proves less damaging – and is sometimes even beneficial.

The lower panels of Fig. 2 illustrate how this is projected to play out at the end of the twenty-first century. The middle panel shows how warming in a high-emissions scenario (+4°C in 2100) is expected to increase mortality rates around the world. The bottom panel shows how warming in the same scenario is expected to alter the average GDP per person. In both cases, the fact that wealthier nations have more resources to protect themselves is accounted for in the calculation. However, what really shines through is the non-linear effect of temperature. Hotter locations in the tropics and subtropics suffer more in terms of health and economic opportunities, with annual mortality rates increasing by more than 100 deaths per 100,000 and national income losses of roughly 50 per cent or more. The impacts are milder in temperate regions. Cold places often benefit, since warming can actually improve human health and economic productivity.

Comparing the lower panels of Fig. 2 to the top panel, we can see how, rather than lifting the global poor up, climate change will slow down their progress, deepening the global divide between rich and poor. /

One additional degree of warming does not have the same effect everywhere, and this has profound implications for global inequality.

Water Shortages

Taikan Oki

At the Stockholm World Water Week in August 2019, I asked Johan Rockström whether Stockholm could remain civilized if the mean temperature changed from 7°C to 15°C and the mean annual precipitation changed from 500 millimetres to 1,500 millimetres within this century. He answered, as I expected, that it would be impossible.

It might be. If Stockholm's climate changed so drastically in such a short period of time, adaptation would certainly be very challenging. But it might not be impossible. Tokyo is not far from having a mean temperature of 15°C and a mean annual precipitation level of 1,500 millimetres. And, so far, the people of Tokyo have a modern, safe and comfortable life (although, on a planet where Stockholm had the climate of present-day Tokyo, summers in Tokyo would be unendurably hot and I would be sorely tempted to migrate to Stockholm). In other words, it is not a question of absolute values, such as what temperature or how much rainfall our societies can cope with, but rather a question of how much the climate changes, and how much time we have to adapt to this change. It is the world's most vulnerable communities that will suffer the most from the adverse effects of climate change. Even if future conditions were such that other communities around the world could adapt and thrive, these communities would face serious difficulties, if not intolerable hardship.

Water is the delivery mechanism of climate change's impacts to society. More than a decade ago, my colleague and I contributed a review on global hydrological cycles and world water resources which stated that 'Climate change is expected to accelerate water cycles' and thereby seemingly *increase* available renewable fresh-water resources. This would slow down the increase of people living under water stress. However, our research also showed that changes in seasonal patterns and the increasing probability of extreme events may offset this effect where the precipitation event is more intermittent. We went on to warn, 'If society is not prepared for such changes and fails to monitor variations in the hydrological cycle, large numbers of people run the risk of living under water stress or seeing their livelihoods devastated by hazards such as floods.'

Unfortunately, the number of natural disasters has increased since we wrote this. According to a report compiled by the United Nations Office for Disaster Risk Reduction, the number of reported droughts has increased by 1.29 times, storms by 1.4 times, floods by 2.34 times and heatwaves by 3.32 times in the first two decades of the twenty-first century compared to the last two decades of the twentieth century. These impacts are expected to become more severe as climate change progresses, and it is not just vulnerable communities that will have difficulties. According to the Ecological Threat Register published by the Institute for Economics and Peace, while the developed world has the capacity to deal with resource depletion and natural disasters, it will not be able to escape the aftermath of the influx of distressed migrants driven from their homes by these issues. The 2015 migration crisis, for example, saw an influx of people equivalent to just 0.5 per cent of Europe's population, and this resulted in political tension and social unrest. With current warming trends, 1.2 billion people could be forced to migrate by 2050. The UN High Commissioner for Refugees estimates that about 20 per cent of these people will migrate out of their countries or regions entirely. The numbers on future migration vary based on the source, but are consistently alarming. The World Bank's Groundswell Report Part II estimates that 216 million people could be internally displaced by 2050.

These climate and environmental crises have not been brought about by any single politician, government or company but rather by the aggregate of the moment-to-moment choices we make in our daily lives. We are beginning to wake up to this fact, even if only from a selfish, utilitarian perspective: many businesses now understand that, in the long run, taking steps to avoid climate and ecological crises is the wisest choice for them; many politicians and governments are highly sensitive to public opinion, which is increasingly united behind climate justice. If there had been more of us trying to keep the climate stable by altering our behaviours than those who were changing the climate by not altering their behaviours, decisive climate action for a just transition would have been taken much earlier.

As things stand, we are unable to stop the progress of climate change; instead, the world has agreed to pursue efforts to limit the temperature increase to 1.5°C above pre-industrial levels. This means that, while freshwater resources in some parts of the world may increase in the years to come, many of us will still suffer from the impacts of increased droughts and floods, and the approximately 733 million people that currently live in high and critically water-stressed countries are especially at risk. /

3.19
Climate Conflicts
Marshall Burke

By many measures, humans have grown more peaceful towards one another during our relatively brief time on this Earth. Large countries fight each other with less frequency, fewer people die in battle, and many types of conflicts between individuals such as assaults or homicides have become less common in many societies.

Nevertheless, our world remains a violent place. Hundreds of thousands of people die in homicides every year, and homicide rates are now trending upwards in many countries. The total number of organized armed conflicts around the world is also trending up and is now at the highest level seen in nearly a century (Fig. 1), leading to record numbers of internally displaced people and alarming levels of global hunger. And growing evidence suggests that climate change could exacerbate these trends towards violence.

Scholars and writers have long suggested that climate might play a role in how humans behave towards one another. In Shakespeare's *Romeo and Juliet*, Benvolio tells his friend Mercutio that they should return home because the heat of the day makes it likely that there will be a fight; they fail to do so, and tragedy ensues. In Camus's *The Stranger*, the protagonist,

Number of state-based armed conflicts per year since 1946

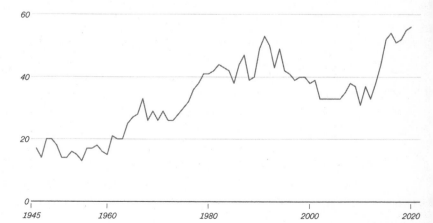

Figure 1: Conflicts where at least one party is the government of a state.

Meursault, becomes overheated while on an Algerian beach and shoots a man. More than a century ago, articles in mainstream economics journals argued that climatic shifts in the centuries around the birth of Christ led to the eventual violent fall of the Roman Empire. And in the last decade, a large number of researchers – enabled by much better data on the timing and location of human conflicts around the world – have shown how a changing climate, in some contexts, can raise the likelihood of human conflict.

Why might a changing climate play any role in human conflict? Any individual conflict – one person harming another in an altercation, or a guerilla group taking up arms against a government – is a complex event, likely with many interacting causes. While climate is never the only cause of a given conflict, a large body of research across many academic disciplines shows that it can be a 'finger on the scale', amplifying individuals' or groups' willingness, ability or incentives to fight one another. This is why the US Department of Defense has long called climate change a 'threat multiplier' – a factor that can exacerbate and amplify the myriad other reasons humans choose to fight.

For decades, researchers in psychology have shown in laboratory studies that humans are more irritable and can be made to act more aggressively by turning up the temperature in the room they're in. This physiological response is evident outside the lab as well: in studies from around the world, hotter temperatures have been shown to lead to increases in aggressive driving, to increased violence in professional sports and to increases in an array of violent crime, from domestic violence and aggravated assault to murder.

Warming temperatures and more extreme variations in rainfall have also been shown to exacerbate the likelihood of group-level conflict, which includes events ranging from gang violence and riots to civil war. Hotter temperatures inflame gang violence in Mexico; drought and hot temperatures increase civil conflict in Africa; and El Niños provoke more civil conflict around the world. It's important to note that these findings are not just correlations. They are from carefully designed studies whose entire purpose is to carefully isolate the role of climatic variables from the myriad other factors that might cause conflict.

There are enough of these high-quality studies that we can do a 'meta-analysis' – a study of studies – to summarize overall findings across dozens of published works. In doing so we again find consistent evidence that warming temperatures in particular can amplify the risk of various types of conflict, with the risk of damaging group-level conflicts going up by as much as 10–20 per cent for every degree Celsius increase in temperature.

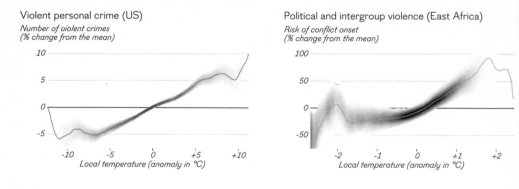

Violent personal crime (US)
Number of violent crimes
(% change from the mean)

Local temperature (anomaly in °C)

Political and intergroup violence (East Africa)
Risk of conflict onset
(% change from the mean)

Local temperature (anomaly in °C)

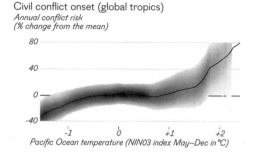

Figure 2:
Many types
of human
conflict become
more likely when
temperatures
increase, from
individual violence
(top left) to
violence between
social groups
(top right). El Niño
has an impact on
global conflict
(bottom).

Civil conflict onset (global tropics)
Annual conflict risk
(% change from the mean)

Pacific Ocean temperature (NINO3 index May–Dec in °C)

These are large effects and imply a substantial increase in the possibility of violence as temperatures continue to warm throughout this century.

We ignore these future risks at our own peril. However, climate is not destiny. Human societies can choose how much warming we're willing to tolerate, and we can also choose how to deal with the warming that does occur. Civil conflicts, once common in most of the world, have now been all but eliminated in many countries. In these societies, additional warming is unlikely to spur large-scale conflict. Similarly, other research has shown that even in conflict-prone societies or regions, the expansion of social safety nets can help communities maintain their livelihoods in the face of climatic extremes and, in turn, can effectively break the link between climatic extremes and conflict. Investments in vulnerable communities' abilities to withstand – and thrive in – new climates will be key to avoiding the worst consequences of a changing climate. /

The True Cost of Climate Change

Eugene Linden

What might be the socio-economic cost of climate change? If we continue along our current path and the Earth warms by 3°C above pre-industrial levels, the risk is, simply, the collapse of civilization itself. This will be a global calamity marked by financial collapse, mass starvation, mass migration and the descent of many nations into civil disorder. Had governments recognized the gravity of the risk in the early 1990s, this apocalyptic prospect might have galvanized action to contain greenhouse gas emissions and avert potential catastrophe. Instead, early projections of the socio-economic damage of climate change were absurdly low, providing intellectual cover for those who would delay action (one influential 1993 paper by an economist who later won the Nobel Prize calculated the cost to the US economy of a 3°C warming by 2100 to be a minuscule one quarter of 1 per cent of GDP). Now, economics is catching up to reality, awakening to the fact that climate change, not action to prevent it, poses the greatest threat to future prosperity.

But even if we take action to keep warming below 3°C, climate change will still take a significant toll. It is difficult to forecast what it will cost, primarily because of the nature of 'thresholds', and 'tipping points' in climate change, which can lead to orders of magnitude increases in damage inflicted. Hurricane Sandy provided a vivid example of the importance of thresholds when the New York City subways flooded for the first time in 125 years, causing $5 billion in damage. Had the combination of storm surge, high tide and sea-level rise been just 15 per cent lower, the damage would have been negligible.

The problem of tipping points is even more serious, and it utterly defeats any attempt to make a sound prediction of future damages. For instance, accelerating melting of permafrost in the high north could release vast amounts of greenhouse gases, leading to an unstoppable feedback loop of warming producing more warming far beyond the most pessimistic forecasts derived from climate models. Conversely, fresh water flooding into the north Atlantic could shut down the global current which keeps much of Europe warm. We don't know when these tipping points might be crossed, but we do know that, once crossed, they will be irreversible in any timeframe meaningful to humans.

Then there are the indirect impacts of a warming climate to consider. In the American West, warmer temperatures have led to an explosion of bark beetles, which has in turn led to vast die-offs of evergreens, which the larvae feed on. These dead trees have provided a ready supply of fuel for fires, which have been raging across the region, further enhanced by ultra-low humidity, high temperatures and the more intense dry winds that accompany a warming and drying landscape. This combination of direct and indirect impacts then cascades through human societies, with unpredictable results.

For instance, one tributary to pressures for migration in the Middle East has been that extreme temperatures have made stretches of Iran, Syria, Iraq and other nations uninhabitable. Such forced migration feeds internal and then international instability: as we've seen in recent years, the arrival of refugees into Europe has been met by resistance, xenophobia and the rise of populist, authoritarian rulers.

Some of the possibilities are simply unimaginable. Several billion people depend on grains grown in just a few breadbaskets, all of which are exquisitely balanced to temperature and precipitation regimes that have been relatively stable for thousands of years. The IPCC estimates that global maize yields will lower by 5 per cent with 2°C of warming. As temperatures warm further, precipitation patterns change, the soil dries out faster and there's a point where staple crops won't grow at all – there's a reason that there are no breadbaskets in the tropics.

All these impacts interact in unpredictable ways and make it exceedingly difficult to predict exactly how much economic damage will be associated with what degree of warming.

Still, people try.

In 2021, Moody Analytics estimated that the global economic toll of 2°C of warming would be $69 trillion. A study undertaken by Swiss Re estimated that 2.6°C of warming by 2050 would inflict three times more economic damage than the Covid-19 pandemic. Unlike the Covid pandemic, however, the damage inflicted by warming will only grow worse every year. Three degrees of warming would produce a world that hasn't existed since humans emerged as a species. There was plenty of life back then, but no humans. It's certain that such a world could not support 7.8 billion people.

The world might well suffer a climate-related global financial crisis long before temperatures rise by 3°C or even 2°C. In fact, economic damage from

Next pages:
In August 2017,
Hurricane Harvey
caused widespread
flooding in
Houston, Texas,
submerging
Interstate
highway 45.

climate change may already have reached the trillions. According to the insurance giant Aon, the world suffered $1.8 trillion in weather-related losses during the first decade of the new millennium. In the second decade, that number rose to $3 trillion. Recent wildfires in the American West and floods and wind storms on the East Coast have given the world a preview of how a climate financial crisis might unfold, even in a rich country.

Here's how: as floods and fires proliferate, storms intensify and become more frequent, and as temperatures rise, so too will insurance rates for homeowner and business protection against natural disasters. Where they can, insurers will pull out of the most at-risk areas. Without such insurance, most homebuyers can't get a mortgage, and in those fire and flood zones where insurance rates skyrocket, many owners will try to sell – but to whom, and who will finance the purchase? This would set the stage for panic selling and a housing market collapse more serious than the crisis of 2008 because it would not be a one-off event. As we saw in 2008, a housing crisis can quickly morph into a systemic financial crisis because banks own most of the value, and thus the risk, in housing and commercial real estate.

Our global economy is a tightly coupled system – that's the message of 2008 and, more recently, of the supply-chain disruptions resulting from the pandemic. In such systems, even minor disruptions can lead to devastating repercussions. The disruptions coming from climate change are anything but minor and will progressively get worse. The message that should lodge with policymakers, politicians and the public is that climate change must be averted at any price because its ultimate cost can be neither imagined nor calculated. /

The message that should lodge with politicians and the public is that climate change must be averted at any price because its ultimate cost can be neither imagined nor calculated.

PART FOUR /

What We've Done About It

'We don't speak the same language as the planet'

How can we undo our failures if we are unable to admit that we have failed?

Greta Thunberg

Saving the world is voluntary. You could certainly argue against that statement from a moral point of view, but the fact remains: there are no laws or restrictions in place that will force anyone to take the necessary steps towards safeguarding our future living conditions on planet Earth. This is troublesome from many perspectives, not least because – as much as I hate to admit it – Beyoncé was wrong. It is not girls who run the world. It is run by politicians, corporations and financial interests – mainly represented by white, privileged, middle-aged, straight cis-men. And it turns out most of them are – under the current circumstances – terribly ill suited for the job. This may not come as a big surprise. After all, the purpose of a company is not to save the world – it is to make profit. Or rather, it is to make as much profit as it possibly can in order to keep shareholders and market interests happy. The same goes for the financial interests that drive the economy in pursuit of further profit and growth.

This leaves us with our political leaders. They do have great opportunities to improve things, but it turns out that saving the world is not their main priority either. It could be, if enough people wanted it – but that is far from the case today. So it appears that their job is simply to remain in power, get re-elected and to stay in tune with public opinion. Many people say that politicians don't plan or think any further than the next election, but I strongly disagree. My experience is that their long-term policies stretch no further than to the next opinion poll – but usually their main focus is not even that long term; often, they only think as far as tomorrow's newspapers or the daily evening news.

Previous pages:
The world's
worst offshore oil
spill. In April 2010,
the BP-run
*Deepwater
Horizon* rig caught
fire, killing eleven
people. Over 130
million gallons of
crude oil spewed
into the Gulf of
Mexico,
devastating
huge swathes of
this richly
biodiverse area.

Approaching the issues of the climate and ecological crisis inevitably involves confronting numerous uncomfortable questions. Taking on the role of being the one who tells the unpleasant truth, and thereby risking one's popularity, is clearly not on any politician's wish list. So they try to stay clear of the subject until they absolutely cannot avoid it any longer – then they turn to communication tactics and PR to make it seem as if real action is being taken, when in fact the exact opposite is happening.

It gives me no pleasure whatsoever to keep calling out the bullshit of our so-called leaders. I want to believe that people are good. But there really seems to be no end to these cynical games. If your objective as a politician truly is to act on the climate crisis, then surely your first step would be to gather accurate figures for our actual emissions to get a complete overview of the problem, and from there start looking at real solutions? That would also give you a rough idea of the changes needed, the scale of them and how quickly they need to be put in place. This, however, has not been done – or even suggested – by any world leader. Or, to my knowledge, even by any one single politician. To me, that would suggest that the sincerity of their ambition to *solve* this crisis is somewhat limited.

Journalist Alexandra Urisman Otto describes how she started investigating Swedish climate policies and found that only one third of our actual emissions of greenhouse gases were included in our climate targets and the official national statistics. The rest was either outsourced or hidden in the loopholes of international climate accounting frameworks. So whenever the climate crisis is debated in my 'progressive' home country, we conveniently leave out two thirds of the problem. A major investigation by the *Washington Post* in November 2021 has shown that this phenomenon is far from unique to Sweden. Though the figures vary from case to case, this process and the over-all mentality of constantly trying to sweep things under the carpet and blame others is the international norm.

So when our politicians say that *we must solve the climate crisis*, we should all ask them which climate crisis they are referring to. Is it the crisis that contains all our emissions or the one that contains only a part of them? When politicians go a step further and accuse the climate movement of not *offering any solutions to our problems*, we should ask them what problems they are talking about. Is it the problem that is caused by all our emissions or just by the ones they didn't manage to outsource or hide in the statistics? Because these are completely different issues.

It will take many things for us to start facing this emergency – but above all it will take honesty, integrity and courage. The longer we wait to start taking the action needed to stay in line with our international targets,

CO$_2$ mitigation for a 67% chance of staying below 1.5°C of warming

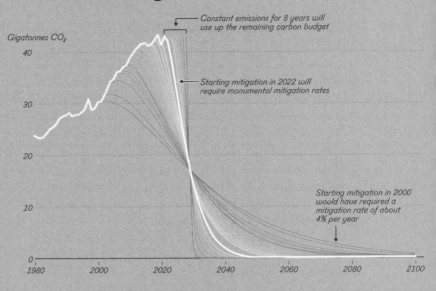

Gigatonnes CO$_2$

Constant emissions for 8 years will use up the remaining carbon budget

Starting mitigation in 2022 will require monumental mitigation rates

Starting mitigation in 2000 would have required a mitigation rate of about 4% per year

the harder and more costly it will get to reach them. The inaction of today and yesterday must be compensated for in the time that lies ahead.

For us to have even a small chance of avoiding setting off irreversible chain reactions far beyond human control we need drastic, immediate, far-reaching emission cuts at the source. When your bathtub is about to overflow, you don't go looking for buckets or start covering the floor with towels – you start by turning off the tap, as soon as you possibly can. Leaving the water running means ignoring or denying the problem, delaying doing anything to resolve it and downplaying its consequences. And when it comes to the climate crisis, no person, group or nation is solely responsible for this level of denial and delay. It takes a whole society to play along, or at least large parts of it. It also takes strong cultural norms and common interests – such as ideological interests, or perhaps more to the point, financial ones – such as the almighty short-term economic policies that currently shape the world.

Leaving capitalist consumerism and market economics as the dominant stewards of the only known civilization in the universe will most likely seem, in retrospect, to have been a terrible idea. But let us keep in mind that when it comes to sustainability, all previous systems have failed too. Just like all current political ideologies – socialism, liberalism, communism, conservatism, centrism, you name it. They have all failed. But, in fairness, some have certainly failed more than others.

One problem that we face today is closely linked to the fact that nearly all those who have dedicated their lives to the public service of politics are firm believers in these ideologies. It was probably this belief that made them go into politics in the first place. It was this belief that made them endure all those endless meetings, campaigns and conferences – the belief that social-ism, conservatism or whatever-ism could deliver the answers to the challenges of our modern, daily lives. This same belief that made them read all those tens of thousands of pages of political reports: the belief that their little niche of current party politics had the keys to unlock all the necessary solutions to society's ills. To give up your beliefs is not an easy thing to do. And yet how will we be able to change if we do not learn from our mistakes? And how can we undo our failures if we are unable to admit that we have failed?

In my experience, most politicians are more or less informed of the situation we are currently facing and yet, for various reasons, they still focus on other things. You could of course argue – and rightly so – that it is the respon-sibility of the media to force them into action. After all, it is public opinion that sets the agenda of the free world, and if enough people cared enough about ecology and sustainability, then our political leaders would have no choice but to start facing up to these matters in a credible way. This is slowly beginning to happen, but so far we are barely scratching the surface.

Nevertheless, our politicians do not need to wait for anyone else in order to start taking action. Nor do they need conferences, treaties, international agreements or outside pressure to start taking real climate action. They could start right away. They also have – and have had for a long time – endless possibilities to speak up and send a clear message about the fact that we must fundamentally change our societies. And yet, with very few exceptions, they actively choose not to. This is a moral decision that will not only cost them dearly in the future, it will put the entire living planet at risk. /

When your bathtub
is about to overflow, you don't
go looking for buckets –
you start by turning off the tap.

4.2

The New Denialism

Kevin Anderson

I'm **sitting above** one of the COP26 pavilions, preparing yet more PowerPoint slides, when there's a sudden crescendo in the hubbub of conversation that fills the air at such events. I look over the barrier to see the corridor below filled with a seething mass of COP delegates desperate for a glimpse of the godlike figure being shepherded to a nearby lectern. Another Obama or Bezos, a celebrity or a royal, there to cast pearls of wisdom before the swine clamouring for selfies. And then there are the journalists not far behind.

Meanwhile, from pavilions just metres away, some Indigenous people are talking about the destruction of their homes; a scientist is explaining the unprecedented melting of Greenland; and a protester, without formal permission to protest, is being 'debadged' and escorted out of the 'Blue Zone'. All go virtually unreported, witnessed only by a few individuals socially distanced around their respective rooms.

Thirty-one years after the first IPCC report on climate change, the Blue Zone – the formal, gated venue where negotiations take place and governments showcase their 'climate action' – is a microcosm of three decades of failure: of rapidly rising emissions, of climate denial, of expedient technical optimism, of 'negative emissions' and, today, of 'net zero, but not in my term of office'. Where is the concern for vulnerable communities already suffering climate impacts, for species extinction, for exchanging rich biodiversity for monocropped wastelands? Where is the concern for our own children's futures?

How did we end up here? At the 1992 UN Earth Summit in Rio de Janeiro hopes were so high. Good people could envisage progressive, low-carbon and sustainable futures.

Back then, financiers were only just waking up to the money-making scams of emissions trading schemes, the financialization of nature and the issuing of catastrophe bonds, all serving to derail meaningful climate action today. The fossil fuel companies, however, were a decade or so ahead of us all. They had known the risks and challenges and lied about them for years. They were prepared. Some issued direct denials; others offered comforting reassurances of future technical salvation. In the years that followed, the unscrupulous operators of Big Finance and Big Oil increasingly came to see the profits to be made from maintaining the status quo behind a veneer of 'decarbonization'. A few even managed to delude themselves that, with elaborate financial bundles of 'offsets', they could actually reconcile the irreconcilable, allowing emissions to continue with no real consequence.

This line-up of high-profile villains is culpable for the fact that, since the IPCC's first report in 1990, we've dumped more carbon dioxide in the atmosphere than throughout all of human history prior to 1990. But climate change is a systemic issue; it has multiple layers of failure. Few of us can hold our heads high – including those actively engaged in the climate issue. Where is the collective chorus of senior academics revealing the white-wash of Big Oil and Big Finance? Where are the CEOs of our established environmental charities, our policymakers and our investigative journalists? We aren't asleep at the wheel; we've been actively steering society towards its own demise. Why? Because we fear rocking the boat and upsetting our paymasters. We enjoy the prestige of hobnobbing with the great and good, and we aspire to honours from the establishment. And because, ultimately, we're afraid of our own conclusions. But we've also convinced ourselves that we deserve our high salaries and accompanying carbon-rich lifestyles. The limelight feels good.

To be clear, I'm not talking here about the climate science itself. Many within the scientific community did a wonderful job of taking the standard tools of science, seasoned with math and statistics, to develop a deep understanding of the climate and of climate change. This was all the more impressive as many scientists had to battle against well-funded, coordinated and powerful forces hell-bent on undermining their credibility – forces driven not by intellectual disagreement (there was none to speak of) but by fear of the policy implications of the science.

In the end, the scientists or, more accurately, the science, won out. While there remain a few enclaves of opposition, most of those previously decrying the science now outwardly accept it. In reality, however, they have transitioned to a second phase of denialism: 'mitigation denial', whereby the need for deep cuts in emissions today is substituted with empty promises of low-carbon technologies tomorrow. But here the net of responsibility needs to be cast more widely, with many climate scientists wriggling in the mesh.

Climate change is a cumulative problem. Our burning of fossil fuels releases carbon dioxide that builds up in the atmosphere, day on day, decade on decade, warming the climate for centuries and even millennia to come.

Each year, we fail to deliver the required reduction in emissions, so the rate of cuts we need to make the following year increases. If we needed a 10 per cent cut in emissions this year to stay within our carbon budget but we only achieved 5 per cent, then to bring us back on track we'd need a cut of over 15 per cent the following year. Put bluntly, when emissions are cut at lower rates than is necessary, this is not a step in the right direction. Rather, it is a step backwards – just not as large a step backwards as it would otherwise have been.

It is this relentless backsliding that has given rise to increasingly elaborate forms of 'mitigation denial', whereby we rely on ever more speculative forms of 'negative emissions'. These range from future carbon-sucking technologies and simplistic 'nature-based-solutions' through to paying poor countries to further cut their emissions on our behalf. All such ruses are designed, in significant part, to 'offset' the responsibility we have to make profound cuts in emissions today. To our deep shame, many of us who work on climate change have bought into this mathematical massaging and, worse still, some enthusiastically peddle such snake oil.

Away from this subterfuge, the science makes clear that for a given chance of staying below a particular temperature (say, 1.5°C) we can emit no more than a particular quantity of carbon dioxide (our carbon budget). There remains some uncertainty as to what the exact quantity is, but science does give us a robust range within which to work.

Our remaining carbon budget range is small and is rapidly shrinking. For a 'likely' chance of not exceeding 1.5°C, we have under eight years at the current rate of emissions. Weakening the commitment to 'well below 2°C' (and thereby accepting more devastating impacts), slows the ticking clock, but still fewer than twenty years' worth of current emissions remain.

To put this into perspective, let's imagine that at 2022's climate jamboree (COP27), world leaders agree to put in place polices to cut emissions in line with a likely chance of 1.5°C. Then, globally, by around 2035, we would need to have eliminated the use of all fossil fuels, stopped all deforestation and made rapid and deep cuts in all other greenhouse gas emissions. However, this is a global assessment, and since the 1992 Rio Earth Summit, the international community has agreed that, for poorer nations, cuts in emissions must not unduly impede their development. Accordingly, richer nations, with much greater historical responsibility for climate change, need to mitigate their emissions earlier and faster than those in early or transitional stages of development. Play out the numbers, and this means that wealthy nations must eliminate their use of fossil fuels by around 2030 for a likely chance of 1.5°C, extending only until around 2035 to 2040 for 2°C.

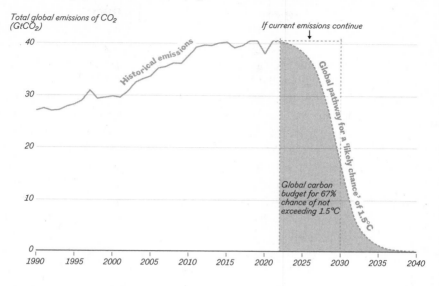

Remaining global carbon budget for 67% chance of 1.5°C

Figure 1:
By around 2035, we would need to have eliminated the use of all fossil fuels, stopped all deforestation and made rapid and deep cuts in all other greenhouse gas emissions.

Total global emissions of CO₂ (GtCO₂)

Historical emissions

If current emissions continue

Global pathway for a 'likely chance' of 1.5°C

Global carbon budget for 67% chance of not exceeding 1.5°C

But this is still far from fair. Even under such demanding conditions, an individual's average annual emissions, from now to the point of zero global emissions, would still be greater for citizens of high-emitting wealthy nations than for those of low-emitting poorer nations. The blunt fact arising from our continued inaction on climate change is that, in terms of quantitative emissions, we have left it far too late for real fairness. The 'least unfair' solution is now the best that can be achieved. It is this unfairness that lends significant credence to arguments for major financial reparations from wealthy high-emitting nations to those poorer nations facing the climate impacts that we have knowingly imposed on them.

All of this may be hard to take in, but we are where we are precisely because for thirty years we've favoured make-believe over real mitigation. We are reaping what we chose to sow or, more accurately, not sow.

Those of us in high-emitting nations cannot hide behind a defence of ignorance. For decades now, science has outlined the repercussions of favouring hedonism over stewardship. Cut through the rhetorical facade of concern, and we have been well aware what the climate impacts of frequent flying, buying SUVs, owning second homes, travelling further and faster and consuming more stuff year on year would be. But the people paying the price for our norms of ever-increasing consumption are not us, they are poorer, more climate-vulnerable communities elsewhere, typically both low emitters and people of colour. And these vulnerable communities extend to our own children. At the same time as we shower them with gifts, drive them to school and take them on foreign holidays, we are hugely discounting their future. When we high-emitting parents are pushing up the daisies, our own offspring and grandchildren will be battling with, and sometimes dying from, our explicit choice to take the easy route, to believe in technical utopias and to point the finger of blame at others.

Recent years have seen a wealth of research demonstrate the huge asymmetry in responsibility for emissions. Disturbingly, the top 1 per cent have lifestyles that give rise to twice the emissions of the bottom 50 per cent of the global population. There is no unified 'we'. Responsibility is not evenly spread; we're not all in this together, neither in terms of mitigation nor impacts. At virtually every level, climate change divides along lines of wealth and income. These divisions are not small – rather, they're tectonic fractures that have been so normalized that, like climate change itself, we appear unable or unwilling to see them.

To give some sense of this emission asymmetry, if the top 10 per cent of global emitters were to cut their carbon footprint to that of the typical EU citizen, and the other 90 per cent made no change to their emissions, that alone would cut global emissions by around one third. Certainly not enough to deliver 1.5°C or even 2°C, but a huge opportunity for rapid and equitable change deliberately omitted from the conversation around mainstream mitigation.

So why does equity still remain such a taboo subject? And who is shaping the climate debate, setting the boundaries, developing the mitigation models and proposing the policies that fail to take it into account? The professors, the policymakers, the journalists, the barristers, the entrepreneurs, the senior civil servants et al., all of whom reside in, or close to, the top 1 per cent of global emitters. For all of us, living within a fair carbon ration would entail profound changes to how 'we' live our lives: the size (and number) of our houses; how often we fly and in which class; how big and how many cars we have and how far we drive them. Even at work, how large is our office, how many foreign meetings and international conferences do we attend and how frequent are our field trips.

The exclusive, high-emitting and high-consuming 'we' has constructed and subsequently hidden itself within the myth of a universal 'we' who must fight climate change together. Once you unpick this false togetherness, the prevailing mitigation narrative explodes.

Mitigation aligned with 1.5–2°C of warming is an issue of major structural change. It means improving the fabric of our homes, rapidly expanding public transport, developing a massive programme of electrification, changing town planning, rolling out e-bikes in cities and shared electric vehicles in the rural environment. All this could be win-win for the majority within our nations. Our cities and urban environments could again be built around people rather than two-tonne metal boxes. High-quality, secure jobs would emerge. Children would be safe to cycle and the air would be better for their lungs.

Such widespread benefits would come with considerable costs, falling primarily on the shoulders of the select 'we': those of us who have thus far shaped the climate debate around speculative tech, electric SUVs, carbon credits that 'we' can afford, heat pumps for our holiday homes and offsets for our flying. It is 'we' who have failed to mitigate warming. We have been unwilling to question our profligate consumption, the allure of infinite economic growth and the hugely disproportionate allocation of society's productive capacity to furnish the relative luxuries of us fortunate few.

But the elite 'we' no longer has such a firm grasp on the wheel. Four years ago, we were given a shake – not by a world leader, or a decorated member of the great and good, but by a fifteen-year-old schoolkid. In the years since, a messy gaggle of youths and grandparents, professional activists, concerned politicians and earlier career academics have found their voices and are beginning to reframe the debate. Climate change has flown its privileged nest and has entered the everyday. Countless people are discussing, testing and refining ideas. Not in the way of a formal experiment or even directly related to the abstract world of carbon molecules, prices or budgets. Rather, climate awareness has infused the collective psyche, enabling the public to see through political rhetoric, question utopian tech and smell a rat, even if they can't put their finger on exactly what's wrong.

Whether a more inclusive 'we' can ultimately usurp the old guard in time to prevent the worst of the climate crisis is up for grabs – emissions are still rising, and spineless policymakers are still in thrall to Big Oil and Big Finance. But, for now, the future is being determined, at least partially, by this newly empowered and much more diverse constituency of the willing and the concerned. /

The Truth about Government Climate Targets

Alexandra Urisman Otto

When I wrote my first article about Greta Thunberg, I didn't have a clue. It was a long interview in my newspaper's weekend magazine in the autumn of 2018, and for most of the conversation I played along, pretending to understand what she was talking about.

I was actually a crime reporter. Loved the thrill of a murder investigation or the tension in a court room. Climate change was, to me, boring, completely uninteresting. Full of dry, hard-to-understand facts, graphs I couldn't interpret. Sure, there was a risk of disaster. But I took comfort in the fact that the situation was somehow under control. Surely the people responsible had to have plans that were being implemented. Above all, I felt grateful that the issue was on somebody else's desk, that it wasn't my job to report on it.

But when Greta Thunberg's journey gained speed, my colleague Roger Turesson and I were invited to follow it. Journalistically, it was impossible to refuse; her story was surreal. So I realized that I had to understand the facts. How else would I be able to check if what she said was true? I started reading.

To get an accurate view of the crisis, I created a new Twitter feed. I started to follow climate scientists, environmental journalists and activists. I read newsletters, climate books and in-depth stories in the international media. In the summer of 2019, I passed a tipping point and went from ignorance and unconcern – straight down into the abyss of despair.

The carbon budget to stay within the targets of the Paris Agreement would be used up within just a few years, I realized. And I had spent my time writing about criminal cases. I had failed, and so had most of my colleagues: the world that was described in the 'usual' daily news reporting in newspapers and on radio and TV was a world where everything was normal, sometimes interrupted by something 'climate related'. There was no crisis. Our readers, listeners and viewers had for decades trusted us as journalists. Yet, in the midst of the biggest-ever crisis for humankind, we kept on giving them the 'news' with the underlying assumption that business as usual was an option. It was an enormous betrayal.

But it wasn't only journalists that had failed. The more I read, the more clearly I could see the real crisis: the fact that nowhere were the political answers to the challenges even close to sufficient.

In the spring of 2021, I started working as a climate reporter. I clambered around Estonian forests to write about the biofuel industry in Sweden and Europe, reported on the climate science studies that over and over again told basically the same, bleak story. I talked to scientists more or less daily. And I started to get a feeling that maybe the political response to the crisis wasn't flawed in the way I had thought – maybe the situation was even worse.

To test my hypothesis, I went to the core of the Swedish climate policy: the country's net zero target for 2045, which supposedly made the country a climate 'frontrunner'. I sat down in a quiet room in the state archive, opening box after box of documents from the praised Parliamentary committee that had negotiated and agreed on the target. And then I compared what I found to the emissions statistics and what the scientists told us was necessary in order to be in line with the Paris Agreement.

It took me months. Understanding the statistics from the authorities, and then structuring them in a way that would be readable. Sweden emits roughly 50 million tonnes of greenhouse gases each year – that's the number always referred to in policy debates and official statistics. But now, together with my colleague, infographic journalist Maria Westholm, I could show that the real figure is way higher. When you add in emissions from consumption and the burning of biomass, the total reaches around 150 million tonnes – three times the official number. And this does not include, for example, emissions from pension funds with fossil capital or the emissions from the state energy company's coal business abroad.

I talked to scientists, to experts on the connection between global justice and the climate transition. Sweden would need double-digit percentage emissions cuts each year to be even close to doing its fair share of the transition, they told me. I asked: if all countries failed to accurately establish their targets to the same extent that Sweden does, what global heating would the world be heading for? The answer: between 2.5°C and 3°C. And that's even if we manage to meet those stated targets, which in itself is rather unlikely: the Swedish Environmental Protection Agency's evaluation of the government's

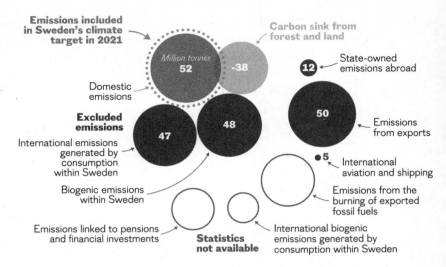

Figure 1:
Less than one
third of Sweden's
total emissions
are included in
its climate
targets; numbers
from 2018

Emissions included in Sweden's climate target in 2021

Million tonnes

52

-38

Carbon sink from forest and land

12 ← **State-owned emissions abroad**

Domestic emissions

Excluded emissions

47

48

50 → Emissions from exports

International emissions generated by consumption within Sweden

Biogenic emissions within Sweden

●5 ← International aviation and shipping

Emissions from the burning of exported fossil fuels

Emissions linked to pensions and financial investments

Statistics not available

International biogenic emissions generated by consumption within Sweden

climate policy showed that the policies in place were only about halfway to the already severely insufficient target.

And there were still more loopholes. The word 'net' in 'net zero by 2045' allowed for up to 10 million tonnes of greenhouse gases to be emitted each year *after* 2045. The government would 'compensate for' these carbon emissions primarily by making widely criticized climate investments abroad – the practice known as carbon offsetting – or by using technical 'solutions' like BECCS (bioenergy with carbon capture and storage) that were not even close to being developed at scale, and which were associated with other problems, including biodiversity loss.

Since most colleagues and readers were uninterested in the climate, as I had been just a few years earlier, the article's publication in the summer went by quite unnoticed. Even when the investigation was awarded the honour of 'best climate journalism of the year' by the newspaper *Aftonbladet*, its findings were still overlooked. That same week, several outlets, including my own newspaper, ran important investigations into climate policy proposals from the Swedish political parties ahead of an upcoming election. None of them used the numbers from my investigation, instead measuring the parties' policies against the 2045 net zero target. None told readers or viewers that the target in itself was insufficient.

Almost simultaneously, during COP26 in Glasgow in November 2021, the *Washington Post* showed that, internationally as well, the existing road-map for navigating the climate crisis landscape was completely insufficient. Its investigation revealed that the gap between the emissions that countries reported to the United Nations and the greenhouse gases actually being

Next pages:
Magnitogorsk steel
works in central
Russia. More than
one in seven of the
420,000 inhabitants
of Magnitogorsk
are employed
in the plant,
which produced
11.3 million tonnes
of steel in 2019 –
almost double the
UK's entire output.

emitted was enormous: between 8.5 and 13.3 billion tonnes a year, leaving a 16–23 per cent gap of unclaimed emissions, the high end equalling almost the total yearly emissions of China.

'In the end, everything becomes a bit of a fantasy,' scientist Philippe Ciais told the newspaper. 'Because between the world of reporting and the real world of emissions, you start to have large discrepancies.'

One of the *Post*'s reporters, Anu Narayanswamy, concluded: 'If we are wrongly calculating the emissions today, the policies that we should be adhering to for the next 50 years are going to be based on these incorrect numbers. So 50 years from now, we are going to be in a much worse place than our models or predictions are foreseeing.'

A journalist's most important task is to give her readers the information they need, not least in order for them to be able to make substantiated democratic decisions. We are decades behind on the 'climate story', and still only a minority of journalists see the climate and ecological crisis as 'their' area. The work of creating a proper roadmap has only just begun. /

The word 'net' in 'net zero by 2045' allowed for Sweden to emit up to 10 million tonnes of greenhouse gases each year *after* 2045.

4.4

We are not moving in the right direction

Greta Thunberg

In the autumn of 2021, the world's biggest direct-air carbon removal plant opened in Iceland. If all goes according to plan and the Climeworks Orca facility operates with no setbacks, it will – according to climate scientist Peter Kalmus's calculations – capture about three seconds' worth of each year's global CO_2 emissions. Carbon capture and storage is a key part of the strategy to which we seem to have blindly entrusted the future living conditions for life on Earth as we know it. Another key part is to cut down trees, forests, crops and other living biological organisms and ship them around the world to be burned for energy while we capture the carbon dioxide in huge chimneys and somehow transport and bury that CO_2 underground or in cavities underneath the ocean floor. This process – known as bioenergy with carbon capture and storage, or BECCS – is, of course, extra-beneficial to policymakers, as the huge emissions from burning wood are excluded from national statistics.

In the coming decades, these three seconds in Iceland will have to be transformed into significantly longer periods of time. And that is *significantly* as in very, very, *very* significantly longer periods of time. We are not just talking about turning seconds into minutes, hours or even days. We are talking about turning them into several weeks, by the middle of this century, or sooner. When our leaders say that *we can still do this*, then scaling up these seconds to weeks is very much included in what they mean. The gap between the carbon removal rhetoric and what is actually being done is so huge that it is almost a joke; and, once again, they more or less get away with it because public interest and the general level of awareness are so painfully low.

If the people in power were the least bit honest about their strategies for staying below 1.5°C or even 2°C of global average temperature rise, they would be pouring money into projects like this plant in Iceland and similar facilities would be popping up in every country, state, province and municipality around the world. All their pathways and pledges are dependent on this technology, and it is not as if this technology is new – it has been around for many, many years. But there are still only twenty or so small carbon capture and storage plants running worldwide, some of which have been shown to actually emit more CO_2 than they capture.

We cannot just buy, invest or build our way out of the climate and environmental crisis. Nevertheless, money is still very much at the heart of the problem. Investment is vital. Financial resources need to be directed to the best available solutions, adaptations and restorations – as much as we can possibly find. But the money seems to be going elsewhere.

The often-used argument that 'we don't have enough money' has been disproven so many times. According to the International Monetary Fund, the production and burning of coal, oil and fossil gas was subsidized by $5.9 trillion in 2020 alone. That is $11 million every minute, earmarked for planetary destruction. During the Covid-19 pandemic, governments around the world launched unprecedented financial rescue packages. These recovery plans were seen as a huge opportunity to set humanity on a brand-new course for a more sustainable economic paradigm. They were called 'our last chance to avert a climate disaster', as the enormous size of the investments would make it impossible for us to undo their consequences in the future if we got that funding even slightly wrong.

However, in June 2021, the International Energy Agency concluded that out of the historic global recovery plan, only a bleak 2 per cent had been invested in green energy, whatever 'green' means in this case. In the European Union, for instance, those 2 per cent might well be spent on fossil gas from Putin's Russia or on burning biomass made from clear-cut forests, as these activities – along with many others – are at the moment considered green in the brand-new EU taxonomy.

So our leaders did not just get it *slightly wrong* – they completely failed. And they continue to fail: despite all the beautiful words and pledges, they are not moving in the right direction. In fact, we are still expanding fossil fuel infrastructure all over the world. In many cases, we are even speeding up the process. China is planning to build forty-three new coal power plants on top of the thousand already in operation. In the US, approvals for companies to drill for oil and fossil methane gas are on schedule to reach their highest level since the presidency of George W. Bush. Oil production is soaring all over the globe: new oil fields are being opened, pipelines are being built, new oil licences are being auctioned and the search for even more production sites is ongoing. Even the use of coal is expanding – the global amount of coal-fired electricity reached an all-time high in 2021. The overall forecast for 2022 tells of further increasing emissions of CO_2.

We are two years – one fifth of the way – into what is called 'the decisive decade'. For even a small chance of staying in line with the 1.5°C target, our emissions must be in an unprecedented decline. But instead, in 2021, we saw the second-biggest emissions rise ever recorded. And it keeps increasing. A United Nations report from September 2021 states that they are expected to rise by 16 per cent by 2030 compared to 2010 levels. Add to this the fact that, at 1.2°C of warming, we are already seeing feedbacks that are not fully accounted for in the scientific pathways. According to the European Union's Copernicus Atmosphere Monitoring Service, global wildfires in 2021 created the equivalent of 6,450 megatonnes of CO_2. That is 148 per cent higher than – way more than double – the total fossil fuel emissions of the entire EU in 2020.

So, all in all, the carbon removal facility in Iceland has some serious scaling up to do – an effort that would dwarf all other human endeavours of the past. Yet that is clearly not happening, which really makes no sense at all. Why foster the idea that this underdeveloped technology could be a substitute for the immediate, drastic mitigation needed? Why bet our entire civilization on it without making the slightest effort to make it work? Why make the world picture a potential solution so vividly that we include it in every possible future scenario and then fail to invest in it? Could it be that it was never even meant to work at scale? That it was just being used – once again – as a way of deflecting attention and delaying any meaningful climate action so that the fossil fuel companies can continue business as usual and keep on making fantasy amounts of money for just a little while longer?

Either way, it is crystal clear that technology alone will unfortunately not save us. And it is still very much the lobbyists, fighting for the interests of short-term economics, who occupy the driver's seat in our society.

In the chapters that follow, scientists and experts show us just how vast the distance is between our actions so far and any real solutions, whether it is the greenwashing of sustainable consumerism, the failure to adapt to renewable energy sources and abandon fossil fuels or our desire to look away from issues of equity and justice. We see in these pages just how bad things have become and how far we still are from embracing the obvious solutions. Companies and politicians have done so much to use false solutions to preserve the status quo. But the real answers are right in front of us. /

4.5

The Persistence of Fossil Fuels

Bill McKibben

Energy lies at the red-hot heart of the climate crisis: our current system of burning fossil fuel drives the temperature ever higher, and replacing that coal, oil and gas with something else is the biggest task humanity has ever taken on. If climate change is, at some level, an arithmetic problem, then our energy sources are the numbers in question – and getting the maths right is our only hope.

Until the eighteenth century, human beings burned only small amounts of fossil fuel; wood was at the centre of our energy economy, such as it was. But, beginning in England, inventors figured out how to run engines with coal, and soon the Industrial Revolution was underway. People noticed the pollution that came with all that combustion, of course – cities choked with smoke, which today kills 8.7 million people every year, more than AIDS, malaria and tuberculosis combined. But they had no idea that the deeper problem lay with what they couldn't see. If you burn, say, a gallon of gas, which weighs about 8 pounds, you emit about 5.5 pounds of carbon; that combines with two oxygen atoms in the air to produce about 22 pounds of carbon dioxide. It's invisible, it's odourless, and it doesn't harm you directly. But since the molecular structure of CO_2 traps heat that would otherwise radiate back out to space, the warming of the Earth had begun.

We've burned enough of this fossil fuel to raise the concentration of CO_2 in our atmosphere from 275 parts per million (ppm) before the Industrial Revolution to about 420 ppm now – and that means we're trapping, each day, the heat equivalent of 500,000 Hiroshima-sized bombs. So it should come as no surprise that ice sheets are melting, oceans are rising and hurricanes are growing in force.

In order to slow or stop climate change we have to stop burning fossil fuel, but there are three reasons that's a difficult job.

One is that fossil fuel is miraculous stuff. Essentially, it represents concentrated sunshine. Over hundreds of millions of years, the sun produced vast forests, seas full of plankton and the plants that fed hundreds of billions of animals. When they died, their carcasses were eventually compressed into coal and gas and oil. In the course of two centuries we've dug up that lineage and set it on fire – it's like living on a planet with many suns, a planet pulsing with energy. A single barrel of oil – about 42 gallons – can do about the same amount of work as a man labouring for 25,000 hours. Put another way, figuring out how to use fossil fuel provided each of us in the western world with the equivalent of dozens of servants. We were able, for the first time, to move ourselves and our belongings easily over long distances; we could extend the day's light long past the setting of the sun; heat and cold were suddenly available at the flick of a switch. Fossil fuel has produced the world we know. It's too bad that it's also wrecking it.

Luckily, just in time, scientists and engineers have come up with a replacement. In the middle of the twentieth century, researchers built the first solar panels – they were designed for use on spacecraft, since it clearly wouldn't work to burn coal in orbit. But those first designs were incredibly expensive – much too expensive to compete with fossil fuel. Over time, however, the price has come steadily down, and in the last decade the cost of solar power has plummeted. The same thing has happened with wind power, as engineers learned to build much bigger turbines and even float them offshore. And now batteries to store the energy when the sun goes down or the wind drops are following the same spiralling downward cost curve. Economists say that each time we double the amount of solar power on the planet, the cost drops another 30 per cent, simply because we get better at making it efficiently.

This is the opposite of fossil fuel: oil, gas and coal do not get cheaper over time, because we've already used up most of the deposits that are easy to get at: where once drillers in the Texas Panhandle would hit 'gushers' shooting crude into the sky, now they have to drill miles beneath the ocean, or heat tar-sand oil until the gunk will flow through pipelines. By this point, renewable energy is the cheapest source of energy almost everywhere on Earth – and that's even before you start to figure in the huge economic cost of overheating the planet.

You would think that would mean we'd quickly convert to renewable energy, and indeed it's starting to happen. But, so far, that transition is going much too slowly to allow us to catch up to the damage of global warming.

Put part of the blame on simple inertia – it's the second reason we're not moving as fast as we need to. Our system is geared to use fossil fuel – there are about 1.446 billion vehicles on the world's roads. My country, the US, has 282 million cars. Almost all those vehicles run on gasoline or diesel; there is an endless network of refineries, pipes and filling stations to keep them going. So it's very good news that engineers have invented electric vehicles, and that they are in most ways superior to the internal combustion machines they replace: quieter, fewer moving parts, and so on. But despite that, it could be decades before gas cars disappear on their own – decades we don't have if we are to address climate change. And cars that run on fossil fuels should be relatively *easy* to phase out, because they last only ten or twelve years on average. Governments are starting to promote electric vehicles, and to subsidize their purchase; car makers are beginning to aggressively market them; victory over inertia seems possible. But think of the furnaces in the basements of homes all over the world – they often last for thirty or forty years before they are replaced. It's going to take much more concerted government action to speed up this process.

But inertia is not the biggest problem. The other trouble – the third reason that we're moving too slowly – is vested interest. Renewable energy obviously makes more sense than fossil fuel: it's cheaper, it's cleaner, it's available everywhere. But these arguments don't hold for one group of humans: the people who own oil wells or coal mines. For them, the advent of renewable energy is a disaster, because if it happens too fast they will never get to dig up and sell their remaining stocks of hydrocarbons.

And these people who own fossil fuels are powerful players in our political world. Until recently, ExxonMobil was the biggest company on Earth. Entire nations – Russia or Saudi Arabia, say – are essentially petro-states, deriving most of their revenue and power from hydrocarbons. The biggest political donors in American history, the Koch brothers, were also the country's biggest oil and gas barons; United States Senator Joe Manchin, who has taken more political donations from the fossil fuel industry than anyone else in Washington and personally has millions invested in coal, was single-handedly able to rewrite climate legislation in 2021. In rich and highly educated nations such as Canada or Australia there are politically powerful regions like Alberta and Queensland that are under the thumb of coal and oil companies.

The fossil fuel industry has single-mindedly used its power to delay action. As Naomi Oreskes points out in Part One, great investigative reporting in recent years has proved that the oil companies knew all about global warming way back in the 1970s – ExxonMobil scientists were able to predict with great accuracy how much the temperature would rise by 2020. And they were believed by company executives – they started building drilling rigs higher, for instance, to compensate for the rise in sea level they knew was coming. But across the industry, instead of explaining the dilemma to the world, they took the opposite tack: they hired a small army of public relations experts, some of whom had worked for the tobacco industry, to try to instil doubt about the science in the public mind. This worked very well: for almost thirty years the world was locked in a sterile debate about whether global warming was 'real', even though both sides in the debate knew there was no question it was. It's just that one of those sides was willing to lie – and the lie cost us the one thing we don't have, which is time.

The fossil fuel industry continue to lobby, to greenwash, to delay. But now they are met by a large citizens' movement that, for instance, has persuaded institutions to divest huge amounts of their stock, making it more difficult for them to raise capital. Other activists have blocked pipelines and coal terminals. So change is coming – the main question is, how fast?

That change won't be perfect, by the way: no way of making energy is without human and environmental cost. It will be important to try to prevent abuses in the mining of the minerals that go into solar panels and batteries. And some people dislike the sight of windmills on the horizon – though others find them beautiful, both because they make the breeze visible and because they represent people taking responsibility for their energy needs closer to home. There are other potential advantages to renewable energy as well: because fossil fuels are concentrated in a few places, the people who control them have undue power – think of the king of Saudi Arabia, say. But sun and wind are available everywhere, which means that there will be at least some reduction in the unfair distribution of power. And consider the nearly billion human beings, mostly in Africa, who still have no access to modern energy at all: the UN now estimates that 90 per cent of them will get their first power from renewable sources, because putting solar panels on the edge of a remote village is so much cheaper and easier than building the conventional grid to these hard-to-reach locations.

If you think about it, it's quite miraculous to live at a moment when the easiest way to produce power is to point a sheet of glass at the sun. I've been in villages where people for the first time ever have small refrigerators to store vaccines (and ice cream), and enough light for kids to be able to study in the evening. This is Hogwarts-scale magic, and if we were smart and kind we would devote ourselves to spreading this new technology everywhere in the next decade. That wouldn't be enough to stop global warming – it's too late for that. But it is our best bet for slowing it down and giving humanity a chance.

And seeing where our energy came from might remind us not to spend it so profligately. Electric cars are a stopgap, in one sense, until the moment when we've built decent systems of electrically powered public transit. If we use cheap renewable energy to build ever bigger homes and stuff them with ever more junk, then we'll still use up the world's farms and forests, still kill off its animals. An energy transition may be our most immediate crisis, but it's far from the only peril we face.

Still, we shouldn't underestimate the potential of this moment. One way to think of it is: we've now reached the point when we need to stop burning things on the surface of the Earth. We shouldn't be digging down to find coal and gas and oil and setting them on fire – that's dirty, dangerous and depressing. Instead, we need to rely on the burning ball of gas 93 million miles up in the sky. Energy from heaven, not from hell! /

The transition to renewable energy is going much too slowly to allow us to catch up to the damage of global warming.

4.6

The Rise of Renewables

Glen Peters

Before 1800, our energy system was dominated by human and animal power; afterwards, by the burning of wood. Since then, fossil fuels have come to dominate, and global CO_2 emissions have grown consistently – with this growth closely linked to increasing prosperity. For 200 years, global CO_2 emissions have grown at a steady rate of 1.6 per cent per year. The recent rapid deployment of non-fossil energy sources – biomass, hydro, nuclear, solar, and wind power – has not been able to keep up with the ever-growing demand for more energy. Consequently, the share of non-fossil energy sources has remained at roughly 22 per cent for several decades, though it has begun to slowly increase in the last few years on the back of growth in wind and solar power and is now at the highest level since the 1950s (Fig. 1).

Beneath these global numbers lies a more complex story. In the world's high-income countries, CO_2 emissions are falling – at a rate of 0.7 per cent per year in the US and 1.4 per cent per year in the EU over the last decade. But this is a story of development, and not a story of climate policy. High-income countries, at least at the aggregate level, have reached a comfortable standard of living. Their energy use has levelled out and in some cases is falling. Their energy infrastructure is old, and energy and climate policy have made solar and wind power cost-competitive. As their coal power plants are nearing their end of life, and since energy use is stable, the deployment of solar and wind power in these countries is largely replacing their ageing energy infrastructure. Meanwhile, they are also taking advantage of global supply chains, whereby the import of consumable goods means there is less pressure placed on the energy system and emissions.

Middle- and low-income countries have a different reality. Living standards are, at the aggregate level, considerably lower than in Europe and the US. To improve them, energy use is growing rapidly. The energy infrastructure of these countries is young. Rapid growth in solar and wind

Primary energy

Figure 1:
The global energy system since 1850 (top) showing the dominance of fossil fuels and the recent emergence of non-fossil sources. CO_2 emissions (bottom) primarily come from fossil sources, but also from biomass and from land-use change (not shown). Biomass is often considered carbon neutral in emissions statistics, with emissions allocated to changes in forest carbon stocks; but when analysing the long-term evolution of the energy system, bioenergy cannot be ignored, as burning wood was the dominant energy source before the broad adoption of fossil fuels. Before 1850, it is also necessary to consider human and animal power.

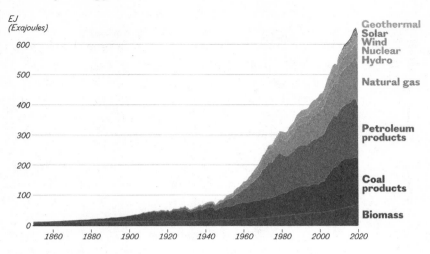

Annual global emissions

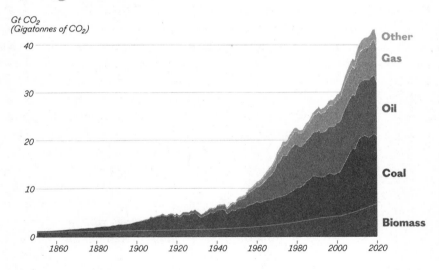

power is not sufficient to capture the increase in energy demand and, consequently, fossil fuel use and CO_2 emissions continue to rise. It is not that these countries are doing a bad job of stabilizing and then reducing their emissions. In many cases, they are world leaders in the deployment of clean technologies. They just have a different context.

In the last decade, global CO_2 emissions have started showing signs of peaking (Fig. 1). This potential peak represents a tug-of-war between the declining emissions in high-income countries and the increasing emissions in middle- and low-income countries. The world could be reaching a point where the opposing forces are beginning to equalize. While in some respects this could be viewed as progress, it is nevertheless progress that would be insufficient to meet our ambitious climate goals.

This tug-of-war is also happening between the decline of fossil fuels and the rise of non-fossil energy sources such as solar and wind power (Fig. 1). Solar and wind power are growing rapidly in many countries, and across broad income levels. The same can be seen with some clean technologies such as electric cars and electric buses. The share of non-fossil energy sources in the global energy system is beginning to increase for the first time in decades – and it is now at the values last seen in the 1950s, when use of biomass (energy from burning wood) was widespread. While this is positive progress, it is nevertheless still far from sufficient.

The progress of the climate crisis does not afford us the time to watch the world slowly transition away from fossil fuels. The simple mathematics of the climate system seen through the 'carbon budget' means we can use fossil fuels only for a few more decades, until around 2050, unless technologies are developed to avoid emitting the CO_2 produced by using fossil fuels (carbon capture and storage) or to take the CO_2 back out of the atmosphere (carbon dioxide removal). Both these technologies have so far failed to deliver. And, since fossil fuel power plants or industrial facilities can have lifetimes of fifty years or more, this means that new fossil fuel infrastructure being built today (or in the recent past) in middle- and low-income countries will need to be retired before it reaches its end of life.

Though energy system transitions have historically been slow, this need not be the case. Policy and social movements can combine with technological development to accelerate the needed energy system transition. The essential tools are readily available – the world knows how to do solar and wind power, how to store energy with batteries or hydrogen and how to decarbonize transportation. It is important to note, however, that decarbonizing the energy system is not just a simple engineering problem; it's not just a question of no longer burning fossil fuels and using renewable energy sources instead. Every aspect of the production system that sustains our lives is intertwined with energy and therefore CO_2 emissions – from simple communication to the provision of essential food and shelter – and even alternative energy sources come with environmental and emissions costs. We can use electricity, hydrogen or biofuels to power our vehicles, for example, but all three can generate significant emissions in their production or provision. Solar and wind power have zero emissions in their use, but they take energy to manufacture too. Solving the climate problem therefore requires taking a system perspective. Eventually, it touches every aspect of our lives. Solving the problem requires us to put less pressure on the system in the first place, and this essentially means shifting to a way of living that is less dependent on material consumption.

With the problem diagnosed, the next question is what to do about it. For too long, the temptation has been to look for the perfect policy mix to efficiently meet climate targets, for example carbon pricing or emissions trading systems. The reality is more complex. Countries have different starting points, different political systems and different contexts. In practice, countries must implement whatever policies and incentives that will work for them, even if they are far from perfect. While such a mosaic of policies and incentives may be the nightmare of economists, the climate system does not allow us the time to find the perfect policy solution that is acceptable to all.

The energy transition is also going to inflict pain on some while bringing rewards to others. This is unavoidable, but the world is full of examples of these transformative shifts. From horses to cars, typewriters to computers, landlines to mobile phones, petrol to electric cars, fossil power to renewable power. Many of these transitions are not driven by policy but by technology and society. The countries and companies that reap the benefits will be those that are ahead of the curve and shape the transition, as opposed to those that seek to hold on to the fading past. It is incumbent on governments to protect and help those who are collateral damage in the energy transition, for example coal miners, but not those that choose to impede it, like some powerful corporations.

The amount of time available to make the energy transition is vanishing. To achieve it, we need all the tools in the toolbox. Technology alone is unlikely to solve the problem, and technologies also come with a variety of risks and challenges of their own. Behavioural change in isolation is unlikely to solve the problem, but ignoring it makes the challenge even harder. Climate-conscious behavioural change can often come with considerable benefits, such as improved health, work–life balance and well-being. Policy is also unlikely to be sufficient, as governments are often constrained by the conflicting goals of the electorate, lobbyists and doing what is best for society. It is at the intersection of all three that progress lies: with a complementary mix of technology and behavioural and political changes, we will drive a transformation of society to avoid the worst risks of the climate crisis. /

Fossil-free Energy Sources

Solar power

The potential of solar power is huge. It is cheap to produce, fast to install and can be scaled up to produce as much electricity as traditional power plants. In Germany, for example, 10 per cent of the country's electricity comes from solar power. The downside? It requires sunshine – or at least a not too cloudy day – to produce large quantities of energy. So, in order to be fully effective, solar power needs to be complemented by electric storage devices such as batteries. That way the energy can be saved for times when its production is lower or non-existent. Another challenge is the fact that large-scale solar power requires a lot of space, so the choice of location must avoid any potential harm to the local environment.. One solution is to place solar panels on existing buildings, where the preservation of biodiversity is not a consideration. Wherever there is a roof or a parking space, there is a potential solar energy source. /

Wind power

Wind power is widely available, relatively fast to install, clean and cheap. And it's rapidly becoming even cheaper. The most frequent criticism of wind power is that the wind doesn't always blow. This is true, but only for small electric grids. On large national or regional grids the wind blows more or less all the time. The real challenge to scaling it up is the risk of disturbing local wildlife and the impact on people living nearby. As with solar power, location is crucial. Wind parks can be built close to motorways and industries, or in places where few people are affected and the consequences for wildlife are minimal, for example in off-shore locations. Technology is also making it increasingly possible to have mobile offshore power plants, which reduce NIMBY (Not in My Backyard) complaints. /

Green hydrogen

Hydrogen is an electricity source and a fuel which leaves only water behind when used in a fuel cell. However, hydrogen largely does not exist by itself in nature so it must be produced from other sources, mainly methane or water. This process requires more energy than the fuel itself gives in return. The gain is that hydrogen can be stored without losing energy over time.

According to the *New Scientist*, 96 per cent of hydrogen is currently being made from fossil fuels, which makes it far from being a renewable or fossil-free energy solution today. But it can also be made from water, using renewable energy such as wind and solar in the process. This is called green hydrogen and it can be used as an alternative to fossil fuels in certain circumstances, for example where the energy source can't be electrified or where the energy needs to be stored for a longer time period than is efficient for a battery.

The problem, however, is that green hydrogen requires an abundance of cheap renewable energy, something that is not likely to be seen in the near future. Making hydrogen through electrolysis using nuclear power is called pink hydrogen, and there is also blue hydrogen, which is made from fossil fuels using carbon capture and storage. But since that technology is also far from being developed at scale, hydrogen as a whole remains a solution with considerable limitations. A Global Witness report from 2022 showed that a 'first of its kind' blue hydrogen plant in Canada was emitting more greenhouse gases than it was capturing. /

Hydro power

Hydro power plants use falling or fast-flowing water to create electricity. According to the International Energy Agency, they provided 17 per cent of the world's electricity in 2020.

Although hydropower is clean, it has a significant impact on the local environment, damaging wildlife and ecosystems as well as affecting people who live near the power stations or the water reservoirs needed to regulate the flow. /

Nuclear power

Electricity can also be produced using nuclear technology, whereby reactors split the atoms of elements such as uranium and plutonium. This reliable, very low-carbon power currently provides around 10 per cent of the world's electricity.

There are, however, many downsides to nuclear power because of its technical complexity. The plants are expensive and take a long time to build. The two most recent plants to be set up in Western Europe, Olkiluoto 3 in Finland and the UK's Hinkley Point, both experienced long delays; when the Finnish reactor finally opened in the winter of 2022, it had taken sixteen years to construct. Even if the construction period could be considerably shortened, it would still be a huge challenge to replace the world's ageing nuclear power stations within the timeframe needed to meet our climate targets.

From a safety perspective, nuclear power has proven to have alarming drawbacks, the disasters at Fukushima in 2011 and Chernobyl in 1986 being two examples. It is also sensitive in terms of security policy, as a nuclear power plant is potentially vulnerable as a target during conflicts or terrorist attacks. Its production and use requires geopolitical stability.

Then there is the question of safely storing the radioactive waste produced – a question that after over seventy years is yet to be resolved globally. Due to its technological intricacies, nuclear power remains a limited global energy source. /

Biomass energy

Biomass energy creates electricity or heat by burning wood, as well as other plant or animal materials such as crops, peat, kelp, rubbish and slaughterhouse waste. It is considered a renewable energy source, but its renewable status depends upon a sustainable agriculture and forest industry, which doesn't exist at any sufficient scale today.

Beyond this, it is renewable only over a vast timescale: it can take over a hundred years for a tree to grow, and many centuries for a forest to fully recover after it has been clear cut – if ever. Also, when we replace forests with tree plantations we lose invaluable biodiversity and resilience. The relative inability of plantations to sequester carbon is another negative aspect, as is the fact that plantations are much more vulnerable to fires and disease.

The fact that biomass is considered renewable has sparked large-scale exploitation of this energy source, which is accelerating deforestation and biodiversity loss. This is a human-made negative feedback loop that is currently spiralling out of control in many places. For biomass to be sustainable and renewable, we need to significantly downscale the entire practice.

Burning wood for energy releases more carbon dioxide into the atmosphere than burning coal does, and the fact that these emissions are excluded from our national statistics – and that they are considered renewable – has created a potentially disastrous loophole. /

Geothermal energy

Geothermal energy comes from inside the Earth's crust. It can be used to generate heat or to produce electricity. To create electric power from geothermal heat, wells are dug deep into the ground to access steam and hot water to drive the turbines that generate electricity.

While geothermal is a low-carbon source of energy – it produces around 17 per cent of the emissions of fossil gas – it does produce other emissions, such as hydrogen sulphide and sulphur dioxide, both of which cause substantial environmental concerns. Geothermal electricity is also geographically limited, as it requires proximity to tectonic fault lines. This is why its technological hotspots are found in places such as Iceland, California, New Zealand, Indonesia, El Salvador and the Philippines. /

There is also huge potential for us to:

- Use less and save energy.
- Make products and buildings more energy efficient.
- Make use of power where and when it gets produced.

4.7

How Can Forests Help Us?

Karl-Heinz Erb and Simone Gingrich

Forests can play an important role in mitigating the climate crisis. They lock away carbon, storing about two times more than the atmosphere, and the wood they provide can replace greenhouse-gas-intensive products and services. Countries in the Global North currently expect that forests will be a substantially larger source of energy and products in the future, while at the same time sequestering additional carbon from the atmosphere. However, there is a risk that this approach could do more harm than good.

It is well known that, globally, deforestation contributes significantly to greenhouse gas emissions, producing around 13.2 gigatonnes of CO_2 equivalent (GtCO_2eq) per year. However, many temperate and boreal forests act as net carbon sinks because, overall, these forests are expanding in area and carbon density. This expansion is often driven by industrial monoculture forests – fast-growing trees destined to be cut down again – not pristine forests. It is paradoxical, but increases in carbon absorption in forests have, in several cases, coincided with simultaneous increases in wood harvesting. Understanding how and why this has occurred is crucial to evaluating the role forests can play in our climate change mitigation strategies.

The riddle can be solved by considering that a forest's ability to absorb carbon does not only result from current management but is strongly affected by the past. Both present-day and historic management practices determine how much carbon a forest holds and thus also how far a forest is from its potential to store carbon. By the early nineteenth century, forests in many regions had become heavily depleted after a long history of intensive land use. Industrialization alleviated this pressure on forested areas, due to the new availability of fossil energy, long-distance trade opportunities and the intensification of agriculture. So forests could recover even if wood harvesting increased, as long as regrowth outpaced harvests.

It is essential to understand that carbon stocks in forests have increased not because of but despite high harvest levels. Harvesting wood causes emissions from soil and withdraws the carbon of trees from the forests. At the global scale, wood products are estimated to be responsible for land-use emissions of 2.4 $GtCO_2eq$ per year, and the carbon stocks of managed forests are considerably lower than those of pristine forests would be – on average, 33 per cent in temperate, 23 per cent in boreal and about 30 per cent in tropical regions.

When we harvest wood we must consider the carbon that would otherwise be sequestered if we left a forest untouched. The climate impact of wood use depends on the average time wood products remain in use. This 'lifetime' of long-lived wood products is typically about fifty years, while trees would remain alive and continue to sequester carbon for decades or even centuries more if they weren't harvested. Meanwhile, burning wood for bioenergy causes more emissions per unit of energy than fossil fuels, and these emissions can only be reabsorbed via forest regrowth. Climate change mitigation is therefore realized only after a forest regrows and enough emissions have been cut to equal the amount of carbon that would have been sequestered had the forest not been harvested in the first place. In temperate and boreal zones, this 'parity time' may take several decades, sometimes even centuries.

What does this mean for forest-based climate change mitigation strategies? It would be wrong to claim that forests should not be harvested at all. Wood can replace many greenhouse-gas-intensive products and help reduce waste problems associated with materials such as plastic. But it should be primarily used for long-lived products, within sustainable limits. The supply of wood must be limited by the amount of timber that is regrowing in forests if deforestation or forest degradation are to be avoided. Biodiversity conservation must also impose an additional limit. But in a world where the mass of human-made artefacts already exceeds the mass of all living biomass, the focus should rather shift towards reducing the use of resources in industrialized countries.

Furthermore, given the urgency of the climate crisis, the parity time of forest bioenergy is prohibitively long. In the EU, currently about a quarter to a third of all wood harvested is directly used for energy. While EU legislation treats forest bioenergy as sustainable and inherently carbon neutral as long as harvests remain less than regrowth, biomass should be considered sustainable only if the material that is burned is strictly limited to manufacturing residues that cannot otherwise be used. In industrial wood-processing chains, residues are not waste streams but resources for the paper and board industry – and here, bioenergy competes with these uses.

The expansion of forested areas in the Global North holds only limited potential for carbon sequestration, because it requires too much time in light of the urgency of the crisis and might lead to land competition. As such, protecting forest carbon sinks by reducing wood harvesting appears to be the optimum strategy: forest carbon sinks currently sequester 10.6 $GtCO_2eq$ per year, compensating for around 30 per cent of total annual emissions. This is the only strategy to sequester atmospheric carbon that is currently readily available at a large scale.

Forest carbon sinks will eventually saturate – not anytime soon, but roughly 50–150 years from now. Natural disturbances will affect and reduce forest carbon stocks, and industrial monocultures are particularly vulnerable. We therefore need a multipronged strategy: harvests should be taken from these monocultures, while efforts should be made to improve forest resilience by increasing the diversity of species and by allowing (at least some) trees to grow old. Resilient, biodiverse forests should be left alone and considered a 'bridging technology' – buying time for other sectors to decarbonize while maximizing the benefits to biodiversity.

Maximizing the climate change mitigation role of forests will probably mean limiting the availability of wood products. In order to prevent this reduced supply from being compensated for by fossil fuels, demand-side strategies need to be explored that allow decreases in material use while safeguarding human well-being and ensuring just access to resources. /

Resilient, biodiverse forests should be left alone and considered a 'bridging technology' – buying time for other sectors to decarbonize.

What about Geoengineering?

Niclas Hällström, Jennie C. Stephens and Isak Stoddard

'Geoengineering' is the intentional technological manipulation of the Earth's atmosphere and ecosystems at scales so large that it would interfere with and alter global climate systems. Most geoengineering technologies are still only speculative ideas, yet they are extremely controversial.

Geoengineering is not intended to reduce production of fossil fuels or emissions of greenhouse gases, the root causes of global warming. Its proponents seek instead to reduce the warming effects of the sun, either by reflecting some of its radiation back into space or by removing carbon dioxide from the atmosphere and somehow storing it. Solar geoengineering includes widely contested proposals to fly fleets of aeroplanes around the globe to continuously spray large quantities of sun-blocking aerosols into the stratosphere, or to cover extensive areas of Arctic ice with glass beads. Carbon dioxide removal at geoengineering scale includes suggestions to fertilize swathes of the ocean to cause massive algal blooms, or to convert enormous land areas to tree plantations with the intention of burning the wood and capturing the CO_2.

All geoengineering approaches involve huge risks, some to the point of threatening both ecosystem and societal breakdown. Many impacts would be irreversible and impossible to predict and would exacerbate existing injustices. This is particularly the case with solar geoengineering, where the injection of aerosols into the stratosphere could disrupt monsoons, intensify droughts and threaten the livelihoods of billions of people. Worse yet, if this process were initiated, and then at some future time the sun-dimming aerosol injections were to stop, the masked heating effect of the CO_2 accumulated in the atmosphere could cause sudden and massive temperature rises, preventing any chance of adaptation and driving a catastrophic 'termination shock'.

Many scholars, experts and activists have concluded that such technologies cannot be managed equitably and safely. Advancing solar geoengineering assumes the existence of stable global systems of governance that could function without failure for hundreds or thousands of years – an impossible requirement. Allowing the development of these technologies also leads to the frightening prospect of powerful states, organizations or even wealthy individuals exerting unilateral control of them, deepening today's inequities in power and financial access, and escalating the risk of wars over attempts to control the Earth's climate systems. Around the world, there are growing calls for an immediate international ban on the advancement of solar geoengineering technology (see www.solargeoeng.org), and many are working to strengthen the existing geoengineering moratorium under the UN Convention on Biological Diversity.

Attempts to advance real-world research and experimentation on solar geoengineering are consistently met with fierce resistance from Indigenous peoples, scientists and civil society organizations, who warn that humanity must not head down the slippery slope of normalization (see www.stopsolargeo.org and www.geoengineeringmonitor.org). Attempts at repackaging the contested term 'geoengineering' into new, less tarnished terms, such as 'climate intervention', 'climate repair' and 'climate protection technologies', shows the ways in which certain actors are attempting to obfuscate the discourse around these controversial technologies.

All geoengineering schemes are attempts to manipulate the Earth with the same domineering mindset that got us into the climate crisis in the first place. The implications of vested interests mainstreaming the idea of geoengineering by discussing it as if it were a viable option may be as dangerous as the impacts of actually deploying geoengineering. Suggesting that geoengineering is a 'plan B' provides convenient excuses for the fossil fuel industry, tech billionaires and other promoters of these ideas to delay and derail the fundamental societal transformations that are urgently needed. Geoengineering is not an option. Intensified climate disruptions and injustices call for something very different: a focus on sufficiency and well-being, curbing emissions at the source and rapidly phasing out fossil fuel production, while prioritizing principles of equity, local livelihood and ecological integrity. /

4.9

Drawdown Technologies

Rob Jackson

The need for 'drawdown', defined here as removing carbon dioxide, methane and other greenhouse gases from the atmosphere *after* their release, arises from failure. We have flooded the atmosphere with 2 trillion tonnes of carbon dioxide pollution – most of it in the last fifty years – even when the danger to life was clear. In fact, annual global fossil carbon dioxide emissions have risen 60 per cent since publication of the first IPCC report in 1990. We have not just failed but failed spectacularly.

Given our failure to act, we've left Greta's generation little choice but to wave a magic wand and fix our emissions retroactively if global temperature increases are to stay below 1.5°C or 2°C thresholds, paying more to remove greenhouse gases from the air later.

Can drawdown technologies actually work? They aren't magic – as we'll see – and they're profoundly expensive.

Under almost every scenario, meeting the 1.5°C temperature target will require removing some previously emitted carbon dioxide (CO_2) from the atmosphere. One recent analysis concluded that if we could keep cumulative global emissions below 750 billion tonnes (about two decades of current emission rates) between 2019 and 2100, about 400 billion tonnes of 'overshoot' carbon dioxide would still need to be removed from the air to keep global temperatures increases below 1.5°C in 2100.

At an aspirational cost of $100 per tonne of carbon dioxide removed, drawing down 400 billion tonnes of CO_2 from the atmosphere would cost $40 trillion – and other analyses suggest this is a conservative estimate. Younger generations are legitimately asking, 'Why should we pay for this?'

The reality is that keeping greenhouse gases out of the atmosphere today is always cheaper than removing them tomorrow. The atmosphere contains about one molecule of carbon dioxide for every 2,500 molecules of other gases, which makes finding and 'removing' carbon dioxide from the atmosphere like pulling needles repeatedly from a haystack. About one in ten molecules emitted by a typical fossil power plant smokestack is carbon dioxide, so it makes no sense for these smokestacks to keep releasing concentrated carbon dioxide into the air and then for us to pay to remove it in diluted form later. Wherever we keep burning fossil fuels, we need to capture the carbon pollution from smokestacks now, before it pollutes the air.

Currently there are only thirty or so carbon capture and storage (CCS) plants running worldwide compared to thousands of fossil power plants globally. If all those fossil plants operate to the end of their lifetimes without carbon capture and storage, their 'committed emissions' will entail hundreds of billions more tonnes of carbon dioxide pollution, more than enough to push us past 1.5°C and possibly 2°C.

If we fail to curb emissions, and fail to capture and store the carbon pollution, then carbon drawdown or removal technologies come into play. Land is one of the most obvious options – especially forests and soils – replacing carbon lost to the atmosphere from deforestation and agricultural activities.

The world lost a billion hectares of forest in the twentieth century, most of this land now used for row crops and cattle ranching. Agricultural activities such as ploughing have also released half a billion tonnes of carbon dioxide into the atmosphere from the world's soils. These carbon losses from soils and forest trees underpin natural climate solutions approaches that stop carbon losses from land and put carbon back into them through conservation, restoration and improved land management. Relatively optimistic estimates suggest such practices could provide one third of the climate mitigation needed up until 2030 to stabilize global warming below 2°C. Natural climate solutions are currently the cheapest way to offset fossil fuel pollution, often at estimated costs of approximately $10 per tonne of CO_2 stored.

We can put billions of tonnes of carbon back into land through natural climate solutions such as forest and wetland restoration, tree planting, no-till farming, and other actions. A more plant-based diet, especially eating less red meat, and a smaller global population would also reduce deforestation and global cattle numbers (reducing methane emissions too), sparing land for other ecosystems and uses.

Can we rely mostly on natural climate solutions, though? No – at least not to offset anywhere near 35–40 billion tonnes of annual fossil carbon pollution.

Without drastic emissions cuts, industrial greenhouse gas removal will be needed to keep global temperature increases below both 1.5°C and 2°C. Scientists have studied atmospheric carbon dioxide removal for more than a decade, the twin steps of capturing carbon dioxide from the air and storing it out of harm's way. Plants, rocks and industrial chemicals can be used to remove CO_2 from air. Plants, including trees, grasses, kelp and phytoplankton, as well as some microbes, take up carbon dioxide through photosynthesis. Beyond the natural climate solutions discussed above, there is a common plant-based approach known as BECCS (bioenergy with carbon capture and storage). In BECCS, you gather or harvest plant biomass, burn it to produce electricity (or convert it into biofuels) and pump the carbon dioxide pollution underground to keep it out of the atmosphere. Of all the drawdown or negative emissions technologies, BECCS is the only one that provides energy rather than requiring it (and, done carefully, can provide it close to carbon free). Like all climate solutions at the billion-tonne scale, BECCS has issues: it's land and water intensive and, as with all underground waste pumping, you have to monitor the reservoir for decades to make sure the CO_2 stays put. Still, BECCS is relatively cheap by negative emission standards (around \$50–\$200 per tonne of CO_2 stored) and facilities already operate commercially today. In 2019, BECCS plants were removing about 1.5 million tonnes of carbon dioxide yearly, the largest of them a corn-ethanol facility in Decatur, Illinois. A study from the US National Academy of Sciences put the potential for BECCS at roughly 3.5–5.2 billion tonnes of carbon dioxide removed per year without large adverse impacts.

Another drawdown technology is enhanced weathering. This approach tries to accelerate the rate at which rocks such as silicates react naturally with atmospheric CO_2. Igneous basalt is one of the most common rocks on Earth and underlies a tenth of all the continental surfaces and most of the ocean floor. Basalt contains lots of the calcium-, magnesium- and iron-rich silicate minerals that react with carbon dioxide and form carbonates and other carbon-rich rocks. Calcium carbonate, or common 'limestone', for instance, combines one calcium atom with carbon dioxide and an extra oxygen atom: $CaCO_3$. The Empire State Building and the Great Pyramid of Giza were built with it.

Imagine mining basalts, crushing them and exposing them to the air to react with carbon dioxide. You might even fertilize an agricultural field with them, enhancing plant growth because of the extra calcium, magnesium and nutrients the rocks release. Alternatively, you might just expose crushed rock to air and rebury it after it reacts fully with atmospheric CO_2. Cost estimates for enhanced weathering are \$75–\$250 per tonne of CO_2 removed. Start-up companies are just forming to try it, but enhanced weathering hasn't scaled to commercial-level projects yet. We know weathering works in nature over thousands of years; the question is whether we can speed it up to a policy-relevant timeframe of years to decades.

Finally, dozens of companies are working on direct-air capture of carbon dioxide using specialized chemicals. Nitrogen-based 'amines' have been used for decades in refineries and petrochemical plants to scrub carbon dioxide from gas streams. Hydroxides are a second family of chemicals used in commercial direct-air capture operations today. In both cases, the original chemicals can be regenerated using heat or by changing the acidity of a solution. Concentrated carbon dioxide is obtained during this chemical regeneration.

In most direct-air capture operations, the carbon dioxide generated must be pressurized and pumped underground, just like with BECCS (the 'carbon capture' and 'storage' part of BECCS). The current cost range for direct-air capture is around \$250–\$600 per tonne of CO_2 removed, far more than for natural climate solutions. Today, companies are removing a couple of million tonnes of carbon dioxide a year from the atmosphere industrially. That's a start, but we're far from the billions of tonnes per year we need.

Beyond carbon dioxide, we'll probably need to remove other greenhouse gases from the atmosphere too. Methane (CH_4) is the second most important greenhouse gas, warming the Earth eighty or ninety times more than an equivalent mass of carbon dioxide over the first twenty years of its release. More than half of global methane emissions come from human actions, including fossil fuel use and agriculture. Global methane concentrations are now 2.6 times higher than they were two centuries ago.

Methane 'removal' (or, more literally, 'oxidation') is difficult. Methane is 200 times less abundant in the atmosphere than carbon dioxide and therefore more difficult to isolate. Its pyramid-like structure also makes it harder to crack open than a CO_2 molecule, except at very high temperatures.

Methane removal has some advantages, though. For one, you don't need to capture it and pump it underground. If you can oxidize it using catalysts or nature's oxidizing agents (atmospheric radicals such as hydroxides and chlorine) you can turn it into CO_2 and release it back into the air.

All methane emitted into the atmosphere eventually oxidizes to CO_2 anyway, so methane removal just speeds up nature's reaction. Because methane is so much more potent than carbon dioxide, the CH_4 to CO_2 trade is a good one for the climate. Another advantage is that we need to remove much smaller quantities of methane compared with carbon dioxide to make a big difference for the climate: 'only' tens or hundreds of millions of tonnes per year instead of billions.

If feasible at scale, methane removal could also help shave tenths of degrees off peak temperatures and delay the time a given temperature threshold is passed. Readying methane removal today could also provide insurance against catastrophic methane release from the Arctic, something many scientists believe is possible, even likely, this century.

As important as I think methane removal is, it needs much more research and investment to be ready commercially. Combining carbon dioxide and methane removal in the same industrial facilities seems especially promising to me, using blowers and air-handling systems to remove multiple gases rather than one gas at a time.

Finally, for all drawdown solutions, we will need a global price on carbon to prompt action. An 'upstream' carbon price adds a fee wherever fossil fuels are extracted, with the extra cost passed on to consumers in the price of products derived from fossil energy. It needs discussions on what to do with the money and how to keep poorer people from paying more for their energy. But this price would better shift the financial burden of emissions on to those responsible for them and would (more closely) reflect the real cost of fossil pollution. None of the options I've discussed above is feasible at large scale without a carbon price or, lacking that, policy mandates for action.

Although the cost of drawdown is high, the cost of doing nothing is staggering. Insurance companies know costs and risk better than anyone. Insurance giant Swiss Re (the world's second-largest reinsurance company – a company that insures insurance companies) recently estimated that the global economy could shrink by 18 per cent if no climate mitigation action is taken, at a cost of as much as $23 trillion annually by 2050. It concluded: 'Our analysis shows the benefit of investing in a net zero economy. For example, adding just 10 per cent to the USD 6.3 trillion of annual global infrastructure investments would limit the average temperature increase to below 2°C. This is just a fraction of the loss in global GDP that we face if we don't take appropriate action.'

To reduce these costs, we need to cut emissions – then cut them some more. We need to implement natural climate solutions, restoring forests and soils wherever possible. We need to lower the cost of drawdown technologies and hope people accept them. We need to discuss personal issues of population, diet, energy use and inequality.

In truth, I'm frustrated writing a chapter about drawdown technologies, because we shouldn't need them. I've watched years of climate inaction roll by like floats in a parade. When will the victory parade finally begin? /

4.10

A whole new way of thinking

Greta Thunberg

'The American way of life is not up for negotiation. Period.'

These are the words of US President George H. W. Bush ahead of the 1992 UN Earth Summit in Rio de Janeiro. In retrospect, it turns out he was speaking on behalf of the entire Global North. And this remains our position to this very day. The solution to this crisis is not exactly rocket science. What we have to do is to halt the emission of greenhouse gases, which, in theory, is a pretty easy thing to do – or at least it was, before we let the problem spiral out of control. It is solving the climate crisis while maximizing economic growth that is the hard part. So hard as to be near impossible.

Since President George H. W. Bush spoke those words, our annual CO_2 emissions have gone up by more than 60 per cent, turning what was then a 'big challenge' into an existential emergency. We have developed impressive creative accounting, loopholes, outsourcing and greenwashing PR narratives that make it seem as if real action is being taken when in fact it is not. Continued economic growth, on the other hand, has been hugely successful . . . At least for a small number of people who boast a carbon footprint the size of entire villages. Nevertheless, economic growth since the 1992 Earth Summit has at least brought us one major advantage – it has proved beyond all reasonable doubt that our ambition was never about saving the climate, it was all about saving our way of life. And it still is.

Until recently, you could argue that it was possible to save the climate without having to change our behaviour. But that is no longer possible. The scientific evidence is crystal clear: our leaders have left it too late for us to avoid major lifestyle and systemic changes. There simply are not enough resources left. If we are to have a chance of minimizing further irreparable damage, we now have to choose. Either we safeguard living conditions for all future generations, or we let a few very fortunate people maintain their constant, destructive search to maximize immediate profits. If we choose the first option and decide to go on as a civilization, then we must start to prioritize. In the years, decades and centuries to come we will no doubt need many

Next pages:
A voluntary
afforestation
campaign at
the edge of the
Badain Jaran
Desert in Linze
County, Gansu
Province of China.

transformations that will have to stretch across our whole societies. And since our resources are limited, we must start getting our priorities straight.

Beyond the very basics, our number-one priority must be to distribute our remaining carbon budgets in a fair and holistic way across the world as well as repay our enormous historical debts. That means those who are most responsible for this crisis must immediately and drastically reduce their emissions. We understand that the world is very complicated and there are countless important variables. That is exactly why we have to start as soon as possible. This will require a whole new way of thinking for our societies, at least in the affluent parts of the world.

People keep asking us climate activists what we should do to *save the climate*. But maybe the question itself is wrong. Maybe, instead, we should start asking what we should stop doing? Sometimes you hear people say that we already have all the solutions to the climate crisis and all we need to do is to implement them. But this is only true if we consider *not doing something* to be a valid solution. If we choose to accept that idea, then we will still be able to get out of this mess.

There is, in reality, absolutely no reason to believe that the necessary changes will make us less happy or less satisfied. If we manage to do this right, then our lives will be given more meaning than selfish, shallow over-consumption can ever give us. Instead, we can make time and space for community, solidarity and love. In no way should this be considered a step back in our development. On the contrary, this would be human evolution – human revolution.

A stable climate and a well-functioning biosphere are basic conditions for life on Earth as we know it. That requires an atmosphere that does not contain too many greenhouse gases. The carbon dioxide safety level for such climate stability is often considered to be around 350 parts per million (ppm) – a level we passed in around 1987. In February 2022, we surpassed 421 ppm. At current emission levels, our remaining carbon budgets for a decent chance of staying below 1.5°C and, by doing so, minimizing the risk of setting off irreversible chain reactions beyond human control, will be gone before the end of this decade. There are no effective policies in place. And there is no silver bullet or magic technological solution in sight. It turns out no one can negotiate with the laws of physics. Not even President George H. W. Bush. /

Our Imprint on the Land

Alexander Popp

Land is not only where we live. Land is also what we live from. We draw our food and fibre, timber and bioenergy from land – a host of everyday services that we take for granted, though it is literally our life-support system. Land is one of the major foundations of human well-being, and we have been using it for innumerable generations past. But misusing it could prove fatal for generations to come. All human activities are embedded and constrained by ecosystem processes and functions, and throughout our existence as a species we have manipulated and transformed land and its natural resources. However, in recent history the extent of human land use has altered those processes and functions with mostly adverse consequences for people and the planet.

When we think of human civilization's imprint on Earth, we tend to think of the world's metropolises and megacities, interconnected by a dense network of roads, power grids and infrastructure. While these do represent pervasive land-use activities, they are almost negligible in terms of their ecological, economic and social influence when compared to agriculture. Globally, farming is the dominant form of land management today, and it is changing the face and the function of our planet. In recent decades, agricultural production has grown much faster than the population. The leading driver of this growth is plain and simple: the growing demand for food. Not only has calorie supply per person increased significantly, diet composition has been changing too – a trend tightly associated with wider economic development and lifestyle shifts. It is evolving from scarce, plant-based diets, with fresh, unprocessed foods, towards affluent diets high in sugar, fat and animal produce, as well as highly processed food products – diets that are also increasing household food waste. As a result, humans and livestock currently account for the great majority of the total mammal biomass on Earth, and the biomass of domestic poultry is nearly three times higher than that of wild birds. This is unprecedented. The production of crops has increased since 1961 by about 3.5 times, the production of animal products by 2.5 times, and forestry by 1.5 times.

Historically, a growing population would meet its increasing need for agricultural commodities primarily by expanding the area used for farming. Consequently, around three quarters of the Earth's ice-free land surface, and most of the highly productive land area, is now under some form of human use. Grazing land for cattle and the like is the largest agricultural land-use category, followed by forestland, with cropland coming third. The total land area used to nurture livestock is mind-blowing — it covers about 37 million square kilometres, roughly four times the size of Brazil; not only does it include all grazing land, but also a significant share of cropland, which is used to produce feed. For woodland, the dominant share of the total forested area around the world is used by humans at different levels of intensity. Less than half of all forest area has been estimated to contain old trees, and large areas of primary forest survive only in the tropics and northern boreal zones. And forests aren't the only 'natural' feature of our planet which are in fact subjected to human use. Of all other, non-forested natural ecosystems – natural grasslands, savannahs, and so on – the majority are under human use. Little nature is left 'natural'.

Besides growth in agricultural production, the volume of agricultural goods being traded internationally has also grown – by a factor of nine in the past half-century – resulting in an increasing disconnect between regions of production and consumption. The result is a net redistribution of agricultural land towards the tropics. While most pastureland expansion has replaced natural grasslands, cropland expansion has mainly replaced forests. Noteworthy large conversions have occurred in tropical dry woodlands and savannahs: for example, about half of the Brazilian Cerrado has been transformed for use in agriculture, while African savannahs are under threat. Natural temperate grasslands are now considered one of the most endangered biomes in the world, while the majority of the world's wetlands have also been lost due to agricultural expansion. While agriculture continues to expand in specific regions such as sub-Saharan Africa and Latin America, most parts of the world have – in a dramatic break with historical trends – seen the expansion of farmland playing a significantly smaller role in increasing agricultural production. Instead, yields have increased due to agricultural intensification. That intensification became dominant due to greater inputs of fertilizer, water and pesticides, new crop strains and other technologies of the 'green revolution': since the early 1960s, the global irrigation area has doubled, total nitrogen fertilizer use has increased by ten times, and now practically all cropland is fertilized. These days, at the global

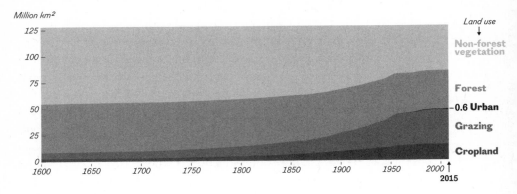

Figure 1

level, around 10 per cent of the ice-free land surface is intensively managed (through practices such as tree plantations, high-density livestock grazing, large agricultural inputs), two thirds moderately and the remainder at low intensities. These increases in agricultural production have supported better living standards for many, reducing the share of people at risk of hunger. However, many countries, particularly in Asia and Africa, still struggle with widespread undernutrition and associated health problems, while others are increasingly faced with the issue of people becoming overweight and, as a consequence, with a rising burden of diseases such as diabetes and cancer.

Beyond these health issues, the fundamental problem is that land is a limited resource which has to fulfil multiple functions. The gains associated with increased agricultural production, either via land expansion or intensification, have been counterbalanced by harmful impacts on the environment, related ecosystem services (such as clean air or water) and, finally, human well-being. For instance, fresh-water use for agricultural irrigation accounts for the great majority of human water withdrawals, but this disturbance degrades many fresh-water ecosystems and has caused their vertebrate populations to decline significantly. Human-made nitrogen fertilizer, which has boosted agricultural yields for over a century, leaks into aquatic ecosystems, leading to groundwater pollution and increased nitrate levels in drinking water, the eutrophication of agro-ecosystems blighted coastal zones, as well as increased frequency and severity of algal blooms. Together with the impacts of climate change and the introduction of invasive species, agro-ecosystems have resulted in the rapid and widespread decline of biodiversity and the degradation of natural ecosystems worldwide. Terrestrial vertebrates have gone extinct at unprecedented rates, and more species are now threatened with extinction than ever before in human history.

Both agricultural expansion – given the associated loss of carbon-rich ecosystems – and agricultural intensification are major contributors to climate change. Today, the Agriculture, Forestry and Other Land Use sector is responsible for about 20 per cent of global anthropogenic (human-caused) greenhouse gas emissions, largely carbon dioxide from tropical deforestation, methane from livestock and rice cultivation, and nitrous oxide from livestock and fertilized soils.

So, this is quite a dilemma. Humanity, by using land to live, is changing it in unprecedented ways – endangering the land's capacity to provide the services we depend on.

Balancing the inherent trade-offs between satisfying immediate human needs and maintaining other ecosystem functions requires holistic and sustainable land-use approaches, especially as different uses of land and associated ecosystem services and their use or protection are not independent of one other. We must find new ways to increase agricultural production while preserving natural habitats and biodiversity. In this context, the conservation of carbon- and biodiversity-rich ecosystems can have crucial effects in reducing biodiversity decline, while also enhancing climate change mitigation. This would be a new way of using nature to support human life – using land by *not* using it. It is about changing land-use change. It would mean further intensifying agriculture so that we can feed a growing world population while protecting land from the ever-increasing expansion of agricultural areas. The big question here is not 'if' but 'how'. How can we sustainably intensify food production around the world?

Agricultural productivity must grow to feed an expanding global population without an increase in the negative environmental impacts associated with it, which means, for example, much more efficient nitrogen, phosphorus and water use. Positive synergies are possible when such measures on the supply side are combined with measures on the demand side, for instance adjusting diets towards an overall healthy and equitable animal-protein intake and reducing food waste. If more people based their diets on plant-based food and avoided waste, there would be reduced pressure on land – thereby supporting health, climate and biodiversity.

This land is our land, for better or worse. We should defend its integrity. Defend it by innovating, defend it by conserving – and defend it against ourselves, if needed. We've undertaken the first 'green revolution', changing the way we use the land in order to feed the world. It is now time to make that change sustainable, leading to what must, this time, be a truly green revolution. /

The Calorie Question

Michael Clark

Where our calories come from is a global issue. Food systems are arguably the single largest driver of environmental degradation. They produce 30 per cent of all greenhouse gas emissions, occupy 40 per cent of Earth's land surface, use at least 70 per cent of Earth's fresh water and are the leading driver of biodiversity loss and nutrient pollution. They are also a major source of poor health and nutrition through the foods we eat and drink and through how they are produced.

The environmental impact of individual foods varies considerably. Broadly speaking, there are three sets of food that range from low impact to high impact per calorie of food produced: plant-based foods have the lowest impact; dairy, eggs, poultry, pork and most fish have five to twenty times the impact of plant-based foods; and some fish, as well as meat that comes from cattle, goats and sheep have upwards of twenty to a hundred times the impact of plant-based foods. This difference is largely driven by the resources that are needed to produce these foods. To consume one plant, you need to produce slightly more than one plant, due to food that is lost or wasted in supply chains. However, to produce dairy, eggs, poultry, pork and fish, you generally need about 2–10 calories of plants to produce 1 calorie of edible food, while producing beef or lamb takes roughly 10 to more than 50 calories of plants to produce 1 calorie of edible meat. Animal-based foods (meat, dairy and eggs) cause additional damage through the manure generated during livestock production, as well as the methane that cows, sheep and goats naturally release during digestion. There are a few exceptions to this general trend. For instance, coffee, tea and cocoa have higher impacts than other plants, mainly because the increasing global demand for these products often causes deforestation in tropical and highly biodiverse regions, which results in large amounts of greenhouse gas emissions and biodiversity loss. Nuts also have much higher overall impacts because they require a relatively large amount of water and are often grown in water-stressed regions (for example California's Central Valley).

The environmental impact of a given food may also vary based on how that food is produced. For instance, a recent analysis found that lower-impact beef producers can have one tenth the impact of less environmentally sustainable producers. However, this variation is typically much smaller than the difference in impacts between foods. A large analysis that drew from evidence across nearly 40,000 farms found that even the most sustainably produced animal-based foods have higher environmental impacts than the least sustainable plant-based foods.

Around the world, diets are transitioning to include more food in total, and more meat, dairy and eggs, as populations become more affluent. At a global scale, average consumption per person per day has increased from roughly 2,200 calories in 1961 to 2,850 calories in 2010, with larger proportional increases in our consumption of animal-based foods and of empty calories (such as sugar, unhealthy oils and alcohol). These transitions are occurring at different rates in different regions: most rapidly in low- to middle-income countries, including many in South East Asia, South and Central America and North Africa; and more slowly in the lowest-income and the highest-income ones. In these richer countries where there is a history of high meat consumption, the environmental impacts of diets can be ten times greater per capita than in poorer countries. These wealthy nations bear the largest responsibility for how our food systems are harming our planet, and they have the greatest imperative to reduce their footprint. This includes

Figure 1:
The averaged relative environmental impact (AREI) is a metric that condenses information from five environmental indicators: greenhouse gas emissions; land use; water use; eutrophication potential; acidification potential. AREI is plotted against the environmental impact of a serving of vegetables: a food with an AREI of 10 has 10 times the environmental impact of vegetables.

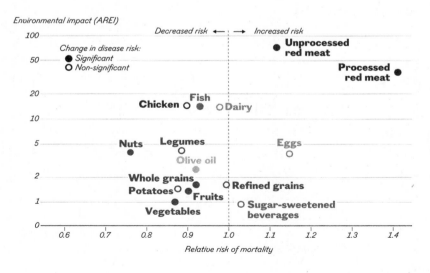

Environmental and health impacts of consuming a serving per day of different foods

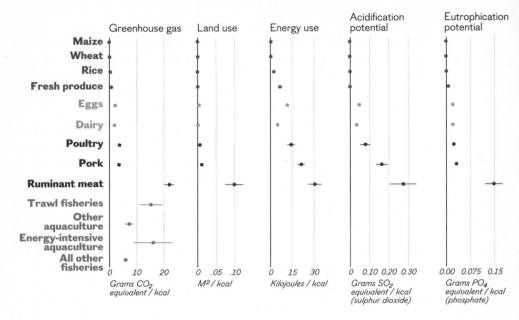

Figure 2:
A dot represents
the mean impact
of the food, while
the bars extending
from it indicate the
plus or minus one
times its standard
error. These five
indicators are used
to calculate AREI.
Fresh produce
includes fruit and
vegetables;
ruminant meat
includes beef,
sheep and goat.

countries such as the US, the UK, much of Europe, Australia, New Zealand and Brazil and Argentina.

As the global population continues to grow, and low- to middle-income countries continue to rapidly change their diets, food systems will surpass environmental sustainability targets in the next several decades. For example, one recent analysis showed that, even if we manage to remove every other source of greenhouse gas emissions, we will pass 1.5°C of warming within the next few decades, and 2°C shortly after the end of century, if we do not change how we produce and consume food. The additional land needed to meet future food production could also lead to 1,280 species of birds, mammals and amphibians losing more than 25 per cent of their remaining habitat in the next several decades.

One of the most effective ways to reduce food-related environmental impacts is to bring meat, dairy and egg consumption in line with health guidelines (Fig. 1). For most countries, this means rapid reductions in consumption of these foods; in the US and the UK, for example, this would be an approximately 80 per cent decrease in pork and beef consumption. However, in lower-income countries, this could mean an increase in consumption to improve health and well-being. Overall, a global transition to a more plant-based diet has been repeatedly estimated to reduce food-related greenhouse gas emissions by 50–70 per cent, and would also bring benefits to other aspects of the environment.

While a shift to plant-based diets is the single largest change we could make to lower the impact food has on the environment, there are many other opportunities to create food systems that contribute to both environmental sustainability and human well-being. These include changing how food is produced, through new fertilizer management regimes or crop rotations, and reducing how much food is lost or wasted throughout the food supply chain – a third of all food produced is not consumed. However, even if many of these strategies are rapidly implemented at a global scale, it is unlikely that we will meet the 1.5°C climate target unless a shift to plant-based diets occur, simultaneously.

Fortunately, many of the same changes that would improve our environment would also improve human health. Many of the most environmentally sustainable foods are among the healthiest and most nutritious, and many of the most environmentally sustainable diets would reduce global premature mortality by 10 per cent. /

Greenhouse gas emissions under different food system trajectories compared to the remaining emissions to meet the 1.5°C or 2°C climate targets

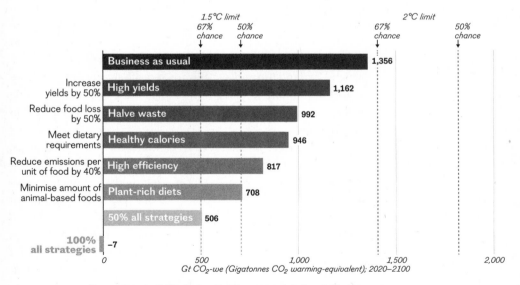

Figure 3: Assuming that food system strategies are properly implemented by 2050, all five strategies simultaneously implemented to half of their potential would give us a 67% chance of limiting warming to 1.5°C.

Designing New Food Systems

Sonja Vermeulen

After fuel, food is the next frontier for solutions to the climate crisis. Going one step better than carbon neutrality, agricultural landscapes and food supply systems have the capacity to flip from carbon sources to carbon sinks. This isn't just a theoretical capacity. For millennia, the landscapes from which we harvest food and on which we raise livestock have been net carbon sinks. Over millions of years, living creatures have been the route by which carbon flows from the atmosphere into the ground – into soils, oil and coal. Our challenge today is to recouple the carbon cycle across earth, biota and atmosphere to restore net flows back into the ground. The nitrogen cycle is tightly linked to carbon, and likewise needs recoupling.

Scientists are in strong consensus on how we could achieve the flip from source to sink for our food systems. The fundamental finding is that to cross the line we need action on all three of the following: diet, food waste and agriculture. The consumption side – diet and waste – is critical because it drives the demand for land use and agriculture. But the supply side – how we farm – also matters. Empirical studies show that the environmental footprint of the same food product can differ fiftyfold or more, depending on agricultural practices.

We have a good collective sense now of the most promising set of actions to shift agriculture on a global scale towards being a carbon sink while also delivering on food security, local livelihoods, biodiversity and other environmental outcomes. But that set can read like a long recipe list of disconnected items: controlling water levels in rice paddies to reduce methane emissions, changing livestock's pasture and feed, mulching fields, growing trees on farms. To understand what connects these, the underlying principles, requires us to think about how we have upset carbon and nitrogen cycles and how we can restore them.

For example, before artificial fertilizers were invented in the early twentieth century, cattle were so valued for their manure – as well as for their draught power, their milk and their cultural significance – that slaughter for meat was rare. This remains true in many African and Asian rural economies. Yet in wealthier regions we have decoupled these nutrient cycles. We take inert nitrogen out of the atmosphere and turn it into fertilizer via energy-intensive manufacture, then feed crops for livestock and finally animal protein for human consumption. Vast quantities of reactive nitrogen drive climate change through nitrous oxide emissions from farms, then food is being transported across continents via complex webs of international trade and end up flushed as urban sewage into inland waterways and coastal areas, damaging biodiversity and ecosystem functions. Clearly, different approaches are needed both locally and globally.

Acknowledging that the current system is broken does not mean that we need to return to the practices of the past, that low-tech cottage-style agriculture is necessarily the better path. We can put in place modern knowledge-intensive farming that brings agriculture into sync faster and better than before. The promise of high-tech solutions certainly has a place – for example, understanding and manipulating the fungal root nodules on crops, or manufacturing fertilizers with new methods – but these need to fit within local ecological and social contexts, rather than be applied as universal silver-bullet fixes.

Thinking ecologically, and looking at the numbers, the biggest bang for carbon in food systems is conserving as many of our biodiverse carbon-rich ecosystems – forests, peatlands, mangroves, and other wetlands – as possible, as well as restoring those we have damaged or destroyed. Agriculture remains the greatest driver of these ecosystems' destruction, through both deliberate clearance and runaway fires. Thus the single most important change needed in agriculture is to reverse, or at the very least reduce, the pressure for greater and greater expansion. This means that we must shrink our demand for agricultural products (the diet and waste side of the story) and make smarter use of the land that we do have – which is technically achievable but presents a complicated political conundrum.

Much of the politically tricky challenge to reverse the expansion of agriculture's footprint centres on growing more with less. Sustainable intensification, as it's called, means increasing agricultural yields without negative environmental impact and without converting additional non-agricultural land. In practice, efforts towards sustainable intensification tend to focus on maximizing yields of single crops, for example through breeding higher-yielding varieties or changing fertilizer application, rather than more creative ways of maximizing the value from farmland over the year, such as rotations, different land use during fallow periods, or diversification of both agricultural products and economic activities.

Alternative approaches have been proposed which overlap with sustainable intensification but suggest a different emphasis. One example is organic farming, which avoids manufactured inputs like fertilizers and pesticides. Another is agroecology, which combines a more holistic approach with a social movement and aims for an agriculture steered by local knowledge and in harmony with local ecologies, with outcomes for both social justice and environmental integrity. But these present their own problems – they are too difficult to scale up, or insufficient to provide 10 billion people with adequate and affordable nutrition.

Discourses around agriculture are highly polarized and contentious, in part because universal solutions are elusive. The ongoing debates around sustainable intensification and agroecology demonstrate the impossibility of prescribing a global recipe to improve farming. Context is everything, and farmers will need to make strategic choices from a set of locally feasible options.

The highest priority for change is the 5 per cent of calories that account for 40 per cent of food's environmental burden. These are predominantly high-input livestock and crops grown to supply urban markets. Key suppliers are the row crop farms of the US, China, South Asia and Europe that provide wheat, rice, maize, soybeans, sunflowers, potatoes, canola (rapeseed) and other crops for animal feed, human food and industry. Their primary concern is to optimize inputs, often meaning a reduction of inorganic fertilizers and pesticides. By contrast, in certain low-productivity but high-potential systems, such as smallholder palm oil, intensification can be useful, preferably coupled with diversification into mixed land uses.

Yet there are also certain 'low-input, low-yield, low-impact' systems for which intensifying does not make sense. For example, a cow on a modern European industrial farm is about a hundred times more carbon efficient per litre of milk or kilogram of meat than a cow in a traditional African pastoral system. But that low-input African cow is far more resilient to climate and disease – and provides irreplaceable value to the cattle owners, such as daily fresh nutritious milk without the need for refrigeration or supply chains, power for ploughing or a reserve of capital savings that can be cashed in or exchanged. Equivalent 'low-input, low-yield, low-impact' systems exist in high-income contexts too, for example in wheat farming in Australia.

As well as intensification to free up land outside farms, options also exist to bring about direct increases to carbon held above and below ground on farms themselves, and in rangelands. We can increase biomass through productive trees and perennial plants (agroforestry), as well as windbreaks, green belts, trees to stabilize slopes and sand dunes, and so on. About 20 per cent cover of bushes and trees would bring impressive benefits to both biodiversity and carbon, with minimal effects on productivity in most places. We should also make deliberate efforts to raise soil carbon by reducing tillage, improving irrigation and moisture conservation techniques (such as harvesting rain-water using Indigenous and local practices, and drip irrigation), maintaining vegetation and mulch cover for as much of the year as possible, and leaving crop residues in situ without burning. Rangeland management systems based on sustainable grazing and re-vegetation can increase productivity of meat and milk, as well as reducing climate-critical methane emissions and increasing soil carbon.

Many of the practices described here support adaptation to climate change as well as mitigation; agriculture offers many such win-wins. Most of the practices are also self-evident to the stewards of our land and water – to farmers, livestock keepers and fishers. But these stewards are up against a set of economic and policy incentives that limit choice and drive contrary behaviours. Many farmers are stuck in supply, management, insurance and loan contracts that leave little wriggle room for more progressive approaches; if rewritten, these contracts could be powerful vehicles for sustainability. Likewise, a crescendo of voices now calls for repurposing the more than $500 billion worth of subsidies that go into agriculture each year, to support more sustainable practices and trajectories. And progress on deeper societal issues, such as women's equality and land tenure, would also accelerate positive change in agriculture.

For change-makers who are not farmers, some modest advice on remod-elling the future of farming: remember that we need action in the three interconnected domains of diet, food waste and agriculture, and at multiple scales. Your individual choices make a difference, so knowing where your food comes from and making conscious choices is the first step; policies and markets matter tremendously, so strategic, collective advocacy and engage-ment can create larger-scale change; and our food system is an outcome of wider issues of social justice, so your activism in spheres such as women's rights, business ethics and law or government transparency will shape the future of our food and our climate. /

Mapping Emissions in an Industrial World

John Barrett and Alice Garvey

The world we live in has, quite literally, been built by 'industry'— a term which describes all the economic activity related to extracting or growing raw materials, processing those materials and transforming them into the infrastructure we inhabit and the products we buy. Today, this complex supply chain involves millions of businesses and exchanges goods and services across a truly global economy, providing jobs and income for a huge number of people around the world. However, it is also responsible for over 30 per cent of global greenhouse gas emissions which are driving climate change and causing other serious human health issues through local air and river pollution.

The global industrial sector is made up of a vast and varied number of businesses, but there are particular activities that dominate emissions. These all come from 'heavy industry', that is, the manufacture of materials and products involving large equipment and complex processes. The largest share of industrial emissions comes from the production of iron and steel, followed by cement. Overall, just three branches of heavy industry (steel, chemicals and cement) contribute 70 per cent of industrial CO_2 emissions.

What's unique about steel and cement is the importance of 'process emissions' – the largely unavoidable result of the chemical reactions required to manufacture the materials. In the production of cement, half of all the CO_2 emitted comes from process emissions.

Globally, industrial emissions are mainly measured using so-called territorial emissions accounting techniques. Territorial emissions are the greenhouse gas emissions that occur within a particular country and are therefore considered the responsibility of that nation. But this fails to account for the fact that, once a material or product has been made, it can be transported all over the world, which allows for a separation between the highly carbon-intensive production of industrial goods and their (less carbon-intensive) consumption. In recent decades, developed nations

Contribution of process emissions to global steel and cement CO_2 production

Figure 1:
A scenario of no additional climate policy is shown for the year 2050, but where current mitigation trends continue.

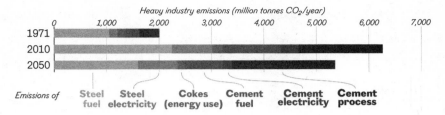

Heavy industry emissions (million tonnes CO_2/year)

| | 0 | 1,000 | 2,000 | 3,000 | 4,000 | 5,000 | 6,000 | 7,000 |

1971
2010
2050

Emissions of | **Steel fuel** | **Steel electricity** | **Cokes (energy use)** | **Cement fuel** | **Cement electricity** | **Cement process**

have predominantly reduced their industrial emissions by offshoring industrial production to emerging economies. While this enables the former nations to meet their own domestic emissions reduction targets, it acts against the global imperative to reduce industrial emissions overall.

A territorial emissions accounting approach does not reflect the fact that there is a significant and growing demand for industrial products from the developed world, and instead allocates the emissions incurred in making those products to the largely developing states where the industrial activity occurs. This process has allowed developed states to defer responsibility for their growing consumption while appearing to be taking climate action.

Consumption-based emissions accounting offers an alternative perspective: in this measure, emissions are allocated to the country of the consumer. For example, in calculating the footprint of manufacturing a car, a consumption-based emissions perspective would allocate the bulk of emissions to the country of final consumption, since this is where the demand is being generated, while a territorial perspective would allocate most of the emissions to the developing countries responsible for manufacturing the component parts. The figure is typically calculated by working out a country's production emissions, minus the emissions from the production of exports, plus the emissions from the production of imports to that country.

A consumption-based emissions accounting approach therefore accounts for the full international impact of final demand for industrial materials and products; this is a vital step towards global equity and acknowledging the UN principle that states have a 'common but differentiated responsibility' to decarbonize, given the disparities in economic development and historic carbon emissions between developed and developing nations.

The task of addressing consumption-based emissions is, therefore, a far more demanding task for developed nations with greater levels of absolute final demand – particularly given that this level is currently expected to remain constant or even increase.

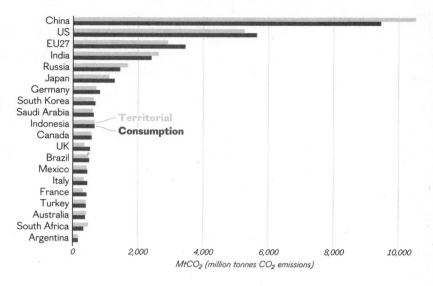

CO$_2$ emissions of G20 countries from territorial- and consumption-based accounting perspectives

Figure 2: Consumption-based accounting allocates the bulk of emissions to the country of final consumption, where the demand is being generated, rather than the country of production; numbers from 2021

Territorial

Consumption

China
US
EU27
India
Russia
Japan
Germany
South Korea
Saudi Arabia
Indonesia
Canada
UK
Brazil
Mexico
Italy
France
Turkey
Australia
South Africa
Argentina

0   2,000   4,000   6,000   8,000   10,000

MtCO$_2$ (million tonnes CO$_2$ emissions)

So far, industry has primarily sought to mitigate emissions through efficiency gains, that is, by making its equipment and processes less energy intensive. But while it has taken advantage of low-hanging fruit – the most cost-effective efficiency improvements – the scale and pace of the sector's response to date has been lacking. Industrial inertia on decarbonization is partly due to the long investment cycles required to replace key equipment and products (such as a blast furnace for steel-making), coupled with global market pressures, which make it harder for firms and governments to make significant investments in low-carbon-production technologies. Even so, the sector overall has managed to become more energy efficient in recent years, reducing the average carbon intensity of industrial production. But these gains have been offset by rising demand for industrial materials and products, particularly in emerging economies.

Demand is expected to at least double by 2050. Emerging markets and developing countries will be the main drivers of growth, as they need to invest in the development of infrastructure and capital stocks, and demand for steel and cement is closely linked to overall patterns of economic activity. Developed states have already benefitted from decades of developing material-intensive infrastructures using steel, cement and other industrial products. Developing states are doing this now. For instance, 61 per cent of the rise in China's greenhouse gas emissions was due to capital investments such as roads, electricity networks and railways between 2005 and 2007. Similarly, by 2050 a fifth of global steel production is expected to be located in India, scaling up from a current 5 per cent. Concerningly, however, estimates suggest that 37 per cent of the carbon we can emit (consistent with global climate ambitions) for steel production to 2050 has already been used.

Proportion of industrial emissions in advanced and developing states in 2020 and 2050, by industrial subsector

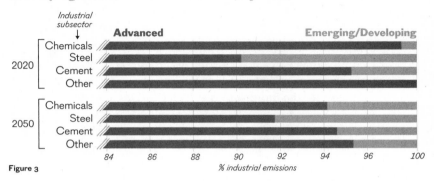

Figure 3

Clearly, as well as adopting low-carbon industrial technologies, action is also needed on the demand side. Some reductions to the level of demand for industrial materials and products can be achieved through 'material efficiency' strategies – that is, by using less material input to deliver the same useful output, or 'doing more with less'. This can involve designing products more efficiently, making them smaller and lighter, and minimizing the amount of scrap or waste generated in the production process. But while levels of demand can fall for some materials and, in some regions, they won't disappear completely, we also require investment in technological solutions.

Transition economies cannot be penalized for their growing demand for industrial products and materials. Given developed states' cumulative historic responsibility for emissions, and the advantages conferred by early opportunities for industrial development, it is they who must do the most to address their levels of demand. But we must all do everything we can to decarbonize industry, taking steps which include using next-to-zero-carbon fuels to replace fossil fuels and implementing more efficient production processes to reduce the impact of everything we make. Above all, we need to consume less, derive economic value in different ways and replace the constant throughput of materials and products with a circular economy. /

The Technical Hitch

Ketan Joshi

Spend even a moment looking over the climate and sustainability documents of the world's high-emitting industries and you will be met with a wall of content that tries to bully you into feeling optimistic. An onslaught of glossy PDFs packed with stock photography of warmly smiling engineers and very serious businesspeople gives you the impression that there is a *plan* for the future and that high-emitting industries are very much *in control*. Over-produced soft-focus promotional videos with slow-motion footage feature a narrator telling you about the suite of ready-to-roll solutions right around the corner. The message is clear: heavy industry is calmly plucking itself out of the deep well of carbon reliance it has found itself in.

Sectors such as power, transport and agriculture are well known for their contribution to global greenhouse gas emissions. But heavy industry is somewhat disconnected from public consciousness, because it functions several steps back from the end products we engage with in our lives. We understand a coal plant producing power, and we get a combustion engine moving a car. But the chain of events behind the concrete in our walls or the steel in the bus on the street or the plastic that wraps a chocolate bar is more opaque.

In 2020 the industry sector was responsible for producing 8,736 megatonnes of CO_2, out of a total of 34,156 megatonnes of CO_2 emitted globally, according to the International Energy Agency's (IEA) World Energy Outlook. Despite the optimistic marketing materials, emissions from this sector are described as 'hard to abate', and for good reason. The machinery it uses is designed to last decades, and to be replaced rarely. The high heat required for processes can currently only be achieved through burning fossil fuels. And public and investor pressure to decarbonize is less on companies in heavy industry. Their high-emissions processes are more hidden in the supply chain than those of coal plants and cars, often at a geographic remove, since emissions-intensive goods are often exported.

There are some technical options that would partially reduce the high emissions associated with heavy industry today. The IEA's 2021 World Energy Outlook and its 2020 Energy Technology Perspectives report both detail methods such as electrifying industrial processes (for example, steel manufacturing or low-heat combustion), improving the efficiency of the supply chains (for example, in cement manufacturing or iron ore production), or replacing fossil fuels with hydrogen (which can itself be produced using zero-carbon electricity and which does not produce carbon dioxide when it is burned for energy).

The World Energy Outlook provides some insight into what heavy industry's future climate impact might actually be. It models a range of scenarios, including sectoral emissions under current policies, under 'announced pledges' (emissions if the promises of governments are truly met), and under a 'net zero by 2050' scenario, in which industry becomes carbon neutral and warming is limited to around 1.5°C relative to pre-industrial temperatures. It then compares our actual technological progress on decarbonizing industry against these three scenarios.

There is a huge gap between the projected emissions under current policies and what's required to hit the best-case scenario of 1.5°C. Even relative to announced pledges, assuming these are actually met with policy and effort, the decarbonization gap remains vast.

The IEA splits this gap by the various technologies and processes available for decarbonizing industry – increased recycling of plastic, for example. Of these, carbon capture, use and storage (CCS) has the biggest job to do. Under currently announced policies, CCS technologies will capture 15 megatonnes of industrial carbon in 2030, but the IEA's net zero scenario would require 220 megatonnes to be captured.

It's not a rare thing to fill in the gaps in climate plans using CCS. But the technology is plagued by a history of failure that suggests it ought to be anything but central to climate plans, particularly for sectors that are extremely difficult to decarbonize. There aren't many places that better present the modern history of CCS than my home, Norway. In 2007 then Prime Minister Jens Stoltenberg spoke of the Mongstad refinery CCS project, designed to capture emissions from a fossil gas power plant, in wildly confident terms. 'It will be an important breakthrough for reducing emissions in Norway, and when we succeed, I think the world will follow,' he said. 'This is a great project for the country. This is our moon landing.'

Six years later, it was dramatically cancelled due to budget overruns. It had gone a blistering 1.7 billion kroner over budget, with a total government spend of 7.2 billion kroner. Following its cancellation, Norway's petroleum minister at the time, Borten Moe, insisted that a full-scale plant would be built on the project site by 2020. At the time of writing it's late 2021, and there is still no operational plant. Norway's trip to the moon never really left the surface of Earth.

Nonetheless, the cycle of over-promise and non-delivery has remained utterly unchanged: CCS is still central to Norway's efforts to mitigate industrial emissions. One of Norway's premier modern climate policies is 'Langskip' ('Longship'), touted as the largest climate project in Norway's history. It's an organized sequence of carbon capture, transport and permanent underground storage of carbon dioxide from industrial and waste sectors, to become operational over the next decade.

The first stage of Longship features the capture of carbon dioxide that would otherwise be released during production at the Norcem cement factory in Porsgrunn. Cement production is responsible for 5–7 per cent of global carbon dioxide emissions – a staggering figure.

In early November 2021 the Norcem CCS project announced a budget breach, showing the required investment had increased by 912 million kroner, up to 4,146 million kroner in total. The future of the project is up in the air. 'Unless the parties agree to continue, or one of the parties undertakes to finance the completion alone, the project will be shelved and each of the parties will bear its own costs,' wrote the Norwegian government. In the event that this project does end up operational (purportedly around 2024), it is predicted to capture around 0.4 megatonnes of carbon dioxide each year, less than half a per cent of the company's total emissions.

Another key project within Longship is the Klemetsrud waste facility's long-delayed CCS plant. Originally slated to begin operation in 2020, the effort to capture the substantial emissions from burning Oslo's 'non-recyclable' waste remains dormant. Oslo's goal of 95 per cent emissions reductions by 2030 cannot be reached without somehow dealing with the emissions from this waste-burning plant, which is the city's single largest source of CO_2 emissions. Despite several successful small-scale tests, the plant remains non-existent because it remains unaffordable without funding from the European Union (an application which was rejected in November 2021).

For Longship's final stage – in which the extracted carbon will be returned deep underground – a consortium of fossil fuel players (Shell, Equinor and Total) will take the reins. The Northern Lights project will purportedly transport the captured CO_2 and inject it into the many depleted oil and gas reservoirs around Norway. For the first twenty-five years, the project will have the capacity to store 1.5 megatonnes of CO_2 per year, and potentially 5 megatonnes of CO_2 after that. In 2019 the combined emissions of Shell, Equinor and Total were 2,350 megatonnes of CO_2. Even when CCS works, it operates on a scale that's barely perceptible relative to the size of the problem it's meant to address.

Currently, the global capacity of CCS to capture CO_2 is around 40 mega-tonnes per annum (mtpa). More than 100 of the 149 CCS projects originally planned to be operational by 2020 have been terminated or placed on indefinite hold. A recent research paper found that CCS fails so frequently because it is expensive to build, the technology is unreliable, and there is no real profit to be earned unless you use the captured CO_2 to – horrifically – enable greater extraction of underground oil and gas wells.

But even if we could trust that these projects would come to fruition, the IEA's 'net zero by 2050' scenario assumes 1,578 mtpa of capacity by 2030.

In the world of glossy climate plans in PDFs and stock photography, CCS is a shining saviour. In the grimy real world, CCS is a failure. This disconnect persists because CCS serves an emotional purpose rather than a technological one. It coats the fantasy of continued, unchanged fossil fuel use with a protective rhetorical magic. It remains persistently 'around the corner', while serving as a justification for the ever-worsening expansion of fossil fuel projects and the cause of delay of real action.

Remove this talisman from climate solutions and the reality hits hard – change must be quick, and deep, and, most importantly, aimed at those most directly and deeply invested in the fossil fuel economy. Reductions in demand must go far beyond efficiencies and shifts in material methodologies.

CCS capture: historical, planned and the IEA's 'net zero by 2050' scenario

Figure 1:
Predicted installed capacity falls short of the required 1,578 megatonnes per year in 2030 to achieve the IEA's 'net zero by 2050' goal. 2020 and 2021 reports from the Global CCS Institute.

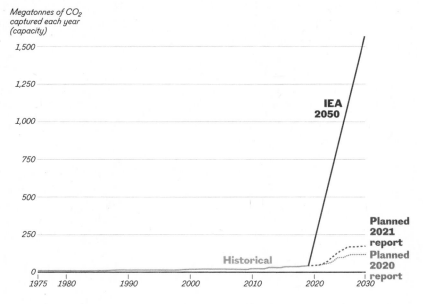

The societal root cause of overconsumption of materials must be faced head on; particularly in the most stuff-hungry, rich, wealthy, white nations. Suddenly, we have to think about the shape of society rather than just the type of machine.

There is no single clearer manifestation of the hazard of tech solutionism than carbon capture and storage. It has become a beacon of false hope for high emitters terrified of fast, deep change. For sectors that are genuinely more difficult to abate, such as heavy industry, technological fantasy serves the most immediate purpose: avoiding any real discussions about whether we ought to simply produce and consume far less stuff.

There is, without doubt, a role for design and technology to play in decarbonization: as the IEA has highlighted, 'increased material efficiency' can help curb demand for industrial products and the energy used to make them. With cement, for example, this means better renovation of existing constructions to extend their lifetimes, or optimizing designs to reduce the quantity of concrete required. But there is no way to arrive at a decarbonization pathway that is fast enough without having a deeper conversation about demand.

We don't need a new smartphone every year. Many of us don't really need to own an entire multi-tonne vehicle, whether it's combustion or electric. Lives can be good and rich even when they don't involve being a source of demand for emissions-intensive *stuff*. But a broad effort to cut demand for the products of industry remains stifled by the stunning dominance of empty promises from the companies supplying and relying on fossil fuels. Until we ditch the glossy, over-produced tech fantasies and come back to reality, we'll continue paying a terrible price. /

CCS serves an emotional purpose rather than a technological one. It coats the fantasy of continued, unchanged fossil fuel use with a protective rhetorical magic.

The Challenge of Transport

Alice Larkin

Travel has always been an essential element of how humans have built relationships, organized communities, traded, and developed as civilizations and societies. Whether we go on foot or use motorized vehicles, we all travel: for work, education and leisure, to move people and goods, and sometimes simply for pleasure – to enhance our mental or physical well-being.

Modes of transport, from bikes to aeroplanes, are continually evolving, as are our travel habits. As people's incomes rise, we tend to travel further – not because we use more of our time to travel; this has remained stable for many years – but because technology increases speed and cuts journey times. For some, this has brought the possibility of commuting long distances to work; for others, opportunities to undertake their education in a different country or go on holiday overseas. Technological advances in vehicles have also transformed trade: while international trade by sea has a long history, the things we consume today have complex global supply chains, and many of us are accustomed to having products arrive from a faraway origin sometimes just hours after we've ordered them.

Yet for all the benefits that transport offers, it has significant and wide-ranging impacts on the environment. Extracting resources to make roads and railways, bikes and heavy goods vehicles, requires energy, generates pollution and often harms biodiversity. Sailing a ship across the ocean creates vibrations that can disturb wildlife, and people living close to airports are subject to noise levels that can impact negatively on their health. Combusting petrol in a car, diesel in a ship or kerosene in a plane releases harmful emissions into the air we breathe and contributes to global warming: overall, the sector represents around a quarter of global carbon dioxide emissions from burning fossil fuels. As our economies grow and transport expands, emissions increase both in absolute terms and, typically, in their proportion compared with other sectors (Fig. 1, inset).

Knowing the benefits brought to people by transport, policymakers are often reluctant in their efforts to mitigate the environmental damage it causes. This cannot continue. Today we are witnessing the climate emergency unfolding around the world. Transport has a key role to play in slowing temperature rises and protecting lives, but this can only happen if we recognize and tackle its environmental impacts.

Transport is a huge and diverse sector, and the most commonly used modes of transport in one part of the world are quite different from those in another. A global study has shown that 47 per cent of road-vehicle kilometres travelled in South Asia were by mopeds and motorcycles and just 15 per cent by car, whereas in North America less than 0.5 per cent were travelled by moped or motorcycle, and 57 per cent were by car. Similarly, it is estimated

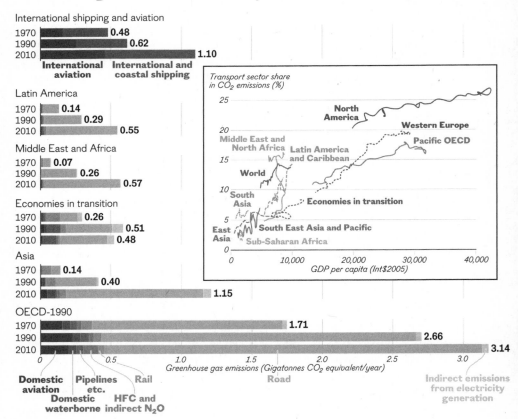

Greenhouse gas emissions from transport subsectors in 1970, 1990 and 2010

Figure 1: Totals of greenhouse gas emissions do not include indirect emissions. Inset: the share of transport emissions tended to increase due to structural changes as countries became richer. CO2 emissions are relative to GDP in the period 1970–2010, the latter being measured in 2005 International dollars, a unit comparing purchasing power of different currencies.

that about a quarter of the global population could have, in theory, taken a flight in 2018; less than 2 per cent in low-income countries compared with 100 per cent in high-income countries. However, through survey data we know that, even in high-income countries, the majority of people do not take a flight in any one year, which means that air travel is practised by only a small percentage of the global population. This variation in transport use (Fig. 2) is important when considering who holds responsibility for environmental impacts and directing policy measures to mitigate damage.

When it comes to the production of warming gases, each type of vehicle emits different amounts for every kilometre travelled (Fig. 3). For example, the emission intensity of a domestic flight in the UK is nearly seven times more than travel by train, even though UK train travel is a mixture of diesel and electric. Travelling on a long-haul first-class flight can be over 130 times worse than travelling by international rail. In addition, once you take account of the 'load factor' (the number of people in a vehicle), the figures change. For instance, if two passengers travel in a car rather than one, the emissions per person are halved. Moreover, while the *emissions per kilometre* of a long-haul economy flight are similar to that of one person in a small petrol car,

Total passenger distance travelled by mode in 2000 and 2010

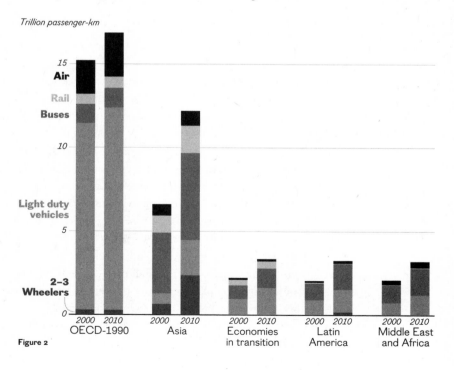

Figure 2

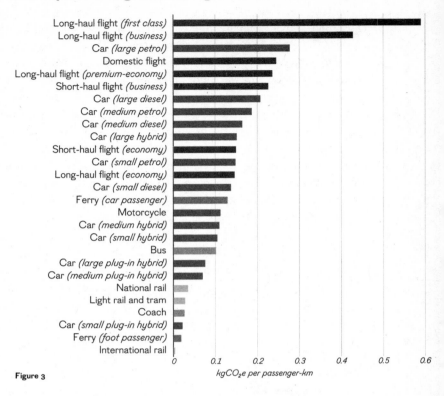

Long-haul flight *(first class)*
Long-haul flight *(business)*
Car *(large petrol)*
Domestic flight
Long-haul flight *(premium-economy)*
Short-haul flight *(business)*
Car *(large diesel)*
Car *(medium petrol)*
Car *(medium diesel)*
Car *(large hybrid)*
Short-haul flight *(economy)*
Car *(small petrol)*
Long-haul flight *(economy)*
Car *(small diesel)*
Ferry *(car passenger)*
Motorcycle
Car *(medium hybrid)*
Car *(small hybrid)*
Bus
Car *(large plug-in hybrid)*
Car *(medium plug-in hybrid)*
National rail
Light rail and tram
Coach
Car *(small plug-in hybrid)*
Ferry *(foot passenger)*
International rail

0   0.1   0.2   0.3   0.4   0.5   0.6

$kgCO_2e$ per passenger-km

Figure 3

it is likely that the *total* number of kilometres travelled by air will be higher, meaning that emissions will, in reality, be much greater.

The picture becomes more complicated when emissions from extraction, conversion and processing are considered – 'well to wheel emissions'. For example, a battery electric vehicle charged using an electricity grid dominated by gas will have higher emissions than if the same vehicle is charged by a grid comprised of nuclear and wind power. And caution must be exercised when assuming that alternative fuels such as biofuel or hydrogen always offer a very low-carbon alternative to fossil fuels, as the production process can be energy and emissions intensive. Only if this energy is *also* supplied by renewable sources, and/or if it is coupled with direct CO_2 air technologies, can the final product be truly 'low-carbon'.

Transport is often considered more difficult to decarbonize than many other sectors. In some wealthy nations, progress has been made to curb emissions overall, but transport emissions have continued to grow (Fig. 1). And there is an added, highly significant complication: international aviation and shipping normally fall outside national emission accounting, targets, policies and carbon budgets (Fig. 1). This is a legacy of the Kyoto Protocol, which charged the International Civil Aviation Organization and the International Maritime Organization with mitigating emissions released in international airspace and waters. Unfortunately, this arrangement has meant that these sectors are yet to develop and implement policies aligned with a 1.5°C goal. This is extremely problematic since international aviation and shipping together contribute the equivalent amount of CO_2 as Japan, the fifth biggest CO_2-emitting nation in the world.

Although the two modes are similar insofar as they both release emissions that can't easily be pinned to a particular country, aviation and shipping are in fact very different in nature and in terms of the challenges they pose. First, one is used mostly for leisure travel, while the other is largely used for transporting goods, including food and material resources. Aviation is also the preserve of a minority of the global population, while shipping delivers goods and resources directly or indirectly to the vast majority – though it should be noted that levels of material consumption are hugely inequitable and skewed to wealthier nations. Both are closely linked to the growth of the global economy, but if we were to ground most aircraft the economic impact would be minor compared to what would happen if ships were prevented from carrying goods around the world.

When it comes to aviation, it is well established that we are many years away from the technological solutions that would allow us to fly en masse without relying on fossil fuel propulsion. This is in part because aircraft, like ships, are typically used for twenty or more years. It is also because producing large quantities of low-carbon fuel of the quality and high density needed to lift a typical passenger aircraft off the ground requires huge technological advances and probably some form of direct CO_2 air capture – a technology that remains, in 2021, at the small-scale demonstration stage. Some are optimistic that the requisite technological change is coming, but there are immense engineering and socio-economic challenges still to be overcome, which even many within the industry recognize will take time, probably decades, to make a material difference.

In the meantime, the aviation sector is seeking to mitigate its impact by using voluntary carbon offsetting. There are deep concerns about whether such a scheme is an effective way of reducing emissions, particularly as it doesn't account for the additional warming caused by the non-CO_2 emissions that aircraft release high into the atmosphere, where they create more warming than they would if released on the ground. This means that constraining demand immediately is essential to reducing aviation's climate impact – otherwise our carbon budgets will be breached too soon. Various mechanisms could be used to do this, including a moratorium on airport expansion in wealthy parts of the world, as well as a frequent-flyer levy.

Ships, however, already have a multitude of available options to curb emissions in the short term, such as retrofitting wind-propulsion or cutting ship speeds; in the longer term, a range of new fuels will help alleviate emissions too. Given that fossil fuels are primarily transported by sea, a wider switch to renewables would also mean a decline in a large chunk of shipped trade, which would likely help to reduce shipping emissions. Of course, rising levels of consumption will increase the demand for freight transport, and it's critical that we ensure we are individually and collectively managing our consumption in order to minimize environmental damage while fairly distributing capacity to support development globally.

We all need to rethink how much we and our goods travel, and why, and the most appropriate mode of transportation for these journeys. Doing so could lead to transformative reductions in resource use and in local and global pollution. When it comes to climate change, we must always keep in mind how difficult it is to ensure that our transport habits are minimizing their overall impact on the environment, and just how long it takes to decarbonize the sector. /

International aviation and shipping contribute the equivalent amount of CO_2 as Japan, the fifth biggest CO_2 emitting nation in the world.

Is the Future Electric?

Jillian Anable and Christian Brand

When it comes to decarbonizing transport, local, national and international policy discussions place overwhelming and sometimes exclusive emphasis on technological solutions. Such solutions are largely focused on the electrification of lighter vehicles (cars and vans) and buses, trams and trains. There is an additional, limited role in the short term for the sustainable production of hydrogen, biofuels and potential synthetic liquid fuels.

The primary limitation of this approach is its inability to keep pace with projected increases in demand for mobility. As economies and populations grow, demand for goods grows, as does the number of people with the desire and means to travel. Globally, total transport activity is expected to more than double by 2050 compared to 2015. This huge rise in the use and ownership of cars, as well as the movements of heavy goods vehicles, aviation and shipping, will more than offset any reductions in emissions from technological change, particularly in the next two, critical decades. It is now widely agreed that there is no way we can meet the decarbonization targets of the Paris Agreement by 2050 without focusing on the amount of movement of people and goods.

The scale of the challenge comes into sharp focus when our near total dependence on oil across all forms of passenger and freight transport is considered. As of 2021, transport is still 95 per cent dependent on oil. Cars, vans and buses tend to stay in circulation for fifteen to twenty years, lorries for around twenty, planes for twenty-five and ships for forty. This means that even if, from tomorrow, 100 per cent of new cars and other modes of transport were fully electric or fuelled by another renewable source, it would take decades for fossil fuels to completely disappear from this sector. Even an optimistic scenario, where global new car sales were 60 per cent electric by the end of this decade, would see global CO_2 emissions from cars drop by only 14 per cent by 2030 compared to 2018.

As well as the timescales involved, electric vehicles (EVs) are not a panacea, as life-cycle emissions depend heavily on the carbon content of electricity, materials used and battery production. Over the past fifty years, increasing vehicle weight and power have been reducing the rate of efficiency improvements for cars and vans across the board. The most popular type of car, the large, heavy sports utility vehicle (SUV), made up 45 per cent of all light-duty vehicle sales worldwide in 2021, outselling electric cars by about five to one. This eroded up to 40 per cent of improvements in fuel economy. Due to the pandemic, the International Energy Agency reported that carbon emissions in 2020 fell across *all* sectors except for one – SUVs. Due to the better margins they make on them, manufacturers benefit financially from selling larger, higher-end cars. The US, for instance, incentivizes large EVs by offering emissions credits. The German government provides direct subsidies for plug-in hybrid electric company cars, many of which are SUVs. Concerningly, many of the statistics published to record the growth in sales of electric vehicles include these plug-in hybrid cars – which make up about a fifth of global EV sales – even though they still rely heavily on, and lock in for a long time to come, fossil fuel combustion. One of the fastest, easiest and most efficient methods of reducing emissions would be to restructure incentives to encourage the sale of lighter electric vehicles and to phase out the use of large SUVs in cities immediately: ban their advertising, and tax their ownership and use. In the UK alone, phasing out the most polluting large vehicles could cumulatively save about 100 million tonnes of CO_2 by 2050.

There is another, even more fundamental issue with our reliance on electrified transport: it requires a reliable electricity supply, which is simply not guaranteed in many parts of the world. EVs do not address social inequality within and between countries, especially in the Global South, where e-cars may well be an option only for the powerful and wealthy. And even if they could be made widely available, EVs do not 'solve' the problems of road traffic congestion, parking pressures, safety or transport poverty. Car dependence gives rise to urban sprawl and sets in motion a vicious cycle whereby places and jobs become less and less accessible by other means, which leads to subsequent falls in the use of public transport and, in turn, reduced revenue and cuts in service levels, resulting in an even greater dependence on cars, and so on. The flipside of the freedom afforded by widespread individual car ownership is that an increasing number of people are being forced to own a car which they struggle to afford and which they maintain at the cost of sacrificing other things in their lives.

So what can we do? In the shorter term, while the majority of vehicles on the road remain fossil fuelled, a low-hanging fruit is lowering traffic speeds on highways. Introducing a motorway speed limit of 130 kilometres per hour (kph) in speed-loving Germany would lower carbon emissions by 1.9 million tonnes annually. This is more than the annual total of sixty of the lowest-emitting countries. Lowering the speed limit to 100 kph would save around 5.4 million tonnes annually – or more than the annual emissions of eighty-six countries, including Nicaragua and Uganda. Yet the debate in Germany has been going on for decades, with no political party in power prepared to implement the policy. A collective sense of entitlement and dislike of limiting 'personal choice' are a lot to blame for this inaction[1].

As this indicates, we need significant behavioural change in addition to technological change, and the two are intimately interlinked: political leaders, town planners, manufacturers and consumers need to adopt, enable and promote new travel habits, as well as adopting new techno-logies. When such change in the transport sector is discussed, it is most typically in terms of 'mode switching', whereby inefficient or polluting modes of travel are swapped for more efficient ones on like-for-like journeys: taking local public transport, walking or cycling, instead of making a short car journey. This is clearly an important step: in the UK, for example, 59 per cent of car trips are less than 5 miles (or 8 km) in length. Walking, cycling and e-biking – or active travel, as it's called – can reduce emissions rela-tively quickly. Alongside active travel, new, lighter forms of e-micromobility are generally becoming cheaper and outselling e-cars in many parts of the world, including sub-Saharan Africa and some parts of Asia; these have the potential to undertake longer journeys, including journeys beyond urban areas. Light rail systems have also long provided vital efficient connectivity in large and smaller towns alike and have mostly been electrified since their inception, while buses are some of the fastest components of the transport system to be making the transition to battery technology.

There are two main limitations, however, to focusing on cycling, walk-ing or high-quality electrified public transport as the solution. Firstly, in the absence of car-restricting policies, car travel will continue to increase alongside any growth in alternative modes. This has been observed in experiments in

---

1 The same principle – simply lowering speed – also has significant potential to reduce emis-sions in the difficult-to-decarbonize maritime shipping sector, which was not part of the Paris Agreement and which is projected to grow as a proportion of global emissions. About 80 per cent of global trade is transported across the oceans. Cargo ships are mostly powered by fossil fuels – largely the most polluting type of diesel. Electrification is not a viable option for deep-sea ocean-going vessels – just as it is not for long-haul aviation – but the sector has significant potential to reduce emissions through a combination of 'slow steaming' and retrofitting to use zero-carbon fuels such as green ammonia. A 20 per cent reduction in ship speeds can save about 24 per cent of CO_2.

European countries, where offering free local bus services to all passengers primarily saw increased uptake among existing users, walkers and cyclists, with a limited effect on reducing car use overall. Similarly, car-restricting policies were crucial for the Netherlands in achieving its world-leading rates of cycling; they made cycling more convenient than driving. Yet even after the success gained by applying a combination of 'carrots and sticks' at the local level, average per capita carbon emissions from personal transport in the Netherlands are as high as those in many neighbouring Western European countries because car-restricting policies have not been applied to longer journeys, which account for most of the mileage and thus carbon emissions.

Mode switching for the shortest journeys within relatively built-up urban areas is important, but it will take us only a fraction of the way to the reductions in car mileage in developed countries by 2030 that have been identified as necessary to stay within carbon budgets. In the UK, for instance, a lack of any reduction in transport CO_2 since 1990 means that the transport sector now has only ten years left to cut back its emissions by two thirds. Various modelling studies have identified that total car mileage will need to decline by between 20 per cent and 50 per cent compared to today's levels even as the uptake of electric vehicle sales is ramped up.

This level of change requires not just mode switching but 'destination shifting' in order to reduce the distances travelled on personal or freight journeys. This is impossible without political approaches that venture far beyond the transport system itself. Regional planning policies to create 'fifteen-' or 'twenty-minute neighbourhoods' will need to be rolled out, locating homes and jobs in closer proximity, putting services such as schools and health care back into the urban and suburban areas from which they have often disappeared, and enabling goods to be sourced from nearer at hand. Some types of trips will need to be removed altogether: the Covid-19 pandemic has accelerated trends towards virtual meetings, for example, eliminating the need to travel for international business meetings and conferences while simultaneously increasing participation. Meanwhile, other journeys should be consolidated, with vehicles being shared by different passengers and loads – a practice which is a large component of less formal modes of transport in many developing countries and which has had a resurgence worldwide due to new technology allowing on-demand 'ride matching' and shared payment services. Car-sharing schemes are seeing increased growth in many countries where parking pressures and the costs of owning and running a car are high. Focusing on car *access* rather than ownership as a means to unlocking flexible and equitable personal mobility would be one way in which

Next pages:
Stacked shipping
containers at Asia's
first fully
automated
container terminal
in Qingdao,
Shandong
Province, China,
in January 2022.

developing countries could avoid the mistakes related to car dependence that developed countries have made.

There is no way round the fact that transport decarbonization means reducing the use of cars, trucks and planes and the simultaneous removal of fossil fuels from them. To reverse current travel trends and end our reliance on high-carbon infrastructure will require a significant shift in how we use our land and how we transform our cities. In already car-dependent societies, successful attempts to reduce vehicle kilometres will need to be followed by a further reallocation of any freed-up road space to the most sustainable modes of transport. Without this, more car and truck travel is likely to simply fill up the new capacity and quickly erode its benefits. This is not rocket science but it does require serious political leadership, major funding and a clear strategy on how to communicate both the benefits and the costs of change to society. Foremost among the benefits is the promise of a fairer society.

Transport systems are inherently unequal, and fewer and fewer people are responsible for the majority of the travel sector's emissions. For example, just 11 per cent of the population in England accounts for nearly 44 per cent of total car mileage. Globally, 50 per cent of aviation emissions in 2018 were caused by 1 per cent of the world's population. About 80 per cent of people in the world have never travelled by aeroplane. If we can change the debate to focus on these current injustices rather than allowing the narrative to be dominated by supposedly 'unfair' policies that ask for a small reduction in motorway speed limits, a redistributive approach to accessibility and mobility could help reduce the resistance to changing our travel habits. Even without the prospect of climate change, it has been known for decades that many benefits arise from cutting traffic. It's good for health, safety and air quality; enables more efficient and equitable use of resources; improves social and economic vitality; and makes for better neighbourhoods. /

4.18

They keep saying one thing while doing another

Greta Thunberg

The first step in solving a crisis is not evaluating the full situation or taking immediate action. That comes later. The first step in solving a crisis is to realize that you are in one. And we are not there yet. We are not aware of the fact that we are in a climate emergency. But that's not the main problem. The main problem is that we are not aware of *the fact that we are not aware*. Recognizing this double lack of awareness is the key to understanding the climate crisis. And yet this is exactly what we fail to comprehend. Not only as a society, but also as individuals. Everyone assumes that everyone else knows, and around and around it goes.

Before COP26 in Glasgow, I was invited to help produce an opinion poll in Sweden. A survey which, among other things, was supposed to investigate the general level of awareness about the climate crisis. This was meant to be a groundbreaking report – the first of its kind. When the poll was finished I was told that they could not make any sense of it. The level of knowledge was apparently so low that the material was completely useless. The answers were either so inaccurate or so far off target that the whole section went in the bin.

To me, this is in accordance with my own personal experiences, not only from the thousands of conversations I have had with people all over the world, but also from meetings with journalists, businesspeople, politicians and even world leaders. Here are some of the comments I hear.

> *'The Paris Agreement . . . well, to be honest, I don't think we had any idea of what we signed up for.'*
> *'I wish the people in power had even half the knowledge about climate change as you and the other children have.'*
> *'Why do I have to know about these facts?'*
> *'Can we please not talk about facts, because I am not really aware of the facts when it comes to climate change.'*

These are real words and sentences spoken to me during private meetings with some of the world's most powerful heads of state and spokespeople for the most powerful institutions. These people have access to every imaginable resource to address their ignorance, and still, a shockingly high number of them have failed to use them. Their level of understanding on climate-related issues is embarrassingly low. In fact, the number of people who are actually aware of the climate crisis is probably much smaller than most of us would ever imagine. And yet, everyone assumes that we all understand the problem. We listen and nod along as various pieces of unfamiliar new information come at us from all directions.

Of course, there are many who would disagree with what I say. So, let's briefly explore the possibility that they are right and I am wrong. Let's for a moment assume that our leaders haven't failed, that the emperors are not naked, but fully dressed. Politicians and the media have succeeded in fulfilling their democratic duties and properly informed us citizens about the nature of our situation. Let's assume that they have explained the full consequences of business as usual, and the fact that historic injustice is at the heart of the problem. OK. What would that actually mean?

It would mean that the people of high-emitting nations, such as mine, are causing this destruction on purpose. That people are knowingly risking the survival of our civilization and life on Earth as we know it. That people are condemning their brothers and sisters in the most affected areas to unimaginable amounts of present and future suffering. Suffering that could cause up to 1.2 billion people to flee their homes by the middle of this century.

If those entrusted with power have indeed performed their educational duties, then surely that means that people like you and me are knowingly causing irreparable damage to our life-supporting systems. That we are willingly setting off a mass extinction that will ultimately threaten the survival of our entire species. Because if it's not the case that they have failed to properly inform us, then we – the people – are causing all this unimaginable destruction on purpose. If this is indeed the case, then we are evil, and it does not matter what we do because we are all screwed anyway – but I refuse to believe that.

Here's the thing. When it comes to the climate crisis, we all know that something is wrong. We just don't know exactly what that *something* is. We are fully aware that many scientists say we are facing an existential crisis which ultimately risks the survival of our civilization. But there are other people who are saying similar things, while adding a sentence or two to suggest that we can still 'fix' this without having to make any systemic or personal lifestyle changes. In fact, some even say that with just a few technical adjustments we will limit global warming to 1.5°C, even though we are currently heading for 3.2°C – at the very least, given that this estimate is based on flawed and severely under-reported numbers. These people also say that many nations have already significantly lowered their emissions during the last three decades, even though that is largely the result of outsourcing production and excluding large swathes of our actual emissions, for example those from biomass. The problem is, the people who say these things are presidents, prime ministers, leading businesspeople and major international newspaper editors.

Yes, we have been told about the fact that we are facing the greatest threat humanity has ever faced. We have been told that we are about to pass points of no return. We have even been told that our entire civilization is threatened unless we take immediate, unprecedented action. But we have not been told in the right way. And we have definitely not been told by the right people. And when we have been told, those same people who have told us just continue as if nothing is wrong. Celebrities who have been appointed as spokespeople for the climate continue flying in private jets; media outlets that put lots of effort into reporting the climate crisis still carry advertisements for fossil-fuel-dependent practices; and so on. As long as they keep saying one thing while doing another, people will not believe them. As long as they keep living like there is no tomorrow, the vast majority of us will aspire to do exactly the same.

We do not have all the solutions to solve this crisis within today's systems. But that cannot stop us from using the ones we do have in every possible way. We must use, evolve and keep pushing for the miracles that are already in place, for example wind and solar power. But it is equally important that, while we are doing everything we can, we constantly remind ourselves and others that this will not be enough, because we need a system change. Our main priority must still be to wake people up – not send them back to sleep with soothing stories of progress from a deeply flawed system. As an example: yes, we should be suing big polluters – governments and energy companies – as often and as much as we can. But we must also remember that there are still no laws to keep the oil in the ground and to keep our civilization safe in the long run. We need new laws, new structures, new frameworks. We must no longer define progress only by economic growth, by GDP or the amount of profit given to shareholders. We need to move beyond compulsive consumerism and redefine growth. We need a whole new way of thinking. /

The Cost of Consumerism

Annie Lowrey

Who is responsible for the climate crisis? Whose behaviour needs to change to stem the catastrophe of climate change?

To answer those questions, we often look to governments. We also look at industries. In the US, the agriculture, construction, electricity, manufacturing and transportation sectors are the predominant emitters, as they are in many countries, and need to switch to renewable energy and sustainable production methods as fast as possible. We look at companies too. Just twenty firms are responsible for one third of all carbon emissions, with Saudi Aramco, Chevron, Gazprom and ExxonMobil at the top of the list.

Surely countries, industries and companies will need to change their practices if we are to save our planet. Yet such analyses absent the households and individuals responsible for buying what those corporations are selling, and for electing such governments into office. They miss the people whose excessive – even wanton – consumption is harming the planet and whose homes and cars and pantries and wardrobes need to change. They miss out consumerism as a root cause.

To be sure, no single person or family, no matter how rapacious and wasteful their lifestyle, is responsible for more than a minuscule share of the excess carbon in the atmosphere or the waste in our landfills or the plastic in our oceans. No household changing its ways will make a measurable difference in terms of combating climate change, either. A wealthy, urban family going vegan, flight-free, and car-free might reduce its carbon emissions by a few tonnes a year, when the world's problems are measured in the tens of billions of tonnes. Moreover, families' carbon budgets and broader environmental impact are in no small part determined by the built environment they live in, the economy they work in and the policy choices made for them by their representatives. Given those dynamics, it will take governmental and corporate action to help heal the planet.

Yet individuals are the ultimate consumers of much of what gets pumped, built, slaughtered, mined, woven, cut, processed and shipped around the world, year in and year out. And it is their materialism and consumerism – *our* materialism and consumerism – that are leading to the destruction of the planet. More than 60 per cent of greenhouse gas emissions and up to four fifths of land, material and water use stem from household demand, with the most affluent bearing the greatest responsibility.

Indeed, the rise of the global middle class might account for much of the growth in greenhouse gas emissions and real resource demands of late, with countries such as China, Nigeria and Indonesia increasing both their absolute and their per capita emissions as families rise out of poverty and living standards improve. But that does not mean that lower-income countries or their citizens are an outsize share of the problem, or that the world needs to accept extreme poverty and cruel inequality if it is to stop climate change.

The rich in rich countries are the ones who hold the most responsibility for using our planet up. An American one-per-center accounts for ten times the greenhouse gas emissions of the average American; the average American is responsible for three times the emissions of the average person in France; the average French person accounts for ten times that of the typical person in Bangladesh. Other measures of resource use follow not dissimilar patterns, whether looking at kilos of factory-farmed meat consumed, kilos of junk sent to the landfill, grams of plastic dumped in the ocean, kilometres of flights taken, litres of gas pumped, or square metres occupied per person.

Our culture values convenience, valorizes excess, encourages keeping-up and hides the true cost of our lifestyles from us – ignoring the fact that the worst consequences will be borne by non-human animals and by people in the generations to come. Too many use too much. Waste too much. Have too much. And care too little. The problem is particularly acute in the United States, where higher education, health care and child care are extremely expensive but stuff is cheap. The average American home has tripled in size in the past half-century, though families have become smaller. A household in the US contains, on average, 300,000 individual items – no wonder one in ten households rents a storage unit and one in four people with a garage say it is too full to house a car.

Surely some of this stuff is necessary, and some of this consumption makes people happier, healthier and more fulfilled. But much of it does not. Indeed, research has shown again and again that once a family has reached a certain middle-class status, spending on things does not bolster its self-reported well-being, though spending money on experiences does.

That is perhaps because, as families become richer, they spend more money on 'positional' goods, ones that fulfil no basic need but instead situate a family among its peers and broadcast its wealth and taste. (Thorstein Veblen, of course, recognized this dynamic more than a century ago.)

One way or another, consumerism is harming a planet already under extraordinary strain. Consider the contemporary obsession with the SUV. The gas-guzzlers' share of the vehicle market has doubled in the past decade, for no other reason than consumer preference. (Family sizes are stable or shrinking and the share of workers in agricultural or industrial professions is falling in the relevant countries.) That shift has overwhelmed the beneficial effect of electric vehicles on energy consumption overall. Or take fast fashion. Garment makers are now producing 100 billion new garments a year, with the average person buying twice as many items of clothing as they were a generation ago. Much of that clothing is never worn, or is worn just a handful of times, and just 1 per cent of fabric is ultimately recycled, the Ellen MacArthur Foundation has found. The fashion industry fills dumps, not wardrobes.

The answer, retailers might have you believe, lies in the world's elite consuming differently. A reusable water bottle, a canvas shopping bag, a silicone straw, an electric vehicle, smart appliances – such purchases are small steps towards a better world, we are told. Except they are not. In terms of resource use and emissions, buying nothing beats buying something virtually all the time: better to keep driving the car you own than shelling out for a brand-new Tesla, or wearing out what is in your wardrobe rather than buying a new capsule wardrobe in the name of ethical fashion. A particularly telling statistic: a person would have to use an organic cotton tote every single day for half a century to offset the impact of its production, the Danish government has estimated.

No, less needs to be less: that is the uncomfortable, essential truth. We as individuals need to stop it with the excessive and the unnecessary, from bags to plane flights to second cars. We need to reside in smaller, greener homes and adopt mass transit. More than that, we need to develop a profound scepticism of the economic ideologies that have brought us to mass extinction and catastrophic warming and begin acting as if this disaster is indeed a disaster.

But didn't I just write that individuals changing their ways won't make a measurable impact, and that it is governments and corporations who are primarily responsible? Sure. But household action is a crucial predicate for broader action. Humans are social creatures, and people influence their friends, families and neighbours in concrete, measurable ways.

Consumer preferences also alter business practices, with companies both trying to convince the market what it should want and delivering what the market demands. Moreover, individuals starting to take the personal initiative to combat the climate crisis would make it easier for governments to do the same. Hefty taxes on gasoline or levies on oversize vehicles would be easier to pass and implement if fewer people were already relying on gas-guzzlers, for instance; people using bicycles to commute to work demand bicycle lanes which tend to induce more people to cycle instead of driving. People rejecting consumerism and voting with the future of the planet in mind would bring more green-friendly politicians into office too.

Everything needs to change, and everything must change. And that change starts at home. /

A household in the US contains, on average, 300,000 individual items – no wonder one in ten households rents a storage unit and one in four people with a garage say it is too full to house a car.

How (Not) to Buy

Mike Berners-Lee

Here we are in a climate and ecological emergency. One critical element of the solution is to rethink and change the way we, the people of the world, consume. Every bit of fossil fuel that comes out of the ground is burned to meet a consumer need or wish. Sometimes the emissions are direct and obvious, for example the exhaust fumes from a car, but just as often they come from chimneys on the other side of the world making things that feed into the production of other products. These in turn become components of another item that somebody buys, without any idea of the emissions that have been released into our collective atmosphere to produce it. For example, a new laptop could have a similar carbon footprint to a thousand-mile car drive, while a new pair of jeans could have about the same climate impact as either a couple of weeks' worth of sustainably chosen food or just one large joint of meat. Most of us, most of the time, have hardly any sense of the scale of the invisible carbon footprints of the many things we do and buy.

As humans have globalized, our supply chains have become ever more complex and opaque. Especially in the developed Global North, we've got used to a world in which items appear on the shelf, as if by magic. We are almost completely protected from having to understand anything about the climate impacts, or other environmental or social impacts, resulting from a complex array of processes that have enabled the production of our purchase.

Why do we buy so much, with so little awareness?

We don't just consume products and services with carbon footprints. We also consume information and misinformation which change how we think about the things we could buy or aspire to buy. There are multibillion-dollar advertising and marketing industries geared entirely to making us want to buy things, whether or not they are in our best interests and whether or not they are good for the planet. L'Oréal's famous 'Because You're Worth It' campaign comes with the implication that if you don't buy their products you will be a less valuable person.

Ninety-seven per cent of Facebook's revenue comes from advertisers paying to influence how its users think and how much they want to buy their products. Film-makers are paid handsomely for product placements that quietly suggest, usually without us even noticing, that we can have the style of 007 if we drink the right beer and buy a new laptop of the right brand. Heineken reportedly paid $45 million for James Bond to take a swig of its beer in a single film. This persuasion to overconsume is everywhere, seeping into all of us from all directions. It can come at us from every corner of our culture, including our workplaces, our politicians, our news media and probably even our family and friends – because they too have been victims of this pervasive manipulation.

So this quick guide to sustainable consumption will look at the physical side of when and how to buy stuff and, just as importantly, how we can protect ourselves from the onslaught of influences that are talking us into planet-wrecking lifestyles with fake promises of happiness.

The consumption of information

Learn to notice and critically evaluate every message. This is a continuous skill we all need to develop as best we can. Ask who is influencing us and how. For every advertisement you see, ask yourself, 'What do they want me to believe? Is it true that I'll be happier or more attractive if I buy that product? What values are implied? Do I subscribe to those values?' This applies to all media, whether it pertains to be fact- or fiction-based. The same questioning principles apply to conversations with our friends, families and workmates. Ask, 'Am I being pushed into believing that I need to buy something unnecessary by a message that is either explicit or implicit?'

If you find yourself rejecting the messaging you are receiving, think about how you can protect yourself from similar influences. Watch different TV channels, subscribe to a different media source, put an ad-blocker on your browser or switch your social media.

For news media, a critical question is who owns and funds the organization: what interests do they have and how do they want you to think? Do their financial and political interests lead you to believe you can trust them to give you the best understanding of what is going on? If not, switch to something else.

How not to buy

- **Learn to pause first.** Much of the world's most planet-trashing and mindless consumption comes down to spontaneous impulse. According to CNBC, the average US consumer spends $5,400 per year on impulse purchases. Ask yourself, 'Why, exactly, am I feeling the need? Could the urge to buy actually be a sign that something else is not right in my life? If so, can I deal with that differently? Have I been influenced by people, adverts or media to think that I need this product in order to have status or to feel good? And are they right, or do I need to make up my own mind?'

- **Repair.** When you get something repaired you show more personal responsibility and you use up far less of the world's resources than if you buy a replacement. On top of that, the money you spend probably goes to someone in your neighbourhood who is earning an honourable livelihood. At the same time, by spending less you will have helped to liberate yourself from the need to earn more in order to buy more. You will have supported a sustainable economic model, and it allows you to continue with a product with which you have a history and a connection.

- **Share.** If you need something, you can reduce the burden on the world by borrowing it (and getting to know a neighbour while you are at it) or renting it (supporting a sustainable economy) or taking part in a sharing scheme.

- **Make do/improvise.** There's a lot of creativity to be had in working with what you've got.

If you have to buy

If you get to the point of knowing you really do need to buy something, the next step is to ask yourself some further questions. What lies behind this product? How do I think it was made? Try to imagine the entire supply-chain network in all its complexity, right back to the materials coming out of the ground. Imagine the emissions, and the chemicals. Think about the people who worked on every stage of production and the land that may have been used. You probably won't know all the answers, but just trying to imagine is still a big step. Make it a routine for every item you purchase.

If you can't research a specific product, research the brand. What do they say about themselves and what do others say about them? What are their values? What is their track record? Can they be trusted? Do they understand the climate emergency and are they pushing for change in every way they can? If an airline has tried to tell you that you can undo the climate impact of a flight for just a few dollars of 'offset', you know you can't trust anything else they want you to believe either. Over time, try to build up knowledge of how things are made and the impacts behind them. One place to start for information is www.ethicalconsumer.org, where, for a small but worth-it subscription fee, you'll find independent research on a huge range of shops, brands and products. For example, if you need clothes and can't buy second-hand, it can quickly help you to understand why you might want to avoid Amazon and Primark but might feel better about other brands.

Consider buying second-hand, thus avoiding the impact of production in the first place. When you have finished with something, try to get it into the second-hand market, whether you sell it yourself or give it to someone else to sell.

If you decide to buy something new, try to buy something that has been made to last and can easily be repaired. This applies especially to clothes, furniture and IT kit such as phones and laptops, where the energy it will use in its lifetime is usually tiny compared to the impact of making it in the first place. For household appliances, it is often the other way around, so energy efficiency is essential. When it comes to vehicles, first think about bikes and electric bikes. To avoid the impact of manufacture, and to help take the status out of cars, keep your current car on the road, rather than buying a new one, unless it is very inefficient or your mileage is very high. If you do need to replace a car, buy a small, efficient one, electric or hydrogen-powered if possible. When buying food, the simple rules of thumb are to eat less meat and dairy (especially less beef and lamb); to make sure you eat everything you buy; and to avoid anything that may have been air-freighted, hothoused or overpackaged.

If you work in marketing or advertising

In today's situation, it is *not* OK to make a living persuading people to think a certain way or to buy things regardless of whether it is in their best interests or in the interests of the planet. Ask yourself carefully whether that is the nature of your work. If so, it has to change. If your company expects it of you, change the company urgently or leave. All those in advertising need to rise to the challenge of entirely reframing this profession.

If you are a producer

Develop a business model that enables people to buy less stuff, less often. Make sure your products are sustainably made, built to last and are repairable. Measure your whole carbon footprint, including your entire supply chains, and set targets and actions to reduce your emissions fast – in line with the world keeping temperature change to 1.5°C. Tell your customers what lies behind the products, prioritizing honesty over greenwash.

If you are a retailer

Buy only from producers who meet the above criteria. Help your customers to be informed. Make repair and second-hand sales part of your business model.

Finally, just note that buying less stuff means we can get off the treadmill of having to earn more to spend more: it gives us more freedom. A world surrounded by sustainable items feels better and is better. Our worth comes not from our possessions but from how we treat others and the environment. Taking the status out of our shiny new high-impact possessions can be psychologically liberating for us, as well as being an essential component of the climate crisis response. /

Most of us, most of the time, have hardly any sense of the scale of the invisible carbon footprints of the many things we do and buy.

Waste around the World

Silpa Kaza

Rubbish or trash – including everything from plastic bags and paper to food waste – is everyone's problem. It is generated daily by households, small businesses and institutions, and is typically managed by local governments. Globally, the solid-waste management sector is usually one of the top three greenhouse gas emitters, contributing approximately 5 per cent of carbon dioxide emissions and up to 20 per cent of methane emissions. It significantly affects our ability to mitigate and adapt to warming, as well as the health, productivity and resilience of local communities. Poor waste management can lead to disease transmission, respiratory problems, water and soil contamination, air pollution, marine pollution and can even hurt local economies (by diminishing tourism, for example). It is a sector that disproportionately harms low-income communities and countries, where waste is primarily burned or dumped. Worldwide, waste is increasing at an alarming rate, and its mismanagement is exacerbating the climate crisis.

When it comes to municipal waste, we have the unusual power to respond locally to a global problem, as our local governments can support national efforts to deliver on global emissions commitments. Seventy-seven per cent of countries have included solutions for reducing waste emissions in their plans to meet the goals of the Paris Agreement. Waste can be a less expensive and less complicated sector to tackle than, for example, industry, where decisions are made at a variety of levels, from federal governments to individual companies, and solutions are sometimes costly. Since municipalities already provide services around solid-waste management for the sake of public health and cleanliness, substantial climate action can enhance their existing efforts.

Some of the key sources of emissions in this sector are from carbon dioxide generated as waste decomposes, methane generated by the mismanagement of organics, and black carbon, or soot, which is generated when trash is poorly incinerated and when it is transported. Some of this can be addressed with basic interventions such as universal waste collection and sanitary landfilling, whereby trash is safely isolated from the environment and the methane it generates can be captured. Other emissions in the sector result indirectly from our 'linear' economies, where virgin materials such as metals and plastics are mined or manufactured, transported, used and then disposed of, rather than being reused or recycled.

Beyond emissions, the mismanagement of waste also directly contributes to flooding and pollution. If waste is not collected properly, it can block drains or canals and exacerbate flooding, which can aid the spread of vector-borne diseases such as malaria. It can leak into waterways and ultimately oceans, threatening aquatic ecosystems. Dumped waste can lead to landslides when combined with heavy rain or flooding. And the uncontrolled burning of waste causes pollution affecting air quality, health and the environment generally.

These problems could worsen significantly in the coming years. Waste generation is rapidly increasing: by 2050, municipal waste is anticipated to outpace population growth by more than 200 per cent. In 2020, an estimated 2.24 billion tonnes of waste were generated; projections for 2050 are for 3.88 billion tonnes, a 73 per cent increase. The waste generated per capita varies drastically by income level – people in low-income countries generate one quarter of what high-income individuals generate daily. This translates to regional variation as well, with people in South Asia and sub-Saharan Africa generating 0.39 kilograms per capita per day and 0.47 kilograms per capita per day respectively, while North Americans generate 2.22 kilograms per capita per day. To prevent a worse waste crisis – and an associated increase in emissions – we must take immediate action to decouple waste generation from income. South Korea illustrates how this can be done: financial incentives, citizen engagement, legislation and accompanying enforcement mechanisms have led to a 50 per cent decrease in waste generated per capita between 1990 and 2000, and it has remained steady since then – despite South Korea nearly tripling its GDP since 2000.

As things stand, however, projected economic development, population growth and urbanization suggest that the total waste generated by sub-Saharan Africa will triple and that of South Asia will double, accounting for more than one third of the waste generated globally by 2050. From a direct emissions perspective, this is concerning, because it is low- and middle-income countries – where waste is more likely to be mismanaged, or unmanaged – that are driving emissions. Particularly in low-income countries, there is a significant disconnect between the budget spent by local

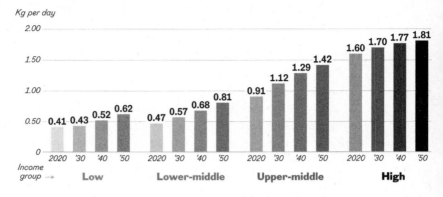

**Projected waste generation per capita
from 2020 to 2050 by income**

Kg per day

| Income group → | Low | Lower-middle | Upper-middle | High |

Low: 2020 0.41, '30 0.43, '40 0.52, '50 0.62
Lower-middle: 2020 0.47, '30 0.57, '40 0.68, '50 0.81
Upper-middle: 2020 0.91, '30 1.12, '40 1.29, '50 1.42
High: 2020 1.60, '30 1.70, '40 1.77, '50 1.81

Figure 1

governments in the provision of solid-waste management services and the quality and coverage of those services. Most of the budget is spent on trash collection and street cleaning, while little is spent on sound management and disposal. Even collected waste is often dumped – usually in areas near poor communities, which frequently do not receive waste collection services in the first place. Meanwhile, informal workers in waste, typically called waste pickers, often work without daily stability in income and in unhealthy conditions without safety gear, dealing with unregulated waste. They tend to be a vulnerable demographic such as women, children, the elderly, the unemployed or migrants, and often face social stigma despite their essential role in reducing emissions, in preventing plastic pollution and in recycling. It is estimated that 1 per cent of the urban population works informally in waste management and is exposed to health and safety risks such as lower life expectancy.

Across low-income countries, 39 per cent of waste is collected yet 93 per cent of waste (both collected and uncollected) is openly burned or dumped. In contrast, high-income countries have nearly universal waste collection and environmentally sound management until disposal. Conservatively, one third of global waste is openly dumped or burned, but the true proportion is likely higher, given the mismanagement of facilities constructed for final disposal. Burning waste releases toxins and particulate matter, which can cause respiratory and neurological diseases, and open dumping results in environmental contamination from the toxic run-off generated. In the fastest-growing regions, sub-Saharan Africa and South Asia, more than two thirds of the waste is currently openly dumped or burned, and this must be addressed immediately.

Projected total waste generation by region from 2020 to 2050

Millions of tonnes

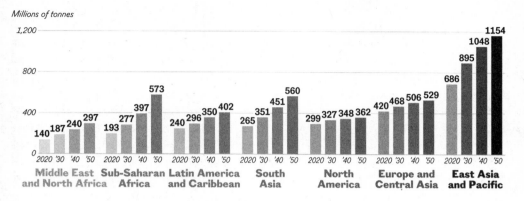

Figure 2

Just 19 per cent of municipal waste is recycled and composted, and leakages of plastic into our oceans are a growing challenge. Annually, 269 million tonnes of plastic are generated in municipal solid waste. It is estimated that 11 million tonnes of plastics entered oceans in 2016 and, if no action is taken, we expect that amount to nearly triple to 29 million tonnes annually by 2040. To visualize this, imagine a trash truck full of plastic bottles dumping its load into the Atlantic Ocean more than once a minute for an entire year. Eighty per cent of the plastic entering the oceans is estimated to be due to a lack of formal waste management systems, that is, from uncollected and dumped waste. Additionally, it is estimated that the 150 million tonnes of plastics accumulated in the oceans to date will more than quadruple to 646 million tonnes by 2040 if nothing is done. Lack of waste management on land, from poor waste collection to dumping of waste, is the largest contributor to marine litter.

Plastics have become an integral material in society, and curbing their use, particularly single-use plastics, will be critical as consumption increases in low-income countries as well as globally. This is a waste stream that cannot be tackled in isolation but must be dealt with as part of an integrated waste management system. We sorely need advocacy and policies designed to reduce consumption, especially of single-use plastics, and increase circularity in plastics and other materials.

To have a circular approach – where waste is reduced, reused, recovered and made back into products with minimal final disposal – the practice of dumping needs to be eliminated, and everyone should receive collection services that allow specific waste materials such as plastics, paper and food waste to be accessed and used productively. There is no single solution, however: local contexts are key. Depending on resources, population density, citizen engagement, land availability, enforcement mechanisms and policies, it may make sense to manage waste locally or more regionally, in order to take advantage of economies of scale. In locations with safe and universal waste management, governments should focus on reducing consumption, optimizing reuse, recovering or treating materials (that is, recycling and composting) and ensuring environmentally sound disposal (in sanitary land-fills and by incineration with energy recovery).

The waste sector can be a part of a low-carbon, resilient world, and its problems are solvable with no-regrets interventions. But cumulative improvements driven by high-income countries are not sufficient to reverse the current trends – it must be a global effort. Massive investment will be needed to prevent increased dumping as the volume of waste surges in the coming years. Moving towards universal waste collection, eliminating dumping, and implementing systems that recover and reuse materials can help mitigate the climate and waste crisis. This means a future without trash surrounding us; it means fewer emissions, cleaner water, more breathable air, and a more resilient world. /

We expect 29 million tonnes of plastics to enter oceans annually by 2040. To visualize this, imagine a trash truck full of plastic bottles dumping its load into the Atlantic Ocean more than once a minute for an entire year.

4.22

The Myth of Recycling

Nina Schrank

1970, the USA: The growing movement against throw-away plastic has led to protests across the country. The big food and drink companies are rightly held responsible. Plastic has been available as a mass consumer product for nearly twenty years, and Coca-Cola has abandoned reusable glass bottles which used to be collected, washed and refilled. By embracing disposable plastic, companies no longer have to pay for washing and refilling operations: instead, they have passed all the costs of dealing with single-use plastic bottles on to local government and taxpayers.

The corporations respond to the protests by releasing what is tipped as one of the most iconic advertisements of all time: the 'Crying Indian' TV advert. An actor dressed in traditional Native American dress paddles through a river awash with plastic packaging, before shedding a tear as litter is thrown out of a moving car.

'People start pollution, people can stop it,' proclaims the strapline. It is designed to deflect attention from the companies and instead blame the public for the deluge of waste. The 'Keep America Beautiful' lobby group behind the advert is made up of the leading drinks and packaging corporations in the US, including Coca-Cola. As it plays on screens across America, they are busy actively opposing legislation that would require them to switch back to reusable bottles.

Today, the Coca-Cola Company produces 100 billion single-use plastic bottles, a quarter of the 470 billion produced by soft-drinks manufacturers every year. The other big polluters of the world – Nestlé, Unilever, Procter & Gamble – collectively churn out billions of tonnes of throw-away plastic packaging every year, despite the fact that no waste system anywhere could possibly handle the volume of disposable plastic being produced.

The mantra that the public is responsible for this global pollution crisis still echoes throughout society. Representatives of leading drinks and packaging companies speak of their 'hatred of littering'. A British MP, recently caught out moonlighting as the chairman of a packaging lobby group, stood in Parliament opposing legislation to curb some of the most damaging single-use plastics. 'It is not the packaging manufacturer that is the polluter – people are,' he said.

The solution to this waste crisis, according to the big consumer goods companies across the world, is recycling. 'Recyclable' is printed on packaging and put front and centre of sustainability initiatives, and governments around the world have followed suit. The message to the Global North is, if we do our bit and put our plastic waste in the right bins, it will be magically swept up, then churned out as new products, in an unstoppable, infinite closed loop.

This narrative is perhaps the greatest example of greenwashing on the planet today. The principle of recycling is a positive one, connected with a sustainable lifestyle, but it has been co-opted as a way to maintain the status quo. Those of us in the Global North have been convinced that our waste is somehow being sustainably managed while, behind the scenes, business as usual continues. And the world's governments and corporations have failed to address the problem of single-use plastic at a systemic level. Some companies have committed to reducing their production of unrecyclable plastic, and some countries are moving to ban some throw-away items, but a recent study has shown that, even if all government and industry commitments to reduce plastic were implemented by 2040, the world would see only a 7 per cent reduction in leakage into the oceans.

The truth is, most plastic packaging never gets recycled. Some of it may be technically recyclable, but the rest of it is so cheaply produced that it is made to be disposed of. The 9 per cent of plastic in the world that is estimated to have made it to a recycling plant is downcycled into other products – doormats or traffic cones – perhaps once or twice, before its chemical make-up makes it impossible to continue and it too meets its final resting place in landfills, incinerators or dumped in the environment.

While some of the worst effects of plastic pollution are happening in the world's oceans, largely out of sight, it's a highly visible problem in countries in Asia and Africa. Here, plastic clogs up beaches and waterways, litters slums and sprawls through cities, towns and villages. The enormous landfill sites and waste dumps in India, the Philippines and Indonesia are a living testament to the flooding of countries with cheap disposable packaging – in greater quantities than these countries' waste systems could ever handle. The global Break Free From Plastic movement has for the last four years conducted global beach clean-ups with more than 11,000 volunteers in 45 countries to identify the most common plastic polluters.

The 2021 audit saw Coca-Cola, Pepsico, Unilever, Nestlé and Procter & Gamble come out on top.

Even when plastic waste is not dumped, it has serious environmental consequences. Landfill stores around a quarter of global plastic, which produces methane and ethylene when it is exposed to solar radiation, breaking down into microplastics that leach out in wind and rain into nearby soil and water bodies. Meanwhile, the energy generated by burning plastic in 'energy from waste' incinerators represents one of the most carbon-intensive sources of power on the planet, second only to coal, and there is nowhere other than landfill sites to put the toxic ash left over.

Yet still the myth of recycling is kept alive, primarily through plastic-waste exports. Countries that produce a lot of plastics, such as the UK, the US, Japan and Germany, lack the ability to manage their own waste; instead, they each export thousands of tonnes of it per year, most notably to Southeast Asia. These exports are conducted under the guise of recycling, even though the countries that receive these imports typically have limited waste management systems and weak or unenforced environmental standards, which means that they are unable to protect their communities and the natural world from the influx. Here, the waste trade is often run using a business model that relies on cherry-picking the most valuable plastics, often using cheap, migrant labour, then dumping the rest.

In 2018, Greenpeace investigators went to Malaysia and found European household waste in dumps 6 metres high. Local activists reported that the waste was being burned at night and that they would wake up struggling to breathe. The health impacts of burning plastic are dire: communities in India and across Southeast Asia have reported respiratory issues, and there is concern that exposure to toxic fumes may also be causing problems with menstruation and higher rates of cancer.

Many countries are now moving to protect themselves from imported plastic waste, as China – once the world's biggest plastic-waste importer – did in 2018. India, Malaysia, Sri Lanka and Thailand are all planning on introducing restrictions. This doesn't stop the industry, however. Shipments are rerouted and adaptations made, and the game of pass the parcel continues. Transit countries are used to disguise the origin of the waste, shipments are mislabelled, and clean, sorted, higher-value plastics are put at the front of shipping containers while dirty, mixed plastics fill the rest. Recyclers import with fake licences and no facilities, leading to the open dumping seen in Malaysia and other parts of the world.

But the governments allowing the export of waste appear to feel little compunction to change this state of affairs. In the UK, the situation is particularly stark. The UK is the second-highest producer of plastic waste per person in the world, behind the US. In 2020, the minister in charge of waste claimed that Britain was recycling 46 per cent of its plastic. In the same year, Greenpeace found that over half of the plastic waste the UK government counted as 'recycled' was sent overseas for other countries to deal with.

The following spring, Greenpeace investigators visited Turkey, the number-one destination for British plastic waste in 2020, taking almost 40 per cent of total exports. We found that half of these exports were either mixed plastic (which is extremely difficult to sort and recycle) or unrecyclable plastic – but the UK still counted it as having been recycled. At ten sites around the outskirts of Adana in southern Turkey, investigators documented piles of plastic waste, much of it from UK households, dumped illegally in fields, near rivers, on train tracks and by the roadside. In many cases, the plastic was on fire or had already been burned.

This is clearly a vast human and environmental tragedy. But one of the very worst impacts of plastic is largely undocumented, though it is happening right under the noses of world leaders: climate change.

Ninety-nine per cent of plastic is made from petrochemical feedstocks, a product of the oil and gas industry. Plastic produces greenhouse gases at every stage of its life cycle, from the moment of extraction, through its transport, up to its disposal.

As the world begins to wean itself off fossil fuels, the biggest oil companies in the world – Saudi Aramco, ExxonMobil, Shell, Total – are investing billions in petrochemicals factories on the assumption that demand for plastic will keep increasing. The International Energy Agency has predicted that petrochemicals will be responsible for more than a third of the growth in world oil demand by 2030, and nearly half of its growth by 2050.

Yet plastics are rarely mentioned when climate change policies are debated, either at national level or on the international stage. If we are to contain emissions, we need to wake up to this – the latest trick by the big oil majors seeking to keep themselves in play.

The solution, of course, is to drastically reduce the amount produced in the first place. The transition from a throw-away society to one that, where possible, either eliminates packaging or adopts reusable packaging, has never been more urgent than it is now. The situation looks set to intensify: by 2040 plastic production is expected to double in capacity, which would triple annual flows of plastic into the ocean.

The world's major brands and plastic packaging producers need a system change, and world governments must step in to enforce it; Greenpeace UK is calling for a 50 per cent reduction in single-use packaging by 2025, at least, and for a minimum of a quarter of this to be met by reusable packaging, rising to a half by 2030. Reuse is how we guarantee a truly closed loop where packaging is used, washed, refilled and used again.

The practice of reuse was embedded for generations in so many cultures across the globe, yet the corporate world has made us forget those traditions and the value we place in objects which have taken natural resources and energy to produce. Our throw-away society doesn't make sense: there has to be a paradigm shift. Business models need to be shaken up, traditions refreshed and innovations embraced so that reuse can flourish in the modern world. /

In 2020, over half of the plastic waste the UK government counted as 'recycled' was sent overseas for other countries to deal with.

Next page:
Waste washed ashore along the coast of Sian Ka'an, a World Heritage Site and federally protected reserve on the Yucatán Peninsula in Mexico. Through his photoseries of items which originated in over sixty countries, artist Alejandro Durán documents a 'new form of colonization by consumerism'.

Curaçao

Russia

Canada

Japan

Colombia

Brazil

Thailand

Morocco

Puerto Rico

Panama

Australia

Philippines

Nicaragua

India

Peru

Ecuador

This is where we draw the line

Greta Thunberg

This is page 301. Make a note of it. Fold over the corner or add a bookmark to your audiobook device. This book contains some stark messages that can be a bit challenging to get your head around. Whenever you are in doubt, or question any of these facts or ideas, come back to this page and read it again.

If we are to stay below the targets set in the 2015 Paris Agreement – and thereby minimize the risk of setting off irreversible chain reactions – we need immediate, drastic, annual emission reductions on a scale unlike anything the world has ever seen. And since we don't have the technological solutions that alone will do anything close to that in the foreseeable future, it means we have to make fundamental changes to our society. This is undeniable. It is also currently the most important piece of information we have when it comes to protecting the well-being of humankind and the only civilization we are aware of in the entire universe. And yet, still, in the year 2022, it is completely absent from every part of the global conversation.

And there is more. According to the United Nations Production Gap Report, the world's planned fossil fuel production by the year 2030 will be more than twice the amount that would be consistent with keeping to the 1.5°C target. This is science's way of telling us that we can no longer reach our targets without a system change. Because meeting our targets would literally require tearing up contracts, valid deals and agreements on an unimaginable scale. This would simply not be possible in the current system.

This should of course be dominating every hour of our everyday newsfeed, every political discussion, every business meeting and every inch of our daily lives. But that is not what is happening. This is not an opinion, or some random report. This is what the current best available science more or less boils down to. And, as you have probably learned from reading this book, it is the nature of science to be far from alarmist or exaggerated. It is cautious and careful.

The media and our political leaders have the opportunity to take drastic and immediate action, and still they choose not to. Perhaps it is because they are still in denial. Maybe it is because they do not care. Maybe it is because they are unaware. Maybe it is because they are more scared of the solutions than of the problem itself. Maybe it is because they are afraid of causing social unrest. Maybe they are afraid of losing their popularity. Maybe they simply did not go into politics or journalism to uproot a system they believe in – a system they have spent all their lives defending. Or maybe the reason for their inaction is a mixture of all these things.

We cannot live sustainably within today's economic system. Yet that is what we are constantly being told we can do. We can buy sustainable cars, travel on sustainable motorways powered by sustainable petroleum. We can eat sustainable meat and drink sustainable soft drinks out of sustainable plastic bottles. We can buy sustainable fast fashion and fly on sustainable aeroplanes using sustainable fuels. And of course we are going to meet our short- and long-term sustainable climate targets too, without making the slightest effort.

'How?' you might ask. How can that be possible when we don't yet have any technical solutions that can fix this crisis alone and the option of stopping doing things is unacceptable from our current economic standpoint? What are we going to do? Well, the answer is the same as always: we will cheat. We will use all those loopholes and all the creative accounting that we have conjured up in our climate frameworks since the very first Conference of the Parties, the 1995 COP1 in Berlin. We will outsource our emissions along with our factories, we will use baseline manipulation and start counting our emission reductions when it suits us best. We will burn trees, forests and biomass, as those have been excluded from the official statistics. We will lock decades of emissions into fossil gas infrastructure and call it *green natural gas*. And then we will offset the rest with vague afforestation projects – trees that might be lost to disease or fire – while we simultaneously cut down the last of our old-growth forests at a much higher speed. Because those emissions are excluded as well. This is the plan. It may not have been the intention of any individual leader or nation. Even so, this is the result of their efforts.

Don't get me wrong. Planting the right trees in the right soil is a great thing to do. It eventually sequesters carbon dioxide from the atmosphere and we should do it wherever it is suitable for the soil and suitable for the people living there who care for that land. But afforestation should not be confused with *offsetting* or *climate compensation*, because that is something completely different. You see, the main problem is that we already have at least forty years of carbon dioxide emissions to 'compensate' for. It is all up there, in the atmosphere, and that is where it will stay, probably for many centuries to come. This historic CO_2 is what we should be focusing on when we are using our present – very limited – ways of removing CO_2 from the atmosphere, in various projects such as planting trees. But offsetting, as we have conceived it, is not meant to do that. It was never created for us to clean up our mess. Far too often it has been used as an excuse for us to continue emitting CO_2, maintain business as usual and meanwhile send a signal that we have a solution and therefore we do not have to change. *We can compensate for our current and future actions, so we can go on like before. Who cares about the past when we have secured the future?* And since public awareness of this disjunction is – again – next to non-existent, the risk of anyone pointing out that 'Hey, this is a cumulative crisis' is pretty slim.

Words matter, and they are being used against us. Just like the idea that we can make sustainable choices and live sustainable lives in an unsustainable world, or that we can compensate our way out of this crisis. These are lies. Dangerous lies that will cause further, disastrous delay. Predictions by the UN conclude that our CO_2 emissions are expected to rise by another 16 per cent by 2030. The time we have left to avoid creating increasing climate catastrophes in many places around the world is rapidly running out.

We are currently on track to have a world that is 3.2°C hotter by the end of the century – and that's if countries fulfil all the policies they have in place, policies that are often based on flawed and under-reported numbers. But in many cases they are nowhere near doing even that. We are 'seemingly light years away from reaching our climate action targets', to quote UN Secretary-General António Guterres in the autumn of 2021. And there is also the matter of our previous track record of failure when it comes to delivering on all those non-binding pledges and promises. Let's just say it is not so impressive or convincing.

Even if we carried out all of our climate action plans, we'd still be in trouble. Even if our leaders all made moral U-turns and managed to fundamentally reorganize their societies in the coming years. Even if, somehow, we miraculously managed to channel all our powers into building the fantasy amounts of negative emissions technologies that our climate plans are completely dependent on. Even if our burning of biomass for BECCS did not create further ecological breakdown. Even if the overshoot – the time we will unavoidably stay above 1.5°C before we somehow pull ourselves back down to safer temperature levels, using technology that does not yet exist – did not set off any severe, irreversible chain reactions. Even if the 0.5°C of additional warming that is already locked in and hidden by the aerosols of

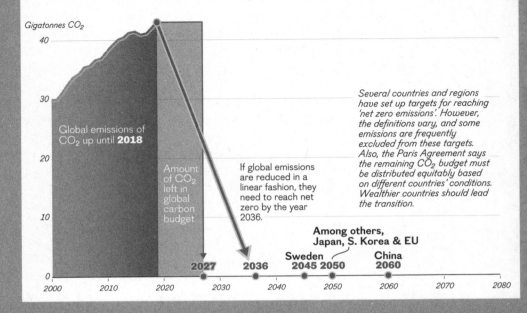

The global carbon budget vs. 'net zero' targets

Gigatonnes CO₂

Global emissions of CO₂ up until **2018**

Amount of CO₂ left in global carbon budget

If global emissions are reduced in a linear fashion, they need to reach net zero by the year 2036.

Several countries and regions have set up targets for reaching 'net zero emissions'. However, the definitions vary, and some emissions are frequently excluded from these targets. Also, the Paris Agreement says the remaining CO₂ budget must be distributed equitably based on different countries' conditions. Wealthier countries should lead the transition.

Among others, Japan, S. Korea & EU

Sweden
2045 2050

China
2060

2027   **2036**

Figure 1: Graph based on 2018 IPCC SR1.5 report.

air pollution described by Bjørn Samset in Part Two of this book was also somehow taken care of ... Even if all these things were done, it still would not be enough.

Net zero by 2050 is simply too little, too late. There is just too much at stake for us to place our destiny in the hands of undeveloped technologies. We need real zero. And we need honesty. At the very least, we need our leaders to start including all our actual emissions in our targets, statistics and policies. Before they do that, any mention of vague, future goals is nothing but a distracting waste of time. They say that we should not let the perfect be the enemy of the good. But what exactly do we do when the 'good' not only fails to keep us safe but is also so far away from what is needed that it can only be described as comedy material. Very dark comedy, but still. What do we do?

The moment we accept their *net zero by 2050* as our goal we not only legitimize the loopholes that threaten the future of the living planet as well as our entire civilization – we also surrender our chance for global equity today and ignore our responsibility for loss and damage and historical emissions. In other words, if we accept *net zero by 2050*, then we forever turn a blind eye to climate justice, to the cumulative crisis that has already happened – and by doing so we close the door to getting the overwhelming majority of the world's population on board. And that will – eventually – kill any idea of a future global climate movement. I agree, the perfect should not be the enemy of the good. But when it comes to the climate and ecological crisis, there is still very little good in sight – let alone perfect.

They say we must be able to compromise. As if the Paris Agreement is not already the world's biggest compromise. A compromise which has already locked in unimaginable amounts of suffering for the most affected people and areas. I say, No more. I say, Stand your ground. Our so-called leaders still think they can bargain with physics and negotiate with the laws of nature. They speak to flowers and forests in the language of US dollars and short-term economics. They hold up their quarterly income reports to impress the wild animals. They read stock-market analysis to the waves of the ocean, like fools.

We are approaching a precipice. And I would strongly suggest that those of us who have not yet been greenwashed out of our senses stand our ground. Do not let them drag us another inch closer to the edge. Not one inch. Right here, right now, is where we draw the line. This is where we stand our ground. /

They say we must be able to compromise. As if the Paris Agreement is not already the world's biggest compromise. A compromise that has already locked in unimaginable amounts of suffering for the most affected people and areas.

4.24

Emissions and Growth

Nicholas Stern

Scientists had long been warning of the risk of climate change when, in 1988, Syukuro Manabe, Michael Oppenheimer and James Hansen testified before the US Congress, awakening the world to the existential threat it posed. In 1992, governments responded by agreeing an international treaty, the United Nations Framework Convention on Climate Change, to limit the growing threat from rising levels of carbon dioxide and other greenhouse gases in the atmosphere.

However, since then, global annual emissions have continued to grow: they were 54 per cent higher in 2019 than in 1990, according to the Netherlands Environmental Assessment Agency. The size of the global economy increased by about 120 per cent over this period, based on World Bank data, and the energy for that growth has been mainly from fossil fuels (the International Energy Agency indicated that 80 per cent of the world's energy in 2019 was from fossil fuels). Fossil-fuelled growth has been the main driver of the increase in emissions.

During this time, many countries have sought to increase their economic productivity while also cutting their annual emissions, with some success. For instance, the annual emissions produced by the UK fell by 44 per cent while its economy grew by 78 per cent between 1990 and 2019. This was achieved primarily by improving energy efficiency and phasing out coal as a source of power. We must note, however, that this calculation excludes important sources of emissions, such as international aviation, and, as the UK Climate Change Committee has pointed out, the reduction would be much smaller (around 15 per cent) if we calculated the emissions embodied in consumption (much of which is imported) rather than production.

Economic decision-making is guided by key indicators, and central among these is GDP, which tries to measure the size of the economy by including all (or at least most) economic activities by companies, governments and individuals. But this does not, of course, measure everything of value, and it excludes the health of both our people and our environment. It does not account for biodiversity loss, environmental degradation and climate change, which constitute losses of real importance to our world and our well-being. In the long term, these losses undermine the economic activities that GDP measures and the health and strength of those who produce. Decision-makers, and all of us, should pay attention to direct measures of the state of our land, seas and atmosphere, as well as our vegetation and wildlife.

It is clearly possible to have economic development across all dimensions, including income, health, education, the environment and social cohesion while tackling climate change. Economic growth of this kind is essential for the nearly 7 billion people who live in developing countries, many of whom suffer from poverty. It can raise their living standards, giving them well-paying jobs and allowing them to access better education and health care. Our challenge is to bring this about in a way that does not damage our environment. This will be possible only if we radically change our ways of producing and consuming, particularly in relation to energy. The next decade is decisive if we are to keep a limit of 1.5°C within reach; we can and we must act swiftly and strongly to create a new form of growth and development which is sustainable, resilient and inclusive.

Unfortunately, much of the economic analysis of climate change has failed to recognize the necessary urgency and scale of action, for three reasons. First, it has failed to capture the immense scale of the risks identified by science. Second, it has underestimated the tremendous potential of alternative sources of energy and associated technologies. Third, it has grossly undervalued our descendants' lives through a misleading and ill-founded approach to discounting: we have discriminated against future generations based on their date of birth.

The people of the world have started to identify and embrace new, exciting and attractive forms of development. And at last the economists are catching up; indeed, some are beginning to contribute to policies and actions that can shape this new world. /

Equity

Sunita Narain

Climate change is an existential threat; we know that. And we know that we need to drastically reduce emissions. But what we continue to deny is that billions still have the right to development in order to live better lives. The most inconvenient truth is not that we have a climate crisis, but that we must build a new economic growth model, one that is accessible and affordable to all while being low-carbon and sustainable.

In my country, India, the poor already living at the margins of survival are severely impacted by extreme weather events. They are the first victims of climate change – and remember: they have not contributed to the stock of greenhouse gases in the atmosphere.

So, as we move ahead, we must recognize the imperative of climate justice. Fossil fuels are still determinants of growth, whatever the rhetoric. And most importantly, billions of people are still waiting to get access to affordable energy, which would bring them the advantages of economic progress. And this at a time when the world has run out of the carbon budget to accommodate their need for development. So the question is, what will this part of the emerging world do? Its growth – tied to its use of fossil fuels – will add to the environmental peril we're all facing. The question, therefore, is, how can growth be reinvented so that it is low-carbon and yet affordable? It is not enough to berate and bully the emerging-world countries into action. There must be supportive policies and real transfer of global finance to enable the transformation.

For far too long, wealthy nations have worked overtime to erase or dilute climate equity in negotiations. This is why the 2015 Paris Agreement was lauded – it got rid of the very concept of historical emissions; consigned climate justice to a postscript. It even removed the idea that the loss and damages that countries are suffering because of climate change ought to be compensated. Worse, it created a weak and meaningless framework of climate action which would depend on what a country could do voluntarily, not what it ought to do, based on its contribution to historical emissions or its fair share. It should not surprise us, then, that the sum of the nationally determined contributions – UN jargon for national reduction targets – takes the world towards a minimum of 3°C temperature rise or more.

Those in power should not dilly-dally around empty promises of net zero targets for 2050. They must envision how countries will frontload emission reductions for 2030. The fact is that the 'old' industrialized countries and new entrant China have appropriated 74 per cent of the carbon budget in the atmosphere up until 2019, and even if they meet their emissions reduction targets, they will still be using up 70 per cent in 2030. This is the carbon budget available to the entire world's population to stay below the guardrail of 1.5°C.

If we do this, then the opportunity of real change opens up – to invest today in the economies of the poorest countries so that they can grow without pollution. There are many possibilities for transformational action. Take, for instance, the energy needs of the poorest in the world, who are without the basic infrastructure of electricity to power their homes or to cook their food – millions of women still use biomass to cook, which is detrimental to their health, as these stoves are extremely polluting. The way ahead would be to use clean renewables to meet the needs of these households that are still outside the fossil fuel energy system. The cost of renewable energy is still beyond the affordability of the poor in the world; those in power must not preach the need for energy transitions but instead pay for them to happen.

This is where the discussions on the use of markets, through instruments like emissions trading, should be put to work. These should be used for transformational action so that projects that will bring 'big bang' carbon reductions can be paid for through financial transfer and carbon credits. For instance, the provision of clean energy through millions of mini-grids in the communities of the poorest. In this way, the market will be led by public policy and intent, and not left susceptible to new scams in the name of carbon offsets.

Similarly, there is an opportunity to use the ecological wealth of poor communities for mitigation, as trees and natural ecosystems sequester carbon dioxide. Their forests and other natural resources should not be viewed as sinks for carbon sequestration but as opportunities to improve the livelihoods and economic well-being of the poor. The rules for carbon offsets for forests must be developed with this in mind – deliberately and with statecraft.

The fact is that we have lost precious time in finding 'smart' ways to do as little as possible to reduce greenhouse emissions, and it is time to take decisive and bold steps. We need to build policies knowing that we live in an interdependent world where cooperation that is driven by fairness and justice is critical. /

4.26

Degrowth

Jason Hickel

People tend to talk about the ecological crisis in terms of 'the Anthropocene', referring to the way that, for the first time in geological history, human activity is dramatically reshaping our planet and our climate. This terminology is useful in certain respects, but it is also incorrect. It is not humans *as such* that are causing the problem but rather a specific economic system – namely, capitalism – which is organized around and dependent on perpetual GDP growth.

This might not be an issue if growth was just plucked out of thin air. But it is not. GDP is tightly coupled to energy and resource use, that is, all the material stuff the global economy extracts, produces and consumes each year. This is a problem, because as our economy grows and our energy use increases it becomes more difficult to decarbonize the energy system fast enough to keep global warming to less than 1.5 or 2 °C. And our resource use – which presently exceeds 100 billion tonnes per year – is already overshooting the maximum sustainable boundary by a factor of two.

Crucially, this crisis is being driven almost entirely by the rich countries of the Global North – and primarily by wealthier classes and corporations within those countries. This is clear when it comes to the climate crisis: the Global North is responsible for 92 per cent of all emissions in excess of the planetary boundary, which scientists have defined as a CO_2 concentration in the atmosphere of 350 parts per million – a level we passed in 1988. Meanwhile, most of the countries in the Global South are still well within their fair share of the boundary, and have therefore not contributed to the crisis at all. And yet the south suffers the vast majority of the damage, including 82–92 per cent of the economic costs of climate breakdown, and 98–99 per cent of climate-related deaths. It would be difficult to overstate the scale of this injustice.

The same is true when it comes to resource use. Rich countries consume on average 28 tonnes of resources per person per year, four times more than the sustainable level and many times higher than the average in the Global South. Moreover, rich countries rely on a large *net* appropriation of resources from the south. This means that the impact of northern consumption is effectively offshored to the south, where the damage occurs, while southern

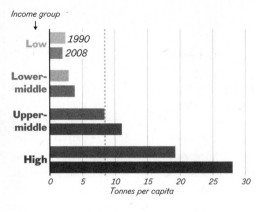

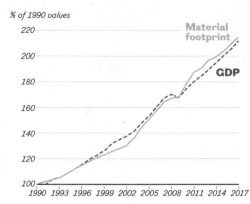

Figure 1 (left): Material footprint of nations, with the sustainable per capita boundary in 2008 as a dashed line.

Figure 2 (right): Global GDP and material footprint in tonnes per capita.

communities are at the same time drained of the resources necessary for development and meeting human needs. This system perpetuates mass poverty and exacerbates global inequality.

In short, the ecological crisis is playing out along colonial lines. Continued growth in the Global North relies upon processes of atmospheric colonization and the appropriation of southern ecosystems. If we are not attentive to the colonial dimensions of the ecological crisis, we are missing the point.

For the past fifty years, many economists and policymakers in the Global North have urged us to press on the accelerator of growth but seek to make it 'green'. The hope is that we can 'decouple' GDP from environmental impact. But scientists reject this narrative as empirically baseless.

First, there is no evidence that growth can be absolutely decoupled from energy and resource use on a global scale, and all existing global models project that it is unlikely to be achieved in the future even under highly optimistic assumptions about efficiency and technological change. These results have been confirmed by scientists several times. A recent study on this issue concludes, 'It is misleading to develop growth-oriented policy around the expectation that decoupling is possible.'

What about emissions? GDP *can* be decoupled from emissions, by replacing fossil fuels with renewable energy, and this is already happening in some nations. The problem is that decarbonization cannot be accomplished fast enough to meet the Paris targets *if high-income economies continue to grow at their current rates*. Remember: more growth means more energy demand, and more energy demand makes it more difficult – and probably impossible – to reduce emissions to zero at a sufficiently rapid rate.

In light of this evidence, ecological economists call for a fundamentally different approach. The first step is to realize that high-income nations do not *need* more growth. In fact, we know that it is possible to meet human needs at a high standard with *much less* energy and resources than rich countries presently use. The key is to scale down less necessary forms of production and organize the economy around human well-being rather than capital accumulation. This is known as degrowth. Degrowth calls for a planned reduction of excess resource and energy use in high-income nations to bring the economy back into balance with the living world in a just and equitable way.

What does this look like in practice? Instead of assuming that *every sector* of the economy must grow, all the time, regardless of whether or not we actually need it, we should decide which sectors of the economy we actually *need* to improve (for example renewable energy, public transportation and health-care), and which are clearly destructive and should be scaled down (SUVs, air travel, fast fashion, industrial beef, advertising, finance, the practice of planned obsolescence, the military industrial complex, and so on). There are huge chunks of the economy that are organized mostly around corporate power and elite consumption, and we would all be better off without them.

Most people would regard this as sensible, except for one thing: what about jobs? Fortunately, there is a simple solution: as the economy requires less labour, we can shorten the working week and share necessary work more evenly. We can also roll out a public jobs programme to ensure that anyone can train to participate in the most important collective projects of our generation: building renewable energy capacity, insulating homes, producing local food and regenerating ecosystems. At the same time, we need to expand universal public services to ensure that all people have access to the resources they need to live good lives (not just health care and education but also housing, public transport, clean energy, water and the internet), while dramatically reducing inequality with progressive taxes on income and wealth.

Taking this approach would ensure good livelihoods and provision for all while at the same time directly reducing energy and resource use, enabling us to decarbonize the economy much more quickly – in a matter of years, not decades – and reverse ecological breakdown. What is more, it also means liberating the countries of the Global South from imperial appropriation so they can mobilize their resources around meeting human needs rather than servicing northern consumption.

This vision may sound utopian, but it is both possible and necessary. This is how we avert ecological breakdown and build a just and equitable civilization for the twenty-first century. Of course, it will require a real struggle against those who benefit so prodigiously from the existing structure of the world economy; it will require organizing, and solidarity, and courage. But so has every struggle for a better world. /

4.27

The Perception Gap

Amitav Ghosh

'**Trees were my teachers,**' wrote the German poet Friedrich Hölderlin, and if there is any place on Earth that could say the same of itself it is Ternate, a tiny island in the archipelago that was once known as the Moluccas, or Spice Islands. It is now part of the province of North Meluku, in the far eastern reaches of Indonesia. The seas here are dotted with volcanic islands, and Ternate is one such; the surface of the island is nothing but the gently sloping cone of a volcano, Mount Gamalama, which rises from the sea floor to a height of over 5,000 feet.

Ternate is a place that would, by most reckonings, be considered very far removed from the pathways of history. But the island was, in fact, a driver of global history for many centuries, as will be evident to anyone who sets eyes on the innumerable colonial forts that line its shores. The reason for this was that a uniquely valuable tree happened to grow on Ternate, and the islands around it: this was *Syzygium aromaticum*, the tree that produces the clove. This spice, which was once immensely valuable, made Ternate prosperous and powerful for hundreds of years. But in the sixteenth century, at the beginning of the era of European colonization, Ternate's 'tree of life' also brought disaster upon the islanders. Various groups of European colonizers fought over Ternate and its surrounding islands in the course of a bloody struggle to establish a monopoly over the trade in cloves. The Dutch eventually prevailed, and in the seventeenth century they turned the island into a colony and decreed that henceforth cloves would only be grown on another island, in the southern Moluccas. The people of Ternate were forced, by the terms of a Dutch-enforced treaty, to 'extirpate' every clove tree on their island. The tree that had been Ternate's teacher would not return to the slopes of Mount Gamalama until the next century, when cloves were already being cultivated elsewhere and had drastically declined in value.

Today, Ternate is a sleepy, quiet place, notable mainly for the ruins of the early Portuguese and Dutch forts that line its shores. But, despite its remoteness from the great centres of contemporary trade, Ternate is by no means a laggard in globalization. Indonesia is one of the fastest-growing economies in the world, and evidence of this is everywhere visible on the island: in the great mass of vehicles, large and small, that throng its streets, and in the fast-rising buildings that dot its villages. Indeed there is no better proof of Indonesia's rapid acceleration than its ability to deliver an abundance of goods and services to this distant corner of its territories.

But Ternate's landscape possesses an additional marker of this era of acceleration. This too is etched upon its landscape, by the island's tree of destiny. Across the island, clove trees are dying; in orchard after orchard they stand in drooping clumps, their branches leafless, their trunks ashen. On the slopes of the volcano, clusters of dead trees can be seen, their leaden colours contrasting vividly with the greenery.

The farmers who tend to the trees are unanimous about the cause of their demise: the climate has changed in recent years, they say; there is less rain and it falls more erratically. This in turn has led to the spread of blights and disease. The lack of rain has been accompanied by another unprecedented phenomenon: wildfires. In March 2016 a fire raged for three days on the slopes of Mount Gamalama. Forest fires of this intensity are new to the islanders' experience.

The ongoing changes in the world's climate have thus placed the people of Ternate once again on the leading edge of history: the trees that guided their first steps in the world are now dying before their eyes as they watch helplessly.

This is a tragic predicament, considering that Ternate's volcanic environment created an especially intimate, sacralized relationship between the island's ecology and its people, who have long seen themselves as custodians of their closely interconnected world. And this is particularly the case for the descendants of the dynasty of sultans that has presided over the island since the fourteenth century. Some members of this dynasty still live on the island, and during my visit, in 2016, I was able to interview one of them, a prince who is the son of the late ruler and the current occupant of the Sultan's Palace.

We sat in a courtyard that faced Mount Gamalama, so it was inevitable that our conversation should touch upon the dying clove trees that I had seen on the volcano's slopes. Like so many others on the island, the prince attributed the death of the trees to climate change – this was, for him, a deeply troubling matter since these trees had sustained his family's fortunes for 700 years.

This being the case, I thought I should ask the prince a question that I had already put to several clove farmers: 'Given the seriousness of the situation, do you think the people of Ternate should make an effort to cut back their carbon emissions?'

Considering his family's special relationship with the clove tree, I thought the prince would see the matter differently from the working clove farmers I had spoken with. But the answer he gave me was more or less the same one I had heard from others on the island. It could be paraphrased as 'Why should *we* cut back? That would be unjust to us. The West had its turn when we were weak and powerless and they were our rulers. It's our turn now.'

The prince's response came as no surprise to me because I had heard its like many times, not just in Indonesia, but also in India, China and many other places. For the farmers, as for the prince, the burden of history's injustices far outweighed the material realities and imminent threats of climate change. Having to tolerate a disrupted environment was, for them, a sacrifice that had to be endured for the sake of a wider, national aspiration.

It is in much the same spirit that the inhabitants of cities such as New Delhi and Lahore endure toxic levels of pollution, knowing that the air they breathe will shorten their lives by several years. The damage to their health and well-being is seen as a sacrifice that is necessary, on the one hand, to enjoy a certain standard of living, and on the other, to advance a wider collective aspiration to attain a better place in the international order. It is by this route that coping with environmental hazards comes to be blended with some of the notions of sacrifice and suffering that underlie nationalism. By the same token, attempts to impose limitations on the carbon emissions of poor countries are widely seen as a covert means of preserving the economic and geopolitical disparities of the last 200 years, since on a per capita basis the carbon emissions of the Global South are still a fraction of those of affluent countries.

These perceptions are mirrored in the West by the idea, now widely prevalent on the right, that the Global South is trying to deprive affluent nations of the hard-earned fruits of their success. In the US the idea of imposing limits on America's carbon emissions is also perceived by many as an infringement of national sovereignty, which is guaranteed, ultimately, by the country's overwhelming military dominance.

In short, nationalism, military power and geopolitical disparities are fundamental to the dynamics that have repeatedly stymied efforts to reach a global agreement on rapid decarbonization. In that sense, it could be said that conflict and national rivalries are fundamental drivers of climate change. Yet these issues are rarely discussed in conferences on global warming, which have come to be focused rather on technocratic and economistic 'solutions' of various kinds. It is no coincidence that the literature on climate change, which is overwhelmingly produced by western universities and think tanks, is also largely centred on technical and economic issues.

As a result, there is an immense gap between perceptions of climate change in the affluent countries of the Global North, which are almost all beneficiaries of centuries of colonialism, and those of the Global South, most of which were subjected to some form of colonial domination. In the north, global warming is largely framed by technology, economics and science; in the south, the same phenomenon is conceived of in terms of disparities of power and affluence that can all be traced back to the geopolitical inequities established in the era of colonialism.

In the south, issues such as violence, race and geopolitical power are implicit in the perceptions of people like the clove farmers of Ternate. In the north, which is largely secure in its position atop the global pyramid, these issues are rarely discussed and climate change is generally treated as a problem of governance which can be resolved through processes of negotiation within multilateral institutions like the UN.

But there is a very significant contradiction here. Multilateral institutions are mandated to operate on the assumptions that all nations and peoples are equal and that wealth and welfare should be justly distributed between nations. Geopolitics, on the other hand, is founded on completely different assumptions. It is not intended to bring about equality and justice but the opposite. It is explicitly about maintaining a structure of dominance – or inequality, in other words.

The dissonance between these two spheres – that of multilateral global governance on the one hand, and geopolitical power on the other – is so great as to be almost irreconcilable. While the structures of global governance produce seemingly endless streams of 'solutions' and treaties, the repeated breakdown of international negotiations points to a different, largely hidden reality. This unacknowledged dynamic was once summed up by a Singaporean journalist with the following words: 'It is our will to power that will help us cope with one of the big driving forces of the future: climate change.'

In other words, global leaders may speak a certain language during international negotiations, but when we examine what they are actually doing it would seem that their actions are indeed driven by a will to power. That perhaps is why affluent nations felt able to contribute only $10 billion to a fund to help countries which are exceptionally vulnerable but had no difficulty in increasing their defence spending by $1 trillion. This suggests that, contrary to what global leaders may say publicly, many of them are in fact preparing for a future of intensified conflict.

Next pages:
Mangroves are
one of the world's
most threatened
ecosystems. An
important wildlife
habitat, they
protect the
coastline from
flooding, tsunamis
and soil erosion,
and help to
mitigate climate
change by filtering
pollutants,
absorbing carbon
dioxide and
releasing oxygen.

Given the intractable nature of the world's geopolitical disparities, what can be done to address the planetary crisis? How can the aspirations of people in the Global South be met when it is clear that humanity would asphyxiate if everybody were to adopt western lifestyles?

One element of this which provides some encouragement is that the aspirations of the middle classes of the Global South are essentially mimetic. That is to say, when an Indian or an Indonesian says, 'Now it's our turn,' what they are really saying is, 'I will not be wealthy or content until I have what the Other has.' It follows from this that if the supposedly wealthy Others were to change their ways and adopt substantially different lifestyles, then this could have a substantial impact on aspirations across the world.

In this regard, the emphasis that Fridays For Future has placed on finding new ways to live is of vital importance. And the fact that its message has resonated so widely, even in the Global South, is a rare cause for encouragement. /

Nationalism, military power and geopolitical disparities are fundamental to the dynamics that have repeatedly stymied efforts to reach a global agreement on rapid decarbonization.

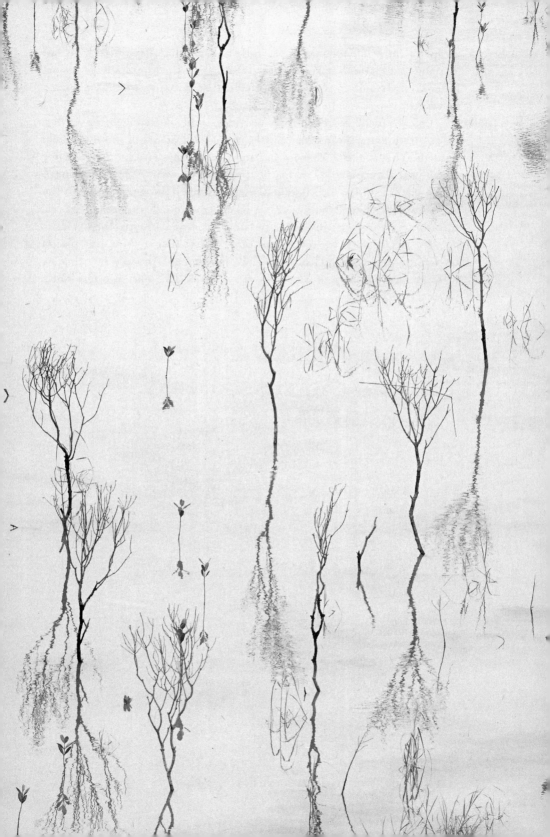

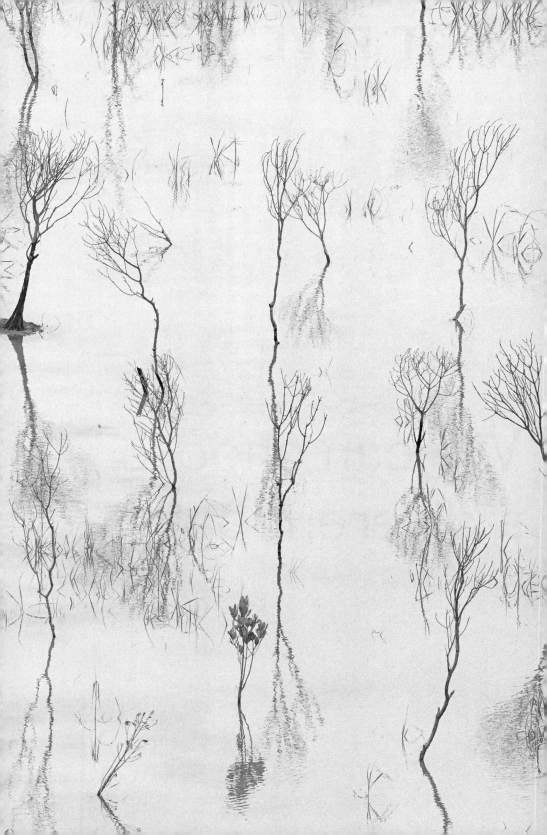

PART FIVE /

What We Must Do Now

'We can choose a different path'

The most effective way to get out of this mess is to educate ourselves

Greta Thunberg

The answer to the question of whether we should be focusing on individual or systemic change is: yes, definitely. We cannot have one without the other. We need both. Solving the climate crisis cannot be left to individuals, nor can it be left to the market. To stay in line with our climate targets – and thereby avoid the worst risks of initiating a climate catastrophe – we need to change our entire societies. To quote the IPCC, 'limiting global warming to 1.5°C will require rapid, far-reaching and unprecedented changes in all aspects of society'. There is no way that such a transformation can be achieved just by individual lifestyle changes, by individual companies finding new ways of manufacturing green cement or by individual governments raising or lowering taxes. Because it will not be enough. But then again, it is equally impossible to bring about such a transformation without individuals; in particular, they need to lead the way at a grassroots level. Individual people, individual movements, individual organizations, individual leaders, individual regions and individual nations need to initiate action.

Throughout history there have been many major societal changes. Some of them have been quite dramatic – for good or for bad. So when we call for unprecedented changes in all aspects of our societies, we do not mean that we should just become vegetarian for one day a week, offset our holiday trips to Thailand or switch our diesel SUV for an electric car. And yet this is what most people in large parts of the world seem to think. And there are understandable reasons for that. We humans are social animals –

herd animals, if you like. As Stuart Capstick and Lorraine Whitmarsh show in the following chapter, we copy the behaviour of others and we follow our leaders. If we do not see anyone else behaving as if we are in a crisis, then very few will understand that we actually are in a crisis.

In other words, it hardly matters if you say that we are facing an emergency if no one is acting as if we are facing an emergency. This is very well understood by the people in power, who have mastered the fine art of saying one thing while doing the exact opposite. This is most likely the reason why we have ended up with a situation where, for example, some of the world's largest oil-producing nations are rapidly expanding their fossil fuel infrastructure and at the same time calling themselves climate leaders.

The Swedish language has produced only a tiny number of words that have achieved international recognition and made it into the global vocabulary, for example 'smörgåsbord' and 'ombudsman'. Recently these words were joined by *flygskam*, or 'flight shame'. It is linked to the international climate movement and the growing number of people who have given up flying, because frequent flying is by far the most climate-destructive individual activity you can engage in – unless you count billionaire-style space travel or owning a large private yacht. The reason *flygskam* took off in Sweden was most likely because a small number of celebrities got behind it. The word itself was created by the media, probably in an attempt to create some click-friendly traction. Hence the addition of the word 'shame'.

I know many people who have given up flying, not just for a year or two, but for good. That is not a decision anyone takes lightly. By doing so, these people have drastically lowered their own carbon footprint. But that is usually not why they did it. Nor did they do it to inflict *shame* on anyone. Most of them did it for the same reason I did – to send a clear message to those around them that we are now at the beginning of a crisis and, in a crisis, you change your behaviour. I certainly did not sail twice across the Atlantic Ocean in order to shame anyone, or to lower my carbon footprint. I did it to point out that there is no way for us as individuals to live sustainably within today's system. And that the solutions needed to enable us to do so will not be even remotely available within the timeframe of our climate targets.

There is, however, another Swedish word that deserves far more attention than *flygskam*, and that word is *folkbildning*. It roughly translates to 'broad, free, voluntary public education' and it has most of its roots in the working-class community that came into being after democracy was introduced to the country in the first decades of the twentieth century – when unions became legal, when workers and women were given the right to vote and Sweden started building its welfare state.

Many probably think that Fridays For Future was initially intended as a protest movement, but that is not the case, or at least this wasn't how it began. Our primary initial aim was to spread information about the crisis – as *folkbildning*, to be more exact. When I sat down outside the Swedish Parliament on August 20, 2018 I not only carried a big white sign reading *Skolstrejk För Klimatet*; most importantly, I also had a huge pile of flyers packed with facts and information about the climate and ecological emergency freely available to anyone passing by. I still have a bundle stuffed away in a desk drawer at my parents' apartment. I guess the flyers were not as effective in putting the point across as the shy girl with the big white sign.

But, to this day, I firmly believe that the most effective way for us to get out of this mess is to educate ourselves and others (a bit ironic, since the idea of school strikes is based on skipping school, but still). Because once you understand the situation we are facing, once you get a sense of the full picture, you will more or less know what to do. And – perhaps just as importantly – you will know what *not* to do. Like focusing on specific details while not taking account of the wider context or, in other words, trying to solve a crisis without treating it like a crisis. I am absolutely convinced that the moment we do go into full crisis mode we will consider every possible individual detail. But until then, debating separate, individual issues will probably be a waste of time, as so many of those separate issues are co-opted to create 'culture wars'. They are often designed to steal everyone's attention and stall all meaningful progress. Like population growth, nuclear power or *what about China?*

Alongside the culture wars, there are many successful strategies to delay, divide and distract. As Naomi Oreskes noted in Part One, the fossil fuel industry 'deflected attention from their role by insisting citizens should take "personal responsibility"' by focusing on their individual carbon footprints. The idea was initially pushed by the oil company BP to deflect attention away from the major destructive industries and on to the individual consumer. It has been very effective. In Part Four, Nina Schrank called attention to a similar effort on the part of beverage companies such as Coca-Cola to shift the blame for skyrocketing plastic pollution on to the consumer, and countless similar campaigns have been injected into the climate debate. A recent, hugely successful one states that a hundred companies are responsible for 70 per cent of the world's emissions. This is the exact opposite argument to that of the carbon footprint narrative, but the result is pretty much the same – namely, inaction. The central message this time is that since it is only a hundred companies that create all these emissions, it doesn't matter what we do as individuals because it would be so much more effective if we somehow just got rid of those companies. How we would get rid of them is unclear, not least since we lack any rules, laws or restrictions to do so other than to boycott their products – which, of course, is an individual action.

Don't get me wrong – I'm all for getting rid of them and having them pay for the indescribable destruction they have caused. It is just that, once those hundred companies are gone, another hundred will no doubt take their place, unless we transform our entire society – a process that requires individual action and systemic change to work hand in hand. So again, we need both. Any suggestion that we can have one without the other – or that any single solution or idea is more important than all the others – is pretty much guaranteed to be aimed at slowing us down.

One thing I should make clear, though, is that when I talk about individual action I do not just mean reducing plastic consumption and eating more plant-based food – even though those things are good methods for generating a sense of urgency. When I talk about individual action I mean that we as individuals should use our voices, and whatever platforms we have, to become activists and communicate the urgency of the situation to those around us. We should all become active citizens and hold the people in power accountable for their actions, and their inaction.

The truth is that if we are to avoid the worst consequences of the climate and ecological crisis, we can no longer pick and choose our actions – we need to do everything we can. And for that we will need everyone: individuals, governments, companies and every other body and institution you can possibly think of. But we have to remember that the time for *small steps in the right direction* is long gone. We do not have the time to bring people along gradually. And 'making some progress' or 'winning slowly' is not good enough. Because when it comes to the climate crisis, to quote the American author Alex Steffen, 'Winning slowly is the same thing as losing.' /

We can no longer pick and choose our actions – we need to do everything we can.

5.2

Individual Action, Social Transformation

Stuart Capstick and Lorraine Whitmarsh

There is a troubling mismatch between the enormity of climate change and the smallness of the response asked of individuals. In the face of an unprecedented existential crisis, we are encouraged to do some recycling, switch off lights and use paper straws, as if these everyday choices could possibly hold back sea-level rise or deadly heatwaves. Even if someone takes every action they can to cut their emissions – becoming vegan, stopping driving and flying, buying as few things as possible – still the nagging sense persists of a drop in the ocean, immaterial in relation to our societies' dependence on fossil fuels and the far-reaching changes needed to move beyond this.

If this is a disheartening viewpoint, the good news is that it also represents a false dichotomy. Focusing attention at two extremes – the individual versus the systemic – overlooks the vast territory in between. It is in this space that we are able to interact with the people around us, helping to bring about change through shaping social expectations and creating shared realities. Exercising our influence in this domain involves far more than being an isolated consumer of products and services. Instead, climate action occurs through the many roles we occupy as human beings in near-constant contact with one another: as people who play a part in the life of communities, families, friendship groups, organizations and workplaces.

One way in which our actions matter in this context is through providing cues and examples to others. Just as each of us is influenced by the opinions and actions of other people – especially those we look up to or care about – so too are other people influenced by us, whether we realize it or not. Many studies have shown that the extent to which people make environmentally friendly choices is affected by their assessment of what others are doing. Other research has illustrated how this interpersonal influence can develop over time and spread throughout a neighbourhood or network of contacts, in a process that has been termed social or behavioural 'contagion'. This can occur when people respond to changes happening around them, as well as through word of mouth. Research examining the diffusion of technology has shown that households installing solar panels have a measurable effect on the likelihood of nearby homes following suit; on average, if two houses within a half-mile radius install a new system, the resulting peer influence prompts one additional household to do likewise. In a similar way, growth in the uptake of electric bikes, scooters and cars has been directly enabled by people discussing their use and encouraging others to try them.

Beyond prompting people towards a particular course of action, patterns of social influence have the potential to set the tone for which ways of life are deemed more or less acceptable. For many years, frequent travel by air has been seen as marker of high social status. More recently, however, awareness of the harmful impacts of flying has begun to shape new social norms set against this and to influence demand for flights: in Sweden, where the phenomenon of *flygskam* (flight shame) took root, domestic air-passenger numbers fell by 9 per cent between 2018 and 2019 as a result. It is out of an interest in influencing others that the Flight Free campaign encourages people to make a commitment to reduce flying, not only to lower one's own emissions (though this matters) but in order to have a wider impact on family and friends and ultimately to change cultural expectations around aviation.

Personal action to tackle climate change has the ability to spark wider transformations of the contexts that underpin our everyday choices, including by influencing business activity and shifting the sense of what represents a normal or desirable way of life. Growing enthusiasm for plant-based diets – which has already led to substantial reductions in greenhouse gas emissions in some parts of the world – has in turn prompted producers to invest in developing new vegan and vegetarian products, with the potential to enable further changes in people's dietary choices as these options become more widespread.

In cases where influential or high-profile public figures commit to personal actions such as reducing flying, this can have a particularly pronounced effect on others. Scientists and campaigners working on climate change can find their credibility enhanced – or undermined – by personal choices that convey a message about the seriousness of the crisis and the relevance of individual action. The capacity to reduce one's emissions and exert influence on others also varies greatly according to socio-economic status and material circumstances. The wealthiest 10 per cent of people globally produce around half of all greenhouse gas emissions; as well as having more to do to achieve a sustainable way of life, their personal resources put them in a better position than most to invest ethically and influence professional practice.

Personal action can also mean activism as part of efforts to push collectively for change. Participation in social movements tackling the climate crisis makes a difference, both for its influence on wider public opinion in favour of climate action and to exert pressure on decision-makers to enact more ambitious policy responses. In many parts of the world, politicians can no longer claim that they do not have the social mandate for taking the climate crisis seriously: citizens are clearly calling for a strong government response, with high levels of public concern about climate change and wide-ranging support for policies to cut emissions. In recognition of this, some senior politicians have actively encouraged citizen activism that pushes them to do more, for example Angela Merkel when she was Chancellor asking young Germans to 'pile on the pressure', and Scottish First Minister Nicola Sturgeon acknowledging that 'our feet do need to be held to the fire'.

In all these ways, our spheres of influence extend from private and personal choices, through persuading and supporting others, to organizing and agitating for change and, ultimately, becoming a part of remaking the very systems and cultures that make up society. Because of the complex interactions between people's actions and social change, the potential exists for domino effects: many separate actions can lead to the overturning of social conventions through disruptive and rapidly spreading tipping points – history shows that such transitions can be sudden and dramatic, and that changes in attitudes and behaviour are a key component of this.

None of this is to argue that the obligation to tackle the climate crisis rests solely with citizens, whose powers are limited and whose choices are often heavily constrained. An emphasis on personal responsibility has been misused by oil companies and others in order to deflect attention from their own deficiencies – a deliberate tactic which deserves to be discredited. It is also essential that governments show leadership in setting the conditions for low-carbon lifestyles and economies without waiting to be pushed hard to do so. But when we reflect upon our own part in tackling the climate crisis, we should remember that there is nothing 'individual' about individual action: it is the vital building block from which social transformation is made possible. /

Towards 1.5°C Lifestyles

Kate Raworth

'**I shop therefore I am,**' declared the artist Barbara Kruger in 1987.

Her iconic words sum up the intensely consumerist lifestyles that, over the course of the twentieth century, came to dominate life in so many high-income cities and nations – while simultaneously degrading the health of the living planet.

This critical decade of climate action calls for a radical rebalancing of consumption between the Global North and South so that it becomes possible to meet the needs of all people within the means of the living planet. The scale and speed required for this rebalancing are unprecedented. According to Oxfam, if humanity is to live well and equitably while keeping global heating within 1.5°C, then, by 2030, the world's richest 10 per cent of people need to reduce their consumption emissions to just one tenth of their 2015 levels – and, in the process, make space for the poorest 50 per cent of people in the world to realize their essential consumption needs.

How, then, can rich communities and countries escape the consumerist lifestyles that have engulfed them for over a hundred years? Let's start by understanding how consumerism was written into the foundational theories and core business models that propelled twentieth-century economic growth.

The founding fathers of economics placed a caricature of humanity at the heart of their theories: a solitary, self-interested individual with an insatiable desire for all the things that money can buy. As Alfred Marshall, the leading economist of his day, put it in 1890, 'human wants and desires are countless in number and very various in kind. The uncivilized man indeed has not many more than the brute animal; but every step in his progress upwards increases the variety of his needs ... he desires a greater choice of things, and things that will satisfy new wants growing up in him.' With such a narrow depiction of humanity as a starting point, no wonder GDP – which measures the total cost of products and services sold in an economy in a year – so readily came to be seen as a reasonable measure of a nation's success.

Although economic theory had already imagined people as insatiable consumers, real people still had to be persuaded of it; indeed the future profitability of the twentieth century's most powerful corporations depended upon it. 'Mass production is profitable only if its rhythm can be maintained,' wrote Edward Bernays in his classic 1928 book *Propaganda*, arguing that business 'cannot afford to wait until the public asks for its product; it must maintain constant touch, through advertising and propaganda . . . to assure itself the continuous demand which alone will make its costly plant profitable'.

Fascinatingly, Bernays – who invented the 'public relations' industry – was the nephew of Sigmund Freud, and he realized that the ideas underlying psychotherapy could be turned into very lucrative retail therapy if he could connect people's deepest desires to the latest products on sale. In the 1920s he convinced women (on behalf of the American Tobacco Corporation) that cigarettes were their 'torches of freedom', while persuading the nation (on behalf of the Beech-Nut Packing Company's pork department) that bacon and eggs were the 'hearty' all-American breakfast. He certainly knew the power of this advertising. 'We are governed, our minds are molded, our tastes formed, our ideas suggested, largely by men we have never heard of,' he wrote. 'It is they who pull the wires which control the public mind.'

The advertising industry rapidly grew and soon embedded consumerism as an aspirational way of life. As the media theorist John Berger put it in his 1972 book, *Ways of Seeing*, 'publicity is not merely an assembly of competing messages: it is a language in itself which is always being used to make the same general proposal . . . it proposes to each of us that we transform ourselves, and our lives, by buying something more'.

If one industry epitomizes the frenetic attempt to transform ourselves by buying something more, it is the fashion industry. Recent decades have seen major fashion retailers increasing their number of annual collections from just four to twelve or even fifty-two 'micro-seasons', holding out the promise of a 'new you' every week of the year. This ever faster cycle of cheaply made garments is echoed in customer habits: between 2000 and 2014 the average consumer bought 60 per cent more clothing but kept each item for just half as long.

The business model behind fast fashion is exploitative of people and planet alike. Pressured to deliver large orders of low-cost clothes under very tight deadlines, factories worldwide often push garment workers into an intense schedule of long hours with low wages, insecure contracts and a ban on organizing the workforce. Add to this the destructive impacts of the industry's use of materials, water, chemicals and energy. Out of all textile fibres currently produced, 12 per cent are discarded or lost in the production process, 73 per cent end up in landfill or incinerated after use, and less than 1 per cent are reused or recycled for new clothing. Additionally, the global fashion industry produces around 2 per cent of all greenhouse gas emissions – these must be almost halved by 2030, but they are still rising. Fashion is clearly wearing out the planet.

Recovering from consumerism

How can societies escape the exploitative dynamics of consumerism – in fashion and far beyond? Can we replace Marshall's caricature with an understanding that we are motivated by so much more than a desire for more things? Can we recover from a hundred years of consumerist propaganda unleashed by Bernays and find a new basis for the relationships we have with each other, with the things we need and use, and with the rest of the living world?

If we are to recover from consumerism – and at the speed required – let's look at what has been learned so far about the most effective ways to rapidly reduce the consumption-intensive lifestyles of high-income nations. A major new analysis of what it would take to achieve '1.5°C lifestyles' explores key sectors including food, housing, personal transport, consumer products, leisure and services. To reduce ecological impacts on the scale required, it recommends ambitious government action to drive systemic change, including 'choice editing' and providing universal basic services.

Policymakers can do much more with regulation, taxes and incentives to 'edit out' harmful consumption options that are not compatible with 1.5°C lifestyles. In the realm of transport, for example, this could include phasing out private jets, mega-yachts, fossil-fuelled cars, short flights and frequent-flyer rewards. At the same time, policymakers must, of course, 'edit in' far better alternatives – from excellent rail networks and electric-car-sharing schemes to dedicated bike and bus lanes – so that the sustainable choice becomes the everyday easy option that is accessible and affordable to all. Such 'choice editing' has long been practised for the sake of health and safety of workers and customers – now it must also be practised for the sake of the health of the planet.

In the realm of transport, this has already begun in some consumption-intensive cities and countries. In 2019 Amsterdam pledged that it would ban fossil-fuelled boats from 2025 and fossil-fuelled motorbikes and cars from 2030. In 2021 the Welsh government announced a freeze on all new road projects, redirecting funding to public transport instead, while the government of France banned short-haul domestic flights for journeys that could be made in under two and a half hours, promoting train travel in their place.

Amsterdam is also leading in editing out the throw-away economy, committing to be 50 per cent circular in the city's use of materials by 2030 and fully circular by 2050 – starting now, with construction, food and textiles. Such policies send a long, loud, legal message to companies: if you want to stay in business here, get circular. The policy has already spurred local innovation, including clothing companies now repairing, reusing and upcycling fabrics. Meanwhile, from Grenoble and Geneva to São Paolo and Chennai, city governments are banning the 'visual pollution' of advertising billboards, literally editing out-of-sight the lure of the advertiser's message.

Eliminating excessive consumption is essential, but so too is ensuring a foundational level of consumption for all. This recognition has led to growing support for universal basic services to ensure the essentials of life – health care, education, housing, nutrition, digital access and transport – for all. In Vienna, for example, over 60 per cent of people live in social housing that is owned by the city or not-for-profit cooperatives because the local government decided, decades ago, that housing was a human right and so ought to be kept affordable for all – and the rents are just a fraction of those in comparable European cities. Public provision of essential services can be achieved at a far lower cost than privately funded alternatives, but also with a far lower ecological footprint. Health-care spending per person in the United States, for example, is almost double that of many comparable European countries, while the carbon footprint of US health care is over three times as high.

These examples of system-changing policies – expanding sustainable options for all while editing out the excessive options of the few – point towards a societal lifestyle that the writer George Monbiot deftly describes as 'public luxury and private sufficiency'. With ambitious policies focused on regulations, infrastructure and public provisioning, 1.5°C lifestyles can indeed become possible.

Exploring a 1.5°C lifestyle

If we want to escape the legacy of consumerism now, rather than waiting for systemic change, then a good place to start may be to examine where we think our own excesses begin. 'Wherever and whenever we are excessive in our lives it is the sign of an as yet unknown deprivation,' writes the psychoanalyst Adam Phillips. 'Our excesses are the best clue we have to our own poverty, and our best way of concealing it from ourselves.' When it comes to consumerism, perhaps the poverty that we aim to conceal lies in our neglected relationships with each other and the rest of the living world. The psychotherapist Sue Gerhardt would certainly agree. 'Although we have relative material abundance, we do not in fact have emotional abundance,' she writes in *The Selfish Society*. 'Many people are deprived of what really matters.'

There are many views on what really matters to us in life – from using our talents and helping others, to standing up for what we believe in. Drawing on a wide array of psychological research, the New Economics Foundation distilled the findings down to five simple acts that are proven to promote well-being: connecting to the people around us, being active in our bodies, taking notice of the living world, learning new skills and giving to others. Take note, Alfred Marshall: people desire so much more than the possession of yet more things – and it turns out that our personal and collective well-being depend upon it.

If connecting to those around us is a key source of well-being, then the momentum created by community-led action makes a great deal of sense. Since 2005 the Transition Network has been connecting and mobilizing community groups that are growing more food locally, installing solar panels on community-run buildings and their own homes, insulating their houses, travelling lighter, and inspiring each other to keep on imagining possible new ways of speeding the transformation that's needed. What started in Totnes in the UK is now a growing network of over a thousand groups worldwide, proving the power of locally led action.

For anyone who is curious to try moving towards a 1.5°C lifestyle, the grassroots citizens movement Take The Jump offers six principles for getting there:

- **End clutter:** keep electronic products for at least seven years.
- **Holiday local:** take short-haul flights only once in three years.
- **Eat green:** adopt a plant-based diet and leave no waste.
- **Dress retro:** buy at most three new items of clothing a year.
- **Travel fresh:** don't make use of private cars, if possible.
- **Change the system:** act to nudge and shift the wider system.

Making changes like these can at first seem intimidating, out of reach or socially impossible – and that's little surprise, since consumerist propaganda has spent over a hundred years convincing whole societies not to be satisfied with a lifestyle of sufficiency. So takethejump.org simply invites people to join its growing community and try these shifts out for a month or more, supporting and inspiring them along the way.

From personal experience, it's been surprisingly positive. In my family, the biggest jump was to give up the convenience of having a car. But we quickly realized that better choices had already been edited into our neighbourhood, with a car-sharing club up and running in the surrounding streets. So we jumped – and we haven't looked back. Possessing less and sharing more can turn out to be liberating. It simply feels good. What I learned in the process, in a very personal way, is that change is often hardest just before you make it. We too easily focus on what we think we are losing, finding it so much harder to imagine what we might gain.

Perhaps this will turn out to be true at the level of society too. Changing the systems that shape our lifestyles may seem so much harder before we do it. But within a decade we might just look back and wonder why we resisted so hard, doubted so much and took so long to adopt the lifestyles that will literally enable all of us to thrive. /

Change is often hardest just before you make it. We too easily focus on what we think we are losing, finding it so much harder to imagine what we might gain.

Overcoming Climate Apathy

Per Espen Stoknes

On reading the latest IPCC reports, it's hard not to think: 'This is bad. It's time for people to wake up. We've got to sound the alarm.' It's not that climate scientists are alarmists. They're generally a cautious bunch. It's that climate science itself is alarming – both for humans and for all living things – so it's natural to respond by declaring an emergency.

I did as much after the first IPCC report, launched in the early 1990s, but I noticed that my anxiety was far from shared by my friends and colleagues – to put it mildly. By the 2000s, I'd grown curious. Why *weren't* people waking up, even as the science got clearer, more confident and more alarming? In December 2009, I travelled to the UN climate summit in Copenhagen and joined what was then the world's largest climate rally. Around 100,000 of us marched through the chilly streets towards the conference centre. 'The time for action is now!' we shouted at the top of our voices. In vain. The talks collapsed. There was no agreement – again. And so my questioning turned to the psychology of effective climate action. Were we – those of us declaring a crisis – achieving what we wanted to achieve? It was evident that sufficient action was still not being taken, so maybe something other than shouting was required. What could that be?

I spent seven years searching for an answer, poring over experiments, books and peer-reviewed papers and the ideas of philosophers and focus groups. I found that, when it comes to climate change, humans tend to put up mental barriers that prevent us from engaging. I summarized these as the 'Five Ds' of psychological defence: Distancing, Doom, Dissonance, Denial and iDentity.

Psychological *Distancing* means that the human brain tends to see climate change as something abstract, invisible, slow-moving, and far away in terms of both space and time. This minimizes our sense of the risk we face. *Doom* refers to the way we frame climate change as a looming disaster that threatens great loss and sacrifice. This framing induces fear and guilt which, after a while, leads to habituation and avoidance of the issue.

Meanwhile, cognitive *Dissonance* between what we do (drive cars, eat beef, fly) and what we know (that carbon emissions wreak havoc on Earth's climate) tempts us to justify ourselves rather than actually change our behaviours. Then, there's *Denial*, which isn't just about rejecting climate science: it's to do with the way we suppress our daily awareness of it, so we can go on living as if we hadn't heard the inconvenient facts.

Finally, the *iDentity* barrier refers to the way that climate policies, by calling for changes of lifestyle, more government and higher taxes, may threaten one's sense of self, freedom and values. If I feel personally attacked by climate activists, I'll push back against them. Together, the five Ds explain why people aren't taking action despite having been exposed repeatedly to the facts. These five shortcomings of the human brain explain why it's so hard for us to go from climate alarm to climate action.

Happily, the five keys for more brain-friendly communications are also clear: make climate action more Social, Simple, Supportive, with Stories and Signals. We can make it feel personal and urgent by centring it around friends and community – our *social* peers. We can make it *simpler* to make climate-friendly choices in our everyday lives by using 'nudging' techniques – for example making a plant-based meal the 'dish of the day' in school canteens. We can apply *supportive* framings to climate action, viewing it as an opportunity to improve our health and prosperity. Also, rather than envisioning endless doom, we can develop better, more vivid *stories* about where we want to go. Finally, in order to keep up our motivation, we need feedback in the sense of frequent and tailored *signals* that tell us whether we are making real social progress on renewables, diets and green jobs, not just planetary data on temperatures or gigatonnes.

The Norwegian ecophilosopher Arne Naess was once asked, 'What kind of environmental activism is best? Should we confront or collaborate with industry in order to achieve systems change?' He answered, 'People are needed along a vast front,' and 'Let everyone perceive *their* particular work as crucial.' What he meant was that a variety of approaches are required to achieve progress.

We need Fridays For Future and school strikes. We need Extinction Rebellion, the Citizens' Climate Lobby, Concerned Scientists United, 350.org and Conservatives for Climate. We need scientists, economists, sociologists and engineers. We need people in finance and administration, particularly those with international networks, to help us all invest in tomorrow's economy. We also need designers, electricians, architects and windmill maintenance crews. We need ecologists, regenerative farmers and gourmet vegan cooks; and we need musicians, sculptors, influencers, artists and fashionistas. When a majority have joined, the politicians will follow (because it will mean they can gain votes, not lose them, by acting ambitiously on climate).

To complain that 'nobody is taking the crisis seriously' or that 'nothing is happening' does little to accelerate systems change. Besides, as a global G20 survey showed, three quarters of people *are* deeply concerned. And in fact, people everywhere are already responding to the challenge. All around, men and women are stepping up. Most media regularly ignore them – as well as the climate crisis itself. But we should all speak more about those heroes and heroines, small and large, who are already taking the lead. Where can we find them? Try drawdown.org, goexplorer.org, wedonthavetime.org, or iclimatechange.org, to mention just four.

Of course, there are good reasons to feel fear, grief and anger. When climate disruption brings those feelings into our hearts we should honour them. We should share and listen without judgement or impatience. Once this has been done, there often comes a shift in the soul. Arne Naess again: 'By interacting with extreme misery, one gains cheerfulness.' By acknowledging the emotions inside us, we eventually find the energy to act again. And there are good reasons, too, to feel deep joy, enthusiasm and gratitude. We're still here – along with the trees, the bees and the other beauties of the living world. Every other second, we breathe in the vibrant, living air. Notice the wind. Notice the life in your breath. Sustenance. /

To complain that 'nobody is taking the crisis seriously' or that 'nothing is happening' does little to accelerate systems change.

5.5

Changing Our Diets

Gidon Eshel

I'm writing these words on a balmy November day in New England, surrounded by a confused forest whose trees are shedding their coloured leaves in a spell of hot, muggy weather.

The Glasgow COP26 notwithstanding, by next fall Earth's atmospheric CO_2 concentration will be 2–3 parts per million (ppm) higher, warming Earth's surface by about 0.01–0.04°C on average. By that time, just under a billion kilograms of nitrogen will have flowed down the Mississippi River into the Gulf of Mexico, predictably stimulating a summer algal bloom along the US Gulf coast which will have robbed seawater of dissolved oxygen, decimating shrimp, oyster and fish populations. Since most of this excess nitrogen originates as farm run-off, fuelled by unutilized fertilizer, this process pits Midwestern commodity farmers against Louisiana fishers, declaring the latter the loser.

Because modern farming relies on regularly mechanically and chemically disturbing the topsoil, croplands lose soil two to five times faster than they would naturally. By next fall, Earth's roughly 1.9 billion hectares of cropland will have thus lost 10–20 trillion kilograms of topsoil, exacerbating an already grave global food security threat.

By next fall, at least several animal species, and quite possibly tens or more, will have bid their final adieu and vacated the world stage forever. Some of these losses will be natural, some will be driven by climate change, but many others will arise from water pollution or shortages, among myriad other environmental stressors dominated by our choices over how to use our resources.

For the environmentally aware and informed, remaining optimistic in the face of these discouraging environmental trends isn't easy. Yet there is a silver lining. Apart from the (forlorn remaining) elephant in the room, anthropogenic climate change, the above challenges are primarily driven by agriculture, and the way we farm dominates how dire their impacts will be. Climate change is thus qualitatively different from the other environmental challenges, because nearly *all* aspects of modern life result in greenhouse gas emissions yet none outright *dominates* emissions, which means that addressing climate change requires a wholesale societal reorganization. Conversely, when it comes to extinctions, topsoil losses that jeopardize food supplies, water pollution by eutrophication (excessive algal growth due to elevated nutrient availability in the water provided by unused fertilizer in farm run-off), and overconsumption of scarce fresh-water resources, food production dominates completely.

This offers a tantalizing possibility. For the solid majority of today's mostly urban population, agriculture means only one thing: food. To significantly ameliorate the series of interconnected major environmental challenges enumerated above, therefore, you need only tweak one thing: diet. Granted, wilful diet modifications are notoriously difficult for individuals, as the world's burgeoning ranks of perennial dieters know painfully well. But because individual diets partly reflect governmental policies that favour the production of some foods over others, and the way food is priced, marketed and taxed, tweaking national- to global-scale diets is vastly more simple to achieve than controlling one's waistline, let alone effectively solving climate change.

So how should we modify our diets, and what positive outcomes can we reasonably expect of those modifications? The most impactful dietary change, unquestionably and peerlessly, is eliminating or drastically reducing our consumption of the most resource-intensive food item: beef.

To illustrate the impact, let's imagine eating a burger as our yard-stick, and then consider alternatives. Producing the roughly 10 grams of protein in a burger results in the emission of 2–10 kilograms of CO_2eq (CO_2 equivalent), and requires the use of 5–35 square metres of total land. The lower numbers in these ranges characterize beef produced from dairy herds or highly intensive beef operations in which steers graze minimally and reach market weight far more quickly, while the higher ones represent beef from cattle reared on extensive grass-based ranches. These sprawling ranches also undermine biodiversity the most, because they disproportionately occupy vast tracts of land in the relatively wilder regions where remaining biodiversity is more likely to be found. Producing those 10 grams of beef protein also requires about 100–600 litres of water for irrigation, and about 40–80 grams of nitrogen fertilizer.

Now, suppose the burger eater wished to reconsider their culinary choice; if these resources were reallocated, what would they yield instead?

Fig. 1a shows that the cropland currently being used to meet the beef protein requirements of one person can deliver the protein replacement needs of four to twenty-eight people (depending on the plant). Addressing the environmental consequences of cropland reallocation,

Fig. 1b shows that greenhouse gas emissions and nitrogenous fertilizer needs for the production of these plant alternatives are merely 2–12 per cent of those required for producing beef.

The graphs have an additional, more subtle message. Water is clearly the limiting resource, with some plant alternatives needing almost as much water as beef, and five out of the twelve plant alternatives requiring even more water per gram of protein than beef. But water needs are fairly easily modified by exploiting geographical variations in climactic conditions. Oats, for example – which are not that distinct from wheat – are often irrigated in the US because they are primarily grown in the fairly dry northern Great Plains. In contrast, a lot of winter wheat is rain-fed. By relocating these crops to suitable rainier locations, such as western New York or Pennsylvania, irrigation needs can be reduced significantly. Further environmental progress can thus be made by redesigning the food system in addition to replacing beef with more resource-efficient plant alternatives, but this initial substitution is the key, because it dramatically enhances protein supplies at far smaller environmental costs.

Replacing beef in one's diet with plant-based foods can significantly reduce our land needs and our use of other resources. Taken with the expected 35 per cent reduction in fresh-water and coastal ocean pollution, this substitution would thoroughly reshape the rural landscapes of developed, wealthy nations, greatly enhancing their biodiversity and environmental integrity. This dietary transition also confers significant nutritional benefits, and would substantially reduce the risk of several currently ubiquitous degenerative diseases, notably cardiovascular conditions and stroke, as well as several cancers. Logistically speaking, replacing beef with plant-based alternatives could be quite easily achieved at the national level within a short timeframe. For some individuals, this might clash with cultural or culinary preferences. But short of such radical options as completely swearing off air travel or car driving, or foregoing all electronics, there are precious few actions that each of us could adopt on our own that would rival shunning beef in expected impact. Replacing US beef with diverse, rigorously nutritious plant-based diets which deliver exactly the same protein mass results in an emission reduction of about 350 million tonnes CO_2eq per year nationwide. As a yardstick, these savings are over 90 per cent of the full emissions of the entire US residential sector. Take it in: replacing beef with plant alternatives would not only vastly improve our health but it would also reduce greenhouse gas emissions by nearly the same amount as is used today in all our energy-intensive dwellings. /

Nutritional and environmental consequences of reallocating high-quality cropland used for beef production

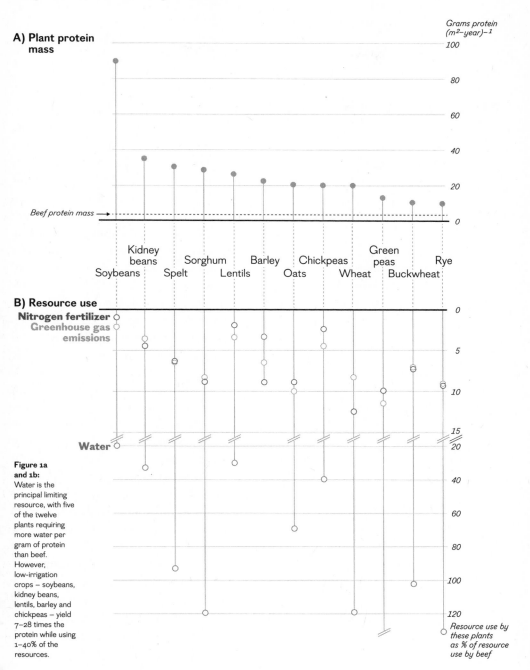

A) Plant protein mass

Grams protein (m²-year)⁻¹

Beef protein mass →

Soybeans | Kidney beans | Spelt | Sorghum | Lentils | Barley | Oats | Chickpeas | Wheat | Green peas | Buckwheat | Rye

B) Resource use

Nitrogen fertilizer
Greenhouse gas emissions

Water

Figure 1a and 1b:
Water is the principal limiting resource, with five of the twelve plants requiring more water per gram of protein than beef. However, low-irrigation crops – soybeans, kidney beans, lentils, barley and chickpeas – yield 7–28 times the protein while using 1–40% of the resources.

Resource use by these plants as % of resource use by beef

Remembering the Ocean

Ayana Elizabeth Johnson

I love the ocean. Maybe you do too. It's easy to love – octopuses, kelp forests, nudibranchs, waves and pufferfish exist! It's also often taken for granted. And the ocean plays a major, under-appreciated role in regulating our global climate.

To wit, the ocean has absorbed about 30 per cent of the carbon dioxide emitted by burning fossil fuels. That has changed the pH of seawater – it has become 30 per cent more acidic since the Industrial Revolution. Ninety-three per cent of the excess heat trapped by greenhouse gases has been soaked up by the ocean – if it weren't for that, the planet would be 36°C hotter. (Also, there are now heatwaves *in the ocean.*) As a result, the surface of the ocean has warmed by 0.88°C since 1900. That additional heat means more evaporation, which in turn is fuelling stronger and wetter storms. This warmer ocean (plus melting ice) changes seawater density and salinity, which alters ocean currents. For example, the Atlantic Meridional Overturning Circulation (AMOC) – the currents that drive the Gulf Stream and keep Europe from freezing – has slowed by approximately 15 per cent since 1950.

And yet, somehow, the ocean is often left out of conversations about climate. I'm frequently raising my hand in discussions about the climate crisis to say, 'Hey, don't forget about the ocean!'

So, credit where it's due, the ocean has been doing us an enormous favour, buffering the impacts of greenhouse gas pollution. (Thank you, ocean.) And here's the good news: the ocean can do us an even bigger favour, because there are plenty of salty climate solutions.

First, though, a reality check: ocean ecosystems and biodiversity are getting pummelled, including by climate change. One third to a half of coastal ecosystems have been lost. Biodiversity is declining at a faster rate than at any other time in human history – around 33 per cent of reef-forming corals, sharks and marine mammals are threatened with extinction. And this biodiversity is essential for human well-being. Approximately 3 billion people depend on ocean ecosystems for their food security, economies and cultures. Ocean changes affect us all, but not equally: low-income communities and communities of colour are the most impacted.

A hotter ocean is causing fish to flee towards the poles and corals to fry in place, upturning food webs and fisheries. Coral bleaching – which happens when seawater is too warm for too long and corals expel the colourful (normally symbiotic) photosynthetic algae that live in their tissues – has become five times more frequent in the last four decades. With 2°C of global warming (which we are currently on track to exceed before 2100), 99 per cent of coral reefs may be lost. As waters warm, metabolically, fish require more oxygen – but warmer waters hold *less* oxygen. Meanwhile, phytoplankton produces over half the oxygen we breathe, but that production is declining by about 1 per cent a year due to climate change.

Beyond warming, we have changed the very chemistry of the entire, enormous ocean by burning fossil fuels. As the ocean becomes increasingly acidic, animals such as oysters (a sustainable seafood) are having a harder time building their shells and making babies. Perhaps more surprisingly, because fish smell via seawater, when the pH changes they can lose their ability to find prey, escape predators or even find their way home.

Given the above, and with nearly 94 per cent of global fish stocks maximally exploited or overfished, we can no longer rely on wild fish to feed the world. At the same time, industrial aquaculture has been largely unsustainable, often focused on carnivorous fish that require a lot of feed, feed which to date is often smaller, wild fish. Globally, industrial seafood is often a disaster both for ecosystems (trawling destroys sea-floor habitats and aquaculture destroys mangroves, both resulting in carbon emissions) and for human rights (dangerous conditions, poverty wages or even slavery), while also using tons of fossil fuel. Fishing emits over 200 million tons of CO_2 annually, as more boats chase fewer fish. Much of this overfishing is fuelled by an annual $20 billion in subsidies, which the United Nations says must be eliminated. I agree.

And yet! Despite these threats – plus all the pollution we throw at it – the ocean is not merely a victim, it's also a hero. We need to reframe the narrative, value the ways in which the ocean is buffering global climate change and learn to turn to the ocean as a key source of climate *solutions*.

Renewable energy

Imagine if homes and businesses along our coasts were powered by the ocean. This doesn't have to remain a dream. Offshore, the wind blows more strongly and consistently than it does over land – a reliable source of energy near population centres. By 2030, the offshore wind industry could generate over 200 gigawatts of electricity globally. There's also burgeoning technology for harnessing the energy of waves and currents, and for floating solar panels.

Regenerative ocean farming

Similar to regenerative agriculture on land, which aims to rebuild soil organic matter, absorb carbon and promote biodiversity, we can farm the ocean back to health. Specifically, farming seaweed and shellfish (oysters, mussels, clams, scallops) is highly sustainable since these organisms live simply off sunlight and nutrients already in seawater – no fertilizer, fresh water, or feed required. These are some of the lowest-carbon sources of food around. Farming seaweed can provide a range of benefits, from helping to reduce local ocean acidification, to harbouring biodiversity, to buffering against the impacts of storms on coastlines, and it has the potential to grow into an industry supporting tens of millions of jobs.

Blue carbon

We hear so much talk about planting trees – billions of them – with little acknowledgement that about 50 per cent of global photosynthesis happens in the ocean. That land-centric myopia misses the carbon drawdown potential of wetlands and seagrasses and coral reefs, of kelp forests and mangrove forests. Per hectare, marine ecosystems can hold up to five times more carbon than a terrestrial forest. Seaweed is particularly promising – seaweed that naturally sinks to the deep sea sequesters around 200 million tons of carbon per year globally; farming, then deliberately sinking, kelp is a promising carbon dioxide removal opportunity. But instead, what we currently have is up to 1 billion tons (1 gigaton) of carbon dioxide being released annually from degraded and destroyed coastal ecosystems – not to mention the release of methane.

Shoreline protection

Protecting and restoring coastal ecosystems is key not only for carbon sequestration but also to protect coastal communities. Loss of coastal ecosystems is putting up to 300 million people at increased risk from floods and storms. Coastal ecosystems serve as the first line of defence against storm surges and rising sea levels. In many cases, they offer cheaper and more effective shoreline protection than sea walls.

Marine protected areas

Scientists recommend we protect *at least* 30 per cent of nature, and fast – by 2030. Protected areas are a triple win because they can safeguard biodiversity and ecosystems, help boost fisheries and sequester blue carbon. Climate impacts are clobbering marine ecosystems and, as they degrade, that releases even more greenhouse gas, contributing to a vicious cycle. The value of protecting them is clear, but right now only 2.8 per cent of the ocean is highly protected. Let's give nature some space to regenerate, shall we?

In sum, marine species and entire marine ecosystems are in peril. But their demise can be constrained, if not halted. Ocean-based climate solutions have the potential to provide an estimated 21 per cent of the reductions in greenhouse gas emissions necessary to limit global temperature rise to 1.5°C – and staying below that threshold is the most important thing we can do for the ocean, marine life, coastal communities and all oxygen-breathing beings. Every bit of habitat we preserve, every tenth of a degree of warming we prevent, really does matter. /

We need to reframe the narrative, and learn to turn to the ocean as a key source of climate *solutions*.

Rewilding

George Monbiot and
Rebecca Wrigley

How do we sustain ourselves in a broken world? How do we prevent ourselves from succumbing to despair, when so much of what we love is disappearing before our eyes, and when the prospect of systemic environmental collapse threatens every hope and ambition we might have entertained? How can we look our children in the eye when we know they might live to witness the toppling of our life support systems?

These are questions that almost everyone seeking to protect life on Earth now confronts. Not only must we contend with the enormous political, economic and technical challenge of seeking to prevent this existential disaster, we must simultaneously navigate the psychological impacts of understanding what we face. Somehow, we must keep finding the energy, the determination, the joy required to carry on. But how?

We need, even when facing the most frightening aspects of this multi-faceted crisis, to sustain in our minds the prospect not just of preventing catastrophe but also of creating a better world. Perhaps our best hope of psychic survival and our best hope of planetary survival can be found in the same place: by seeking the mass restoration of damaged ecosystems, and of our relationship with them.

Anyone who has taken a party of children to the countryside or the seaside for the first time in their lives will testify to something wonderful: a thrilling and spontaneous engagement with these unfamiliar places. Children who have never entered a forest before, or never stepped on to a rocky shore, immediately and instinctively begin to explore them: curiosity and wonder overtake them. They appear to possess an innate desire to engage with the living world.

Almost all of us have a great capacity for delight and enchantment. But most of us live, most of the time, in circumstances in which we can scarcely exercise them. As we disengage from the natural world, we tend to forget the joy that nature has to offer: its spontaneity and serendipity, its capacity to shake us out of our frustrations and humiliations. Unfortunately, even when we do step into what we call 'nature', we often find ourselves in places as disciplined, managed and dismal as the daily grind we might be trying to escape. It is hard to have magnificent experiences in nature, to leave ourselves and our troubles behind, if scarcely any of it is left.

But there is a way that we can begin to mend the living planet and our relationship with it. It is a variety of positive environmentalism which offers the hope of recovery, of re-enchantment with a world that often seems crushingly bleak. It is 'rewilding': the mass restoration of the planet's ecosystems. In essence, rewilding means allowing natural processes to resume. It involves, where people agree, reintroducing missing species, removing fences, blocking drainage ditches and controlling especially virulent invasive exotic species, but otherwise, to the greatest extent possible, allowing nature to find its own way. It means allowing forests and other depleted ecosystems to regenerate. At sea, it means creating meaningful reserves from which extractive industries, especially trawling and dredging, are excluded. Because marine animals tend to be highly mobile during at least one stage of their lives, ocean ecosystems, if left alone, can quickly restore themselves.

To understand what we could restore, we need to see what we are missing. Some countries, such as the UK, have lost almost all the large 'keystone' species – the ecological engineers – that create habitats and drive the dynamic processes which other lifeforms need to flourish. Once, like almost everywhere on Earth, ecosystems here were dominated by enormous beasts: elephants, rhinos, hippos, lions and hyenas. But we have lost not only our megafauna but also most of the mid-sized creatures that used to be abundant here, such as wolves, lynx, moose, boar, beavers, white-tailed eagles, pelicans, cranes and storks. Some of these species are now, slowly and tentatively, being reintroduced, and while their restoration is sometimes controversial, many people respond to them with delight and awe. We have begun to see how simplified, depleted ecosystems can spring back to life when the ecological engineers return.

It's easy to forget that even the emptiest seas once swarmed with living creatures. The waters around the UK were among the most abundant on Earth. Armies of bluefin tuna stormed our coasts, harrying shoals of mackerel and herring many miles long. Halibut the size of barn doors and turbot like tabletops came into shallow water to feed. Cod commonly reached a length of almost 2 metres; haddock grew to a metre. Pods of fin whales and sperm whales could be seen from the shore, while Atlantic grey whales, now extinct, sifted the mud in our estuaries. Gigantic sturgeon poured up the rivers to spawn, pushing through packed shoals of salmon, sea trout, lampreys and shad. On some parts of the seabed the eggs of the herring lay a metre and a half deep.

Almost everywhere on Earth, living systems were so rich and abundant that if we encountered them today we would scarcely believe what we were seeing. A recent scientific paper estimates that only 3 per cent of the Earth's land surface should now be considered 'ecologically intact'. The disappearance of so many of our natural wonders diminishes not only ecosystems but also our own lives. We live in a shadowland, a dim, flattened relic of what there once was, of what there could be again.

As living systems recover, some of them, particularly forests, peat bogs, salt marshes, mangroves and the seabed, could draw down vast amounts of carbon from the atmosphere. While such natural climate solutions should never be used as a substitute for decarbonizing our economies, we now know that a green industrial and economic transition is not enough: even if we cut our emissions almost to zero very quickly, we are likely still to exceed the temperature limits proposed in the Paris Agreement. So we also need to recapture some of the carbon we have already released. The restoration of living systems is a surer, cheaper and less damaging means of doing so than any of the technological alternatives. It enables us to tackle two of our existential crises at once: climate breakdown and ecological breakdown.

The recovery of certain animal populations could radically change the carbon balance. For example, forest elephants and rhinos in Africa and Asia and tapirs in Brazil are natural foresters, maintaining and extending their habitats as they swallow the seeds of trees and spread them, sometimes across many miles, in their dung. If wolves were allowed to reach their natural populations in North America, one paper suggests, their suppression of herbivore populations would store as much carbon every year as between 30 and 70 million cars produce. Healthy populations of predatory crabs and fish protect the carbon in salt marshes, as they prevent herbivorous crabs and snails from wiping out the plants that hold the marshes together. Protecting and rewilding the world's living systems is not just a delightful thing to do. It is an essential survival strategy.

It's important to remember that rewilding is not a replacement for the conservation of existing, rich habitats but a supplement to it. There's no substitute for old-growth forests, long-established reefs of coral, oysters or honeycomb worms; braiding, meandering rivers full of snags and islands; or undisturbed soils reamed by roots and holes. 'Replacing' an old tree is no more meaningful than replacing an Old Master painting. When a trawler ploughs through biological structures on the seabed, they can take hundreds of years fully to recover. When a river is dredged and straightened, it becomes, by comparison to what it once was, an empty shell. The loss of these ancient habitats is one of the forces driving the global shift from

Next pages:
One of the oldest
living organisms on
Earth: a meadow
of Neptune
seagrass
(*Posidonia
oceanica*) in the
Mediterranean
Sea near Ibiza.

large, slow-growing creatures to the small, short-lived species able to survive our onslaughts.

Rewilding seeks to allow our complex natural architectures to recover. It attempts to build a new and deeper respect for the entanglements of nature. It seeks to create the ancient ecosystems that only our grandchildren will see. It does not try to restore the living world to any prior state, but simply to allow it to become as rich, diverse, dynamic and functional as possible.

But it is also about us and the improvement of our lives. It's about people coming together to find ways to live and work within healthy, flourishing ecosystems. Local communities need to be at the heart of any decisions about land- and marine-use change. Nothing should be done without the involvement and consent of Indigenous peoples and other local people. By using a localized people-led approach we can help to create economies that are regenerative and restorative by design, which support human prosperity within nature's flourishing web of life.

To do so, we must start working with nature instead of against it. We would like to see governments, public bodies, businesses, farmers, foresters, fishers and local communities coming together to develop collaborative place-based visions for the ecological restoration of our land and seas, which catalyse the economic restoration of communities. We believe that a new and thriving ecosystem of employment can be built around the healing and rewilding of nature. For example, recent analysis by Rewilding Britain reveals that, across England, rewilding projects have resulted in a 54 per cent increase in full-time-equivalent jobs. Not only has the number of jobs increased, so has their diversity. Rewilding can enrich lives and help us to reconnect with wild nature while providing a sustainable future for local communities.

Rewilding enables us to begin to heal some of the great damage we have inflicted on the living world and, with it, the wounds we have inflicted on ourselves. And this could be our best defence against despair. We can replace our silent spring with a raucous summer. /

We can replace our
silent spring with
a raucous summer.

5.8

We now have to do the seemingly impossible

Greta Thunberg

The fact that our societies are in many ways governed by social norms is a great source of hope, because social norms can be changed. Real change creates real hope, and real hope creates real change. It is a positive feedback loop. But it does not appear out of nowhere. Societal changes are the results of our collective efforts and actions. So instead of asking others if there is still hope, ask yourself, are you prepared to change? Are you prepared to step outside your comfort zone and become part of a movement that will bring about the necessary systemic transformations? Sure, it might feel a bit uncomfortable at first. But then again, the future of our entire civilization is on the line, so it just might be worth it. Instead of looking for hope, we have to go out and create that hope ourselves.

When I sat down outside the Swedish Parliament in August 2018 I was suffering from selective mutism and I could not eat in the company of strangers. It was difficult in the beginning to do around ten interviews a day, five days a week. Sometimes, when young people came up to me, I had to hide and cry because I was so scared of other children. I had been so mistreated that I just naturally assumed that all children were mean. But it was more than worth the effort. I saw that people were listening, despite the fact that all I had to offer them were facts and moral imperatives – or guilt, if you like. I did not have any knowledge about communication tactics. Later, the Norwegian psychologist Per Espen Stoknes told me that, according to psychological research and behavioural studies, I – and the Fridays For Future movement – did everything wrong. But one year later, in the week around the UN climate summit in New York, over 7.5 million people in more than 180 countries flooded the streets of the world, demanding climate justice. 'It wasn't supposed to work,' Stoknes said to me with a smile, 'but it worked.'

The school strike movement is based on climate justice. We seek to shine a light on the intergenerational impacts of climate change and the need for equity for the most affected people in the most affected areas. There is nothing new about this. It is one of the main pillars of the Paris Agreement. All the words that we say have been spoken by others. All our speeches, books and articles follow in the footsteps of those who pioneered the climate and environmental movement. It would be easy to assume that all those before us have failed and that we now are failing too. After all, our emissions are still rising and the necessary action and commitment are still nowhere in sight. But that is not true. We are creating change. Massive change. We are winning. We are just not winning fast enough. We are not a political organization, we are a grassroots movement dedicated to spreading awareness and information. We are not interested in compromises or deals. We have nothing to offer. We tell it like it is.

For this, we are receiving unimaginable amounts of hate and threats. We are being mocked, bullied and ridiculed. And for pointing out that our political leaders have spent thirty years debating this while our emission levels have done nothing but keep on rising, there are elected officials that have called us *a threat to democracy*. Perhaps this level of political desperation should not come as a surprise, since over one third of our anthropogenic CO_2 emissions have occurred since 2005. There are leaders who have been in charge of some of our major emitting nations for large periods of that time. Imagine how their historic responsibilities will be viewed in the future.

Many people say that the actions required for us to stay below 1.5°C or even 2°C of warming are politically impossible today, and I agree. But, as Erica Chenoweth writes, changing what is considered politically possible can most definitely be done. As a matter of fact, it happens all the time. At the beginning of the Covid-19 pandemic we saw it everywhere, often daily. Who created this shift in thinking? The media did. And they did it simply by objectively telling the reality as it was. It turns out they did not have to offer any *inspiration* for people to change their behaviour – contrary to what all the communication experts have been saying for years and years. Nor was it hopeful stories about ninety-five-year-olds who miraculously survived the disease that got us moving. The media just told us the facts, and we reacted to them. We did not become paralysed. We did not give in to apathy. We simply reacted to the information and changed our norms and our behaviour – as you do in a crisis. And we did not do it because we saw financial opportunities. We did not do it to *create new jobs* in the health sector or to benefit the face-mask-manufacturing industry. We changed because others did.

We changed because we got scared, because we became afraid of losing our loved ones, our friends and our livelihoods.

As I am finishing the final edits to this book, Russia has commenced an unprovoked invasion of Ukraine. This horrific violation of all international law has brought about a growing demand for the EU to completely stop all imports of oil and gas from Russia, despite the fact that this action would most likely initiate an unparalleled European energy crisis. This act would severely defund Putin's fascist war machine, yet it was completely unthinkable even a few days ago.

We know what it means to treat something like a crisis, and we know – way beyond any doubt – that the climate crisis has never even once in any way been treated as such. This is the heart of the problem, and it is not the fault of the oil companies. It's not the fault of the logging industries, the airlines, the car manufacturers, the manufacturers of fast fashion or the meat and dairy producers either. They are very much guilty, but their purpose is unfortunately to make money, not to inform citizens about the state of the biosphere or to safeguard democracy.

Our inability to stop the climate and ecological crisis is a result of an ongoing failure in the media, as George Monbiot points out. A crisis of information not getting through – because that information has not been told, packaged or delivered as it should be. And way more importantly, it has been drowned out by other stories. During the week of COP26 in Glasgow, the environmental media coverage was at its peak. But it still struggled to compete with the airtime dedicated to Britney Spears as she regained control over her life. This is one of countless examples of how we are constantly being indirectly told that *we are fine*. After all, if a newspaper dedicates most of its space to sport, celebrities, diets and crime, then surely all that talk of an existential crisis must be blown way out of proportion? And the credibility of all those scientists might not be considered very great if they say all those things about *extinction* and a *code red for humanity* and still get bumped from the front page by Kim Kardashian or Manchester United.

Melting glaciers, wildfires, droughts, deadly heatwaves, floods, hurricanes, loss of biodiversity – these are all starting to make headlines on the front pages and in the evening news. But this is still not reporting about the climate crisis. This is reporting on the *symptoms* of a much larger problem. These stories alone will not explain the challenges we face. To communicate the crisis, you first of all need to convey the fact that the clock is ticking. The climate crisis is about time. If you leave out the aspect of time, then it is just one topic among other topics. If you take away the countdown, then a collapsing glacier, a forest fire or a record heatwave is nothing more than three independent news events – a series of isolated natural disasters. If you fail to include the aspect of time, the climate crisis is not a crisis. Then it is just another story that can be dealt with down the road – in 2030 or 2050; who really cares? Remove the countdown and you lose sight of all the most important details, for example that it might not really matter if we develop technological solutions in the decades to come if we fail to take the necessary action here and now. And that we do not primarily need climate targets for 2030 or 2050. We need them now, for 2022, and for every month and year that follows.

If the media is going to tell the truth about our situation, it must also start to focus on climate justice. The people who are on the front lines of the climate emergency belong on the front pages, as Ugandan climate activist Vanessa Nakate has said. But that is yet to happen. The most affected people in the most affected areas have been erased from the mainstream western media. Yet they are the ones suffering the consequences of our affluence – a way of life that was built with stolen natural resources and forced labour from low-income nations, as Olúfẹ́mi O. Táíwò writes.

Justice means morality – and morality includes guilt and shame. But guilt and shame have been officially banished from the western climate discourse by the media, by the communication experts and by the entire greenwashing community – conveniently closing the door on our historic responsibilities and the losses and damages caused. This is the social and cultural equivalent to what Saleemul Huq described in Part Three, where he explained that low-income nations are not allowed to talk about loss and damage and that words such as 'liability' and 'compensation' have become taboo in high-level climate talks.

How can we possibly address a crisis that is fundamentally created by injustice and inequalities if we are not allowed to talk about morality, justice, responsibility, shame and guilt? We cannot. Ninety per cent of this cumulative crisis has already been created; it is already up there in the atmosphere, and that has to be accounted for. Therefore we have to fundamentally change our social norms. We have to make it not only politically possible but also socially acceptable to address these issues without most people automatically shutting down and sheltering off into a defensive position. And of course that can be done. Guilt, shame, morality and justice are based on social norms, and social norms can easily be changed.

The Finnish philosopher Elisa Aaltola from the University of Turku has argued that shame can be a highly effective moral and psychological method of persuasion. Guilt is not, in fact, a bad thing in itself. On the contrary, it is a necessary part of a functioning society. We pay our bills and we obey the laws to avoid being guilty of a crime. In a way, our entire society is upheld by our desire to avoid guilt. Guilt may not be a pleasant thing to experience, but once we have recognized our mistake, we have the opportunity to apologize and move on, often with a great sense of relief.

And when it comes to climate guilt, very few of us have anything to fear – unless you happen to be a fossil fuel corporation, an energy company or a leader of a major oil-producing nation. Climate injustice is in no way the fault of ordinary people. The vast majority of us are not even remotely aware of historical emissions or of the wrongdoings of the past. Or even of the very basics of global warming, for that matter . . . Because how could we be? We have never been told, at least not in any official sort of way. And it is hardly the responsibility of ordinary citizens to do the work of governments, international newspapers and major TV networks.

But when something that has previously been considered good and desirable – for example, an extremely high-emitting lifestyle – is suddenly revealed to have disastrous consequences for our common society, then we all have a responsibility to find quick ways of making that lifestyle socially unacceptable, just as social norms and laws prohibit theft and violence. And don't get me wrong, it is not guilt that will save us – it is justice. But we cannot have one without the other.

In order to create all the necessary changes, the concepts of climate justice, historical emissions and the mindsets of dominance and inequality that have laid the foundation of the climate and ecological emergency must be explained in the media, over and over again. There are centuries of wrongdoing to acknowledge and make up for. This may seem like an enormous obstacle, but there is no way around it. We cannot continue to create global 'solutions' for just the richest 10 per cent or the wealthiest nations. It simply will not work. To solve global problems, we need a global perspective. And when it comes to climate justice, democracy knows no borders.

None of this will happen unless the people in power are held accountable. Today, our political leaders are allowed to say one thing and do the exact opposite. They can claim to be climate leaders while they rapidly expand their nations' fossil fuel infrastructure. They can say that we are in a climate emergency as they open up new coal mines, new oil fields and new pipelines. It has not only become socially accepted for our leaders to lie, it is more or less what we expect them to do. It is hard to imagine that exemption being given to any other group in society. But this privilege has to end.

You can say that none of this, realistically, is going to happen, and you are most likely right. But I assure you, achieving all this is far more realistic than the notion that our civilization will be able to endure the stress it will face in a 3°C or even a 2°C warmer world. This late in the day, doing well is unacceptable. In fact, even doing our best is no longer good enough. We now have to do the seemingly impossible. The changes we need are enormous, and we need more time to bring people along, to adapt and develop. But we do not have any more time, so all our solutions from here on have to be holistic, sustainable and taken in full consciousness of the ticking clock. I believe that the main reason we have reached this point – the reason we are facing this catastrophe – is because the media has allowed the people in power to create a gigantic greenwashing machine designed to maintain business as usual for the benefit of short-term economic policies. They have failed to hold those responsible for the destruction of our biosphere accountable, effectively acting as gatekeepers of the status quo.

But – and here is the great news – that great failure can be undone. There is still a way out of this. Science has delivered the data. Grassroots movements and non-governmental organizations have brought those facts into our societies. But in order to turn all this into political action, we need to drastically scale up the process. Given the size of our mission and the time we have left to act, there is, frankly, no entity other than the media that has the opportunity to create the necessary transformation of our global society. In order for that to happen they must start treating the climate, ecological and sustainability crisis like the existential crisis it is. It has to dominate the news.

Our safety as a species is on a collision course with our current system. The longer you pretend this is not the case, and the longer you pretend that we can solve this catastrophe within a global societal structure which has no laws or restrictions whatsoever protecting us long term from the ongoing self-destructive greed that has brought us to the very edge of the precipice, the more time we will waste. Time that we no longer have.

So, dear media, you are among those in the driver's seat. You have the ability to steer us out of danger. Whether you choose to turn that ability and responsibility into a mission is your decision – yours, and yours alone. /

Social norms can easily be changed.

Practical Utopias

Margaret Atwood

Way back in 2001, I began writing a novel called *Oryx and Crake*. I was with some bird biologists, and they had been discussing extinction – the probable future extinction of several of the bird species we'd just been looking at, including the red-necked crake – but also extinction of species in general. Our own species was included. How long had we got? If we were to go extinct, would our extinction be self-inflicted? How doomed were we?

Biologists had been having such conversations since at least the 1950s. My father was a forest entomologist, and he was very interested in our collective stupidities and also our collective prospects. At the dinner table, when I was a teenager, a sort of cheerful gloom prevailed. Yes, things would get worse. Yes, we would probably pollute ourselves to death, if we didn't blow ourselves up with atomic bombs. No, people didn't want to face facts. They never do until the facts are unavoidable. The *Titanic* was unsinkable, until suddenly it wasn't. Pass the mashed potatoes.

And that was before the cod population crashed, before the sea level had measurably risen, before Insectageddon, before we had even begun tracking global warming in any serious way. It was when we still had a big chance to stave off the worst impacts of carbon emissions. Now we have only a little chance, because we missed the other chances. Will we miss this one too?

The premises of *Oryx and Crake* are that we currently have the capability to bioengineer a virus capable of wiping out humanity very swiftly; and that someone might be tempted to do just that, in order to save the entire biosphere and all life within it from destruction at the hands of our species. Think of what the scientist Crake does in the novel as a sort of triage: if humanity goes, the rest of life stays; but if not, then not.

It's a high probability that if nothing is done to stop the climate crisis and the parallel species extinctions now well underway, a Crake will appear among us with a mission to put us out of our misery. In *Oryx and Crake*, we are to be replaced by an upgrade: humans without the fatal flaws and desires that have led us into our present dire predicament. The new humans don't need clothing – thus no polluting fabric industries – and they are grass-eating vegans, so no use for agriculture. They are non-violent, self-healing and without jealousy. But *Oryx and Crake* is fiction. In real life, the production of such a species is not believable, or not in the short run. Yes, we are already gene-editing, but not on the scale envisaged by the design-a-species scheme in *Oryx and Crake*. If the climate crisis proceeds unchecked, we will disappear before we can execute a succession stratagem, because the oceans will die, and the major part of our oxygen supply along with them.

Crake did not believe we would have the will or the desire to reverse our lethal modes of living. We present-day humans would have to be eliminated just to keep the blue dot planet alive. If I could summarize humanity's most necessary mission today, it would be in three words: *Prove Crake wrong.*

But how can we prove Crake wrong? A complex question. I'm not sure how to answer it. If we reverse global CO_2 emissions and dial back global warming – and that's a big if – we will have at least begun. But then there are the other pieces of the predicament – the toxic chemical contamination of almost everything, the ongoing destruction of ecosystems, the social chaos that's unleashed when famine, fires, floods and droughts strike and governments can no longer cope . . . the problems can seem overwhelming. One thing is certain: if people lose hope, there is indeed no hope.

In a small effort to begin with hope, I've participated in a thought experiment. It's called Practical Utopias, and it has taken place on an online inter-active learning platform called Disco. Why bother? I suppose this project was a response to a question I have often been asked: why have you written only dystopias and not any utopias?

My answer used to go something like this. In the mid- to late nine-teenth century, utopias were thick on the ground. Some were literary, such as William Morris's *News from Nowhere*, in which beautiful people did a lot of arts and crafts in lovely natural settings, W. H. Hudson's *A Crystal Age*, which solved poverty and perceived overpopulation by doing away with sex, and Edward Bellamy's *Looking Backward*, which anticipated credit cards and was a huge bestseller. Some were real-life attempts, such as the Oneida community, with shared sex and silverware; the Shakers, with no sex, but wonderfully designed simple furniture; and Brook Farm and Fruitlands, which were high on idealism and low on practical experience with, for instance, farms and fruit.

Then there were visions of futures populated with many new things and technologies – air travel, submarines, rapid transit vehicles of various kinds. So many transformative things had already been invented – steam trains, sewing machines, photography; why shouldn't there be more, and then more? Criticisms of capitalism were usual in these utopias, both literary and real-life: surely this rapacious system, with its boom-and-bust cycles and its extreme worker exploitation, should be replaced by something more egalitarian, with wealth distribution and sharing of labour. Utopias in general have addressed the problems that haunted their own ages, and in the nineteenth century, poverty and overcrowding, widespread disease, industrial and urban pollution, the condition of workers and 'the woman question' were seen as the problems of those times. Every literary utopia I've come across offered solutions to each.

But then came the twentieth century. Literary utopias disappeared. Why was that? Possibly because that century witnessed several nightmares that began as utopian social visions. The USSR came into being through the dreams of the Old Bolsheviks, but then turned into Stalin's dictatorship, which liquidated the Old Bolsheviks, along with millions of others. Hitler's Third Reich gained absolute power through promises to create jobs for all – all 'real' Germans, that is – with the results we know. There are other examples too numerous to list, but one possible result is that literary utopias became implausible, whereas literary dystopias such as George Orwell's *Nineteen Eighty-Four* – drawing much inspiration from real ones – proliferated. Does that mean we should stop trying to improve things? Not at all – if we stop trying to get better, the result will be worse, and we'll end up in a dystopia anyway. But it does mean we should be aware of the pitfalls.

Which brings me back to the question I was repeatedly being asked: why not write a utopia? Give us a bit of hope!

Literary utopias are challenging, as fiction – they tend to read like lesson plans or government reports. Everything's perfect, so where's the conflict? I wasn't inclined to try. But then, why not attempt a real lesson plan, of sorts? Practical ideas that might actually address the pressing problems of our time – as literary utopias had attempted to do in the past.

Along came a new interactive learning platform called Disco. Would I do something with them? they asked. Yes, I said: Practical Utopias. In a nutshell, could we create a society that sequestered more carbon than it produced, while also creating a fairer, more equal society? We would have to consider the most basic elements. What would we eat? Who or what would produce our food? Where would we live? In dwellings built of what? The materials would have to be new ones. What would we wear? Made out of what – since the clothing industry is a big carbon polluter? How about energy sources? And travel and transport, if any?

In addition to these, we would have to consider how people would govern themselves, and how they would share wealth. Would there be a tax system? Would there be charities? What political structures would we have? What about our health care? What about gender equality? Diversity and inclusion? Wealth and resource distribution? What sort of arts and entertainment would we have, if any? Would we still make books from paper, and what kind of paper? The beauty-product industry is wasteful: would we cook up our own hand lotion? Would we allow an internet, and, if so, how much energy would it burn? Would there be a police force of any kind? A judicial system? An army? And what about waste management, and – come to that – funerals? Cremation is highly productive of carbon dioxide. What are the alternatives to making one's exit in a puff of smoke?

Just assembling the materials for this course led me and the researchers to a hoard of sources about which we'd had no idea. And inviting special guests with deep knowledge of these issues revealed to us that many of them did not know about the work that the others were doing. Raising awareness, sharing discoveries and envisioning ways of joining forces thus became part of our project. The climate crisis is multidimensional; any solution to it will have to be multidimensional as well. And these solutions, to be effective, would have to be adopted by a large section of society. A daunting prospect.

Les Stroud, who created the television series *Survivorman*, names four elements that anyone attempting to escape a life-threatening situation – a plane crash in the Andes, a boat adrift at sea – must have on their side in order to succeed. They are knowledge, appropriate equipment, willpower and luck. These may be present in varying proportions – even with no equipment, you might make it through if you have enough luck – but if you have none of the four, you won't survive.

We as a species are approaching a life-threatening situation. How do we score on each of the magic four? We have a lot of knowledge: we know what the problems are, and we know – more or less – what must be done to solve them. Appropriate equipment is something we already have a lot of, and we're inventing more by the week: new materials, new techniques, new machines and processes. At the level of households, and even towns and cities, we have the know-how to reinvent our way of life.

But what we're lacking at the moment is willpower. Are we up to the challenges? Can we face the tasks ahead? Or do we prefer to drift aimlessly, thinking someone or something will descend from the sky to save us? Willpower and hope are connected: neither is much use without the other: in order for hope to be efficacious you have to act on it, but without any hope at all you lose the will to struggle on.

However, even if we have the knowledge, the equipment and the willpower, we'll still need luck. But what is luck, apart from good weather? 'We make our own luck' is an old saying. So let's make some luck. /

5.10

People Power

Erica Chenoweth

Today, more people than ever before are conscious of a simple fact: fundamental changes in the global system are urgently required to keep our planet habitable. We have promising technologies and traditions that can help to restore a healthy relationship between humanity and Earth's resources, and we have the ability to invest in more of these technologies and amplify these traditions. We have sophisticated legislation and other roadmaps available for understanding how to transform our economy into a more sustainable one. In many ways, we have the answers to solve our climate problems. What we don't seem to have is the political will.

If history is any guide, only massive collective action – by people all around the world from all walks of life – will embolden decision-makers to take the actions required for climate justice. However, we have also learned how skilled activists, organizers and community leaders can mobilize the public and make politicians willing to stand up and take measures to overcome our shared challenge. This is fertile ground from which to address the climate emergency.

So how can societies create sufficient political pressure to push political leaders, big businesses and other stakeholders to change course?

People power – also called non-violent or civil resistance – is one of the most effective ways by which diverse populations have demanded change. Over the past hundred years, students, workers, children, the elderly, people with disabilities and others who have been marginalized in society have used civil resistance to bring down dictators, end colonial occupation, stop legalized discrimination and oppression, secure fair labour practices and the right to vote, protect human rights, end civil wars, and even create new countries. Civil resistance has been a formidable tool for bringing about meaningful change, and it works by ordinary people exerting political and economic pressure on those who hold power. Most of us are familiar with one or two instances of people power in our own countries, but there have also been global, systems-level changes secured through huge non-violent, popular campaigns. Mass mobilization played a key role in the global movement to abolish legalized slavery, which confronted entrenched social, political and economic interests to bring an end to the slave trade, and enslavement economies, throughout the nineteenth century. The anti-colonial campaigns that followed, especially in the twentieth century, represented a coordinated global struggle for political independence, leading to an incredible expansion in the number of independent nations in the world political system.

In general, social movements have won by following four key strategies.

First, they continually expand in size and diversity. Achieving genuinely large-scale participation is a way to signal the popularity of a movement and its capacity to disrupt the normal order of things, making success seem more likely. Mass participation connects wider networks of society through which a movement can access the decision-makers and stakeholders whose support is critical in effecting change.

Second, movements tend to win when they secure key defections from these powerbrokers, effectively ensuring their support. In the climate movement, this includes institutions that benefit from the status quo, especially corporations and shareholders whose pursuit of profit entails environmentally destructive activity, from extraction and deforestation to overconsumption. Movements succeed when they induce people and institutions with access to power and resources to join the struggle and use this access to expand their leverage.

Third, successful civil resistance campaigns tend to deploy a variety of methods to increase leverage and pressure on their opponents. This means they often go beyond street demonstrations, protests and other symbolic actions and pursue sustained coordinated action. Methods with economic impacts such as targeted strikes, boycotts and other forms of economic non-cooperation – can be especially effective in increasing pressure on those who hold power, either politically or financially.

Fourth, successful mobilization often takes years – not weeks or months – to build the pressure necessary for change. Effective movements stay the course and maintain discipline and strategic unity, even as their support base expands. As a result, they avoid the trap of internal setbacks or external public backlashes which can follow incidents of violence. Non-violent discipline has helped campaigns expand their participation base, in turn expanding their coalitions, shifting the loyalties of those in positions of power and, ultimately, winning.

Where does this leave the climate movement today? It has already mobilized people from all walks of life, with leadership roles taken on by children and youths around the world, by Indigenous and small nations groups, and by minoritized communities – those who are most affected by the crisis and who have been historically marginalized from the formal halls of elite power. Tactics which disrupt the status quo, particularly through economic non-cooperation – for example, strikes, walkouts, boycotts, and divestments which end the profitability of the fossil fuel economy – will likely continue to play an important role, especially if they escalate in frequency and scope. But in the end, every impact hinges on a sustained and dramatic increase in the number and diversity of people participating in the campaign for climate justice. The movement needs to massively expand its membership.

How many people are we talking about? Well, we don't know what the precise threshold is for the situation we face today. But social science gives us a few estimates.

One increasingly prominent estimate uses the '3.5 per cent rule'. This figure comes from the historical observation that among mass non-violent movements attempting to overthrow their own governments, none has failed after mobilizing 3.5 per cent of their population to engage in mass demonstrations. Although that's a small proportion of the overall population, it's a huge number of people. For instance, in the US today, that would be over 11 million people; in Nigeria, over 7 million; in China, over 49 million. Worldwide, 3.5 per cent of the population would be over 271 million people.

Some activists cite the 3.5 per cent rule as the critical threshold to create change when it comes to popular movements more generally. This threshold may well be a helpful rule of thumb for influencing change at the national level, although this has not yet been empirically tested. But there are several limitations to applying the 3.5 per cent rule specifically to the climate movement.

First, the rule has been created from looking at historical cases in which people were trying to topple their own governments. These people were not necessarily pursuing policy reform, much less trying to coordinate durable international change. Its 3.5 per cent threshold has not been tested in a global context where systemic change is required. This is an important distinction, because removing a single, hated dictator is much easier than agreeing to and installing wholly new political institutions, social practices and economic markets simultaneously. Genuine change results from an ongoing project of transformation, not from a one-off victory.

Second, when large numbers of people are willing to mobilize to create change, it's safe to assume that a much larger proportion of the population sympathizes with their cause. This means that the 3.5 per cent rule probably underestimates how many people need to support winning movements, even if they aren't actively mobilizing in the campaign.

Third, mass mobilization of an emboldened minority can create counter-mobilization on the opposing side, which can effectively slow or even stymie progress. Without inviting the public into a broader conversation – one that aims to pull in as many people as possible and change norms, behaviours and expectations – any victories achieved by a small but mighty movement may be short-lived.

Fourth, when people think about the 3.5 per cent rule, they often imagine mobilizing massive numbers of people in the streets. However, although street demonstrations do have enormous symbolic power, they do not by themselves necessarily increase pressure on decision-makers and companies or create widespread behavioural change; they must also disrupt 'business as usual'. This requires careful planning, along with a multipronged communication and political strategy to help popular resistance shift the pillars that support vested interests.

Fortunately, there is additional research that provides some insight as to how many people need to be actively involved in climate action for it to lead to sweeping social change. In his research on the impacts of social networks (that is, real relationships between people, not just those on social media), sociologist Damon Centola finds that the critical tipping point for changing everyone's behaviour is a committed minority of 25 per cent. Assuming this figure applies beyond his studies, then if 25 per cent of the population visibly change their practices, norms and behaviours, the climate movement's victories should be more widely accepted, durable and effective.

Of course, this may seem like a much harder task than organizing 3.5 per cent of the population for mass protests. But research shows that the 25 per cent threshold can be reached with surprising speed. In times of crisis, such as a global pandemic, our societies can quickly change our behaviours and practices, for example wearing a mask, washing our hands and physically distancing ourselves from others. As with public health, when it comes to climate justice, we have a solid understanding of what behaviours need to change directly – which industries we support, what types of energy we buy, how we heat and cool our homes, what we eat, where and how we travel, how we process waste, how much we invest in bold, sustainable technologies and programmes, and how often we consider sustainability in our day-to-day choices – on a global scale.

Over the past fifty years, the climate movement has had undeniable global influence, and it continues to grow. Despite the setbacks, the movement's efforts are paying off. We must not succumb to the belief that our actions don't matter and that we have no power. In many countries around the world, a 3.5 per cent tipping point has led to major breakthroughs for protest movements. And studies suggest that there is a 25 per cent tipping point for genuine behavioural change and transformation on a larger scale. Both of these tipping points are within reach. Tens of millions of people have been taking action to transform our relationship to the planet. Formidable obstacles remain, but if we take history as our guide, these can be overcome by good strategy, effective organizing, and people power. /

The critical tipping point for changing everyone's behaviour is a committed minority of 25 per cent.

Changing the Media Narrative

George Monbiot

If you were to ask me which industry is most responsible for the destruction of life on Earth, I would say the media. This might seem like an astonishing answer. When you look at what the oil, gas and coal industries have done, at the devastating impacts of cattle ranching, timber cutting, industrial fishing, mining, roads, the chemicals industry and the companies manufacturing useless consumer junk, you might wonder how I could justify placing a sector with relatively low environmental impacts at the top of my list.

I do so because none of these industries could continue to operate as they do without the support of newspapers, magazines, radio and television. Most of the media, most of the time, provide them with the social licence they need to persist in their current form. Most of the media, most of the time, have resisted the action required to prevent the collapse of our life support systems. They have attacked and vilified people who challenge the economic system that drives us towards catastrophe, and used their great polemical power to enable business as usual to continue. In many cases, they have simply denied the realities of climate and ecological breakdown.

In other words, the media is the engine of persuasion that allows our Earth-destroying system to persist. It has repeatedly misled us about the choices we face. It has distracted us with trivia, and conjured up bogeymen and scapegoats to prevent us from seeing where our real problems lie. On behalf of its wealthy proprietors, it has sought to justify a political economy that allows a few extremely rich people to grab and destroy the natural wealth on which we all depend.

The case against the billionaire press is easy to make. But the problem is almost universal. Arguably, in my country, the UK, public service broadcasters have caused even more harm than the media empire owned by Rupert Murdoch (which, from Fox News to *The Times*, dominates much of the industry in the US, the UK and Australia).

To give an example from the UK that will doubtless sound familiar to film-makers the world over, between 1995 and 2018 the BBC's channel controllers furiously rejected almost every environmental proposal brought to them, sometimes with a stream of expletives. On the very rare occasions when they allowed an environmental documentary to be broadcast, their terror of upsetting powerful interests drove them to make catastrophic mistakes. In my view, the most environmentally damaging item ever carried on any medium in this country was a two-part documentary broadcast in 2006 titled, without irony, *The Truth about Climate Change*.

It was presented by 'the most trusted man in Britain', Sir David Attenborough, whose word was treated as gospel. Somehow it managed not to mention the fossil fuel industry at all, except as part of the solution: 'the people who extract fossil fuels like oil and gas have now come up with a way to put carbon dioxide back underground'. Carbon capture and storage is a classic oil-industry talking point, always promised, never delivered, whose purpose is to justify continued extraction. Instead of fossil fuel interests, another force entirely was blamed for accelerating greenhouse gas emissions: the '1.3 billion Chinese'. No other cause was named in the series. It immediately triggered a new and virulent form of climate denial that quickly spread around the planet and persists to this day: there is no point in taking action here or anywhere else, because the Chinese are killing the planet.

Channel 4's commissioners went a step further: while blocking almost all environmental documentaries, they broadcast films such as *Against Nature* (1997) and *The Great Global Warming Swindle* (2007), which denied global heating and other environmental crises and repeated falsehoods concocted by fossil fuel companies. These too had a major impact. We don't expect public sector broadcasters to mislead us. But Channel 4 did, blatantly and disastrously.

Worldwide, climate deniers were for years given equal or greater standing than climate scientists. 'Think tanks' which refuse to reveal their sources of funding, and often look more like corporate lobby groups, are still invited to attack environmentalists without disclosing their interests. Advertising, on which most of the media rely for their money, helps to sustain levels of consumption that Earth systems cannot bear.

Without the media, governments would have been forced to act. Without the media, the world's most destructive industries would not have been able to fend off demands for change.

There have been some improvements in recent years, but the most important story of all is still pushed to the margins. Even during major climate disasters – heat domes and droughts, fires and floods – most of the news media merely glance up for a moment before returning to the trivia and court gossip that dominate their bulletins. In one day, NBC, ABC and CBS spent almost as much time covering Jeff Bezos's eleven-minute flight in his giant metal phallus as on all climate issues in the preceding year.

So what do we do? A few established outlets have consistently drawn attention to our environmental crisis, for example the *Guardian* (for which I write), Al Jazeera, *El País*, *Der Spiegel*, Deutsche Welle, *The Nation*, Canada's *National Observer*, the *Daily Star News* in Bangladesh, Africa's *The Continent*, and Cambodia's *Southeast Asia Globe*. We urgently need other newspapers and broadcasters to join them, to prioritize coverage of our existential predicament and stop misleading us on behalf of damaging industry. But it's also crucial that we keep building effective alternatives, such as Mongabay, Democracy Now!, CTXT, the Tyee, the Narwhal and Double Down News. I've been involved in some of these since 1993, when I helped make programmes on Undercurrents, a video newsreel produced by activists and, at the time, distributed by hand and in the post.

New technologies greatly enhance the reach of alternative media, and in many nations they have allowed activists and communicators to reach millions of viewers and readers. The digital promise is at last being fulfilled, as young people turn away from established channels and towards those prepared to tell the truth about the greatest crisis humanity has ever faced. This, for me, is where hope lies.

Every effective movement is an ecosystem in which people bring their different skills together to press for change. Communication is among the most important of these skills. By refocusing the world's attention and changing the narrative, good media, alongside campaigners working power-fully in other fields, can force governments to act. They can hold destructive industries to account, ensuring that they can no longer fend off their critics. They can help provoke the systemic social change we need to prevent systemic environmental collapse. /

Resisting the New Denialism

Michael E. Mann

Back in the late 1990s my co-authors and I published the famous 'hockey stick' curve documenting the unprecedented warming of the past century. Our original graph showed the northern hemisphere's mean temperature over the last 600 years. It was soon extended back to a thousand years. The 'shaft' of the hockey stick depicted relatively minor temperature variations up until the last century, when a dramatic upswing in temperatures formed the 'blade' of the stick. The recent spike in temperatures depicted was seen to accompany the Industrial Revolution, signalling the profound impact that human activity – the burning of fossil fuels in particular – was having on our planet. The 'hockey stick' was a breakthrough in visually demonstrating how intimately connected greenhouse gas emissions were with the rapid warming of our planet. This was inconvenient to fossil fuel interests, and that made it – and me – a target of the fossil-fuel-industry-funded attack machine.

Two decades later, the UN's IPCC update to our graph in its latest summary for policymakers no longer appears to be a hockey stick. Instead, the warming from recent years has made the graph resemble the Grim Reaper's scythe (Fig. 1).

Mother Nature is sending us a message. The latest IPCC report coincided with a scourge of devastating extreme weather events that spread out around the northern half of the planet in the summer of 2021. Record-setting wildfires, floods and heatwaves signalled a new era where climate change isn't just something to plan for in the future – it's here now.

As a consequence of this new reality, climate *inactivists* (that is, fossil fuel companies and the front groups and conservative politicians who do their bidding) can no longer claim that climate change is a myth, or a hoax, or something we can ignore.

So they've engaged in a whole new array of tactics, short of flat-out denial, in what I've called 'the new climate war'. The tactics of this new war on climate action include *division* (dividing climate advocates so they

The 'hockey stick' curve as it appeared in the IPCC's 2001 report already showed unprecedented warming . . .

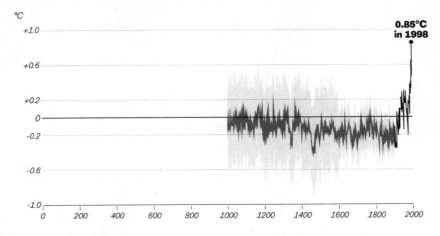

Figure 1: Change in global surface temperature, relative to the 1961–90 average, from 1000 to 1998. Temperatures from 1902 to 1998 are observed; earlier years are reconstructed from records such as tree rings, corals, ice cores, and lake sediments.

. . . the latest version now resembles the Grim Reaper's scythe.

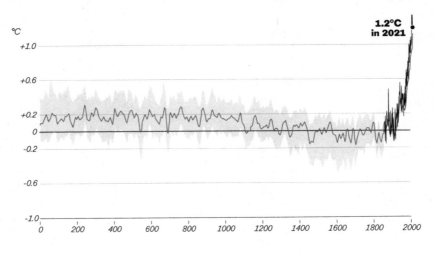

Figure 2: Change in global surface temperature, relative to the 1850–1900 average, from the year 0 to 2021.

do not speak with a single powerful voice), *despair*-mongering (if they can convince us it's too late to act, it potentially leads us down the same path of disengagement as outright denial) and *deflection* (the hyper-focus on the role of individuals to the exclusion of governmental policy).

With regard to this latter tactic, for example, I note in *The New Climate War* how 'the focus on the individual's role in solving climate change was carefully nurtured by industry', explaining how 'the concept of a "personal carbon footprint", was something that the oil company BP promoted in the mid-2000s. Indeed, BP launched one of the first personal carbon footprint calculators.' BP and other fossil fuel companies wanted us so focused on our own individual carbon footprint that we would fail to notice *their* much larger footprint: 70 per cent of total carbon emissions, after all, come from just a hundred polluters. While individuals should do all they can to minimize their environmental impact, we need governmental policies that will prevent them from using our atmosphere as a waste basket.

What can we do to fight back? We can, first of all, call out bad actors, and we must not be afraid to speak truth to power. Climate advocates must resist divisive fights on social media over matters such as our personal lifestyle choices, and instead set positive examples and work together towards the shared goal of holding polluters and those who enable them accountable. We must use our voices and our votes to support and elect politicians who are willing to prioritize meaningful climate action – and vote out those who aren't. And we must be guided by the overriding ethical obligation that we cannot mortgage the lives of future generations by failing to act at this pivotal moment. /

Mother Nature is sending us a message. Climate change isn't just something to plan for in the future – it's here now.

A Genuine Emergency Response

Seth Klein

We've known about global warming for nearly half a century. In response, we've run out the clock with distracting debates about the incremental changes we might make. After so many years of 'blah, blah, blah', how can we know when a government truly gets the climate crisis and has shifted into emergency mode?

I've spent the last few years writing about how my country, Canada, has responded to different emergencies from our past and present. I see in the history of our experience of the Second World War a helpful – and indeed hopeful – reminder that *we have done this before*. We have pivoted at remarkable speed in the face of an emergency. We have mobilized in common cause across class, race and gender to confront an existential threat. And in doing so we have managed to entirely retool our economy – twice, in fact: once to ramp up military production and once to convert back to peacetime – all in the space of six years.

Through my study of Canada's historic mobilizations, I've identified four clear markers which show that a government has genuinely shifted into emergency mode. With respect to the climate emergency – thus far at least – it's clear that our governments are failing on all four counts.

1. Spend what it takes to win

An emergency, once recognized as such, forces governments out of an austerity mindset. Canadian government expenditures during the Second World War were unlike anything before or since. The country's debt-to-GDP ratio at the war's end remains a historic high. When C. D. Howe, then Minister of Munitions and Supply, was pressed about this extraordinary ramp-up in government spending, he famously replied, 'If we lose the war, nothing will matter . . . if we win the war, the cost will still have been of no consequence and will have been forgotten.'

Similarly, during the Covid-19 pandemic, federal government spending increased dramatically, with Canada's debt-to-GDP ratio jumping from about 30 to 50 per cent in a single year. Remarkably, almost all this new debt was taken on by our central bank, which for most of the pandemic's first year was buying $5 billion in federal government securities *per week* to fund the emergency response.

Government spending on climate action and green infrastructure, however, pales in comparison. It currently amounts to about $7 billion *a year*. Former World Bank chief economist Nicholas Stern has said governments should be spending 2 per cent of their GDP on climate mitigation efforts, which in Canadian terms would be about $40 billion per year. Our government isn't merely spending a *little* less than it should in the face of the climate emergency; it is spending less by a massive order of magnitude.

2. Create new economic institutions to get the job done

During the Second World War, starting from a base of virtually nothing, the Canadian economy pumped out planes, military vehicles, ships and armaments at a speed and scale that is simply jaw-dropping. Remarkably, the Canadian government established twenty-eight public corporations to meet the requirements of the war effort.

During the Covid pandemic, we witnessed governments around the world take on a similar role, creating audacious new economic support programmes with a speed that few would have predicted. These programmes provided populations with testing, vaccination and health-care services on an unprecedented scale.

If our governments really saw the climate emergency as an emergency, they would quickly conduct an inventory of our conversion needs to determine how many heat pumps, solar arrays, wind farms, electric buses, and so on, we'd need to electrify virtually everything and end our reliance on fossil fuels. Then they would establish a new generation of public corporations to ensure that those items were manufactured and deployed at the requisite scale. They would also create an audacious new economic programme to catapult climate infrastructure spending and worker retraining.

3. Shift from voluntary and incentive-based policies to mandatory measures

The Second World War saw rationing of core goods and all manner of other individual sacrifices. The Covid pandemic has seen our governments issue health orders and shut down non-essential parts of the economy when needed. But for the climate emergency, we have seen nothing like this.

Virtually all climate policies to date have been voluntary. In Canada, we encourage change. We incentivize change. We offer rebates. We send price signals. But what we have decidedly not done is to *require* change. And our greenhouse gas emissions are not going down, they have merely flatlined.

If we are going to meet the greenhouse gas targets we so urgently need to hit, we must set clear, near-term dates by which certain things will be required. We should declare that it will no longer be legal to sell new fossil fuel-burning vehicles as of 2025. We should mandate that all new buildings will not be permitted to use natural gas or other fossil fuels as of next year. We should ban advertising by fossil-fuel-vehicle-makers and petrol stations. That's how we make it clear that this is serious.

4. Tell the truth about the severity of the crisis

In frequency and tone, in words and in action, emergencies need to look and sound and feel like emergencies. The Second World War leaders we remember best were outstanding communicators who were forthright with the public about the gravity of the crisis yet still managed to impart hope. Their messages were amplified by a news media that knew which side of history it wanted to be on, and by an arts and entertainment sector keen to rally the public.

None of this consistency and coherence, however, is present with respect to the climate emergency. When our governments do not act as if the situation is an emergency – or worse, when they send contradictory messages by approving new fossil fuel infrastructure – they are effectively communicating to the public that it is not an emergency. Where are the regular press briefings on how the climate emergency response is going? Where is the government advertising to boost the level of public 'climate literacy'? Where are the daily media climate emergency reports telling us how this fight for our lives is unfolding at home and abroad? If our current leaders believe we face a climate emergency, then they need to act and speak like it's a damn emergency.

One last wartime lesson: Every great mobilization comes with a pledge to leave no one behind – that life after the fight will be brighter and more just. The climate mobilization must include a jobs guarantee for everyone who wants one, and a just transition for all whose livelihoods are tied to fossil fuels or who live on the front lines of the climate crisis, as part of an overarching commitment to tackle inequality. /

5.14

Lessons from the Pandemic

David Wallace-Wells

In early December 2019, just weeks after 2 million climate strikers gathered around the world to protest the business-as-usual beginning of the COP25 conference in Madrid, the first human case of SARS-CoV-2 was registered in Wuhan. In January, as the World Economic Forum in Davos tried to rebrand itself as a 'climate conference', the first deaths were recorded. In February, as the world outside of China began to panic about the 'novel coronavirus' and the way it might threaten and upend the lives of many millions, 2,718 people died globally of the disease. That same month, approximately 800,000 died globally from the effects of air pollution produced by the burning of fossil fuel.

As the year wore on, the pandemic toll grew gruesomely large, though each new mortality milestone often seemed to produce less horror and outrage than the one before, following the dispiriting if familiar rhythm by which disaster is quickly normalized. By early 2022, after two years of disease, the *Economist* estimated that more than 20 million had died globally, making Covid-19 one of the seven deadliest plagues in human history.

Throughout the pandemic, the climate crisis continued relentlessly, yielding every few weeks – in some cases every few days – what would have once been recognized unmistakably as portents of punishing impacts to come. Across the Horn of Africa swarmed 200 billion locusts, darkening the sky in buzzing clouds as big as whole cities, chewing through as much food as tens of millions of people eat in a day and eventually dying in such agglomerating insect mounds that, when they did fall, they stopped trains in their tracks – all told, 8,000 times as many locusts as would be expected in the absence of climate change.

In California, in 2020, twice as much land burned as had ever burned before in the modern history of the state, with five of the biggest six fires ever recorded there recorded in a single year. Perhaps a quarter of the world's Sequoia trees were incinerated, and more than half of all the air pollution in the western United States was the result of wildfire, meaning that more particulate matter was produced by the burning of forests than by all other industrial and human activity combined. In Siberia, there were 'zombie fires', so-called for burning anomalously all through the Arctic winter, and melting permafrost that buckled a remote power plant, depositing 17,000 tonnes of oil into a local river; in 2021, nearly as much carbon was released by global wildfire as from the whole of the United States, the world's second-biggest emitter. A Category 4 hurricane made landfall in Central America just weeks after and just miles from the landfall of a Category 5. Sixty million Chinese people were relocated away from innocuous-sounding 'river flooding', caused by rains that threatened the world's most intimidating dam and which were yet, by the standards of precipitation and evacuation, only slightly above the recent averages. As the first pandemic year drew to a close, a million people were displaced by floods in South Sudan, one tenth of the country's population. In the pandemic's second year, hundreds were killed by flooding in Western Europe, dozens were killed as rains from Hurricane Ida filled basement apartments in greater New York. The Pacific heat dome so exceeded previous records that climate scientists wondered whether their models and expectations were properly calibrated – in addition to killing several hundred people and several billion marine animals, then setting the stage for wildfire and mudslides from later flooding so intense that Vancouver was effectively blockaded by climate disaster as autumn turned to winter. Just before New Year's Eve, 90 mph winds powered an urban firestorm in suburban Denver, where the warmest and second-driest autumn in a century and a half preceded the most destructive burn in the state's history, the flames surging from house to house through subdivisions and cul-de-sacs that would have looked, the day before, like the very picture of an inflammable modernity.

The world, as a whole, looked away – distracted by the onrushing pandemic, and trained, by the accumulating toll of recent disasters, to see what might once have looked like brutal ruptures in lived reality instead as logical developments in a known pattern. But if we could discern the lessons of the pandemic for the future of climate action, what would we see? Above all, that the pandemic extended an improbable invitation for once unthinkably ambitious action, which the world as a whole then disastrously failed to take. An unprecedented pandemic response might have been directed towards the unprecedented challenge of warming, animated by a truly global spirit and motivated to alleviate the inequitable burdens of those most affected. Instead, that unprecedented response was disgorged in defence of the status quo, the leadership of the Global North hoarding vaccines along with their emissions.

Covid-19 is not as obviously a story of climate change as many of the disasters we looked past in focusing on the more immediate-seeming pandemic threat. But among the many unnerving lessons the two crises share is this one: nature is mighty, and can be scary, and though we call our age the Anthropocene, we have not defeated nature or escaped it but live within it, still subject to its temperamental power, no matter where it is that you live or how protected you may normally feel. We can no longer pretend to draw the rules of reality ourselves, at conferences or in seminar rooms, without consulting the environment first.

And yet for those accustomed to disappointment with global leadership on the climate emergency, even the imperfect initial response to the pandemic was eye-opening – indeed, frankly, exhilarating. In the pandemic's second year, caught in a tangle of ugly Covid nationalism and 'vaccine diplomacy', it can be easy to forget just how shockingly large and immediate that initial reaction was, even among those countries that botched containment – inadequate and even counterproductive in ways, but at a scale and with an urgency that many climate advocates wouldn't have felt comfortable even dreaming of before. In the space of just months, daily life the world over was upended, with more than a billion kids out of school, international travel more or less entirely suspended, and hundreds of millions across dozens of countries sheltering in place out of concern for themselves and those around them. Professional lives were suspended, social, romantic and family lives altered, the business of 'business as usual' entirely rewritten. When climate scientists spoke of the need for a Second World War–scale mobilization to avoid catastrophic warming, this was the kind of action they had in mind – a less punishing transformation, of course, but in other ways just as dramatic. The world's leadership class had effectively dismissed that recommendation when it was first issued, in the autumn of 2018, when the IPCC called to cut carbon emissions in half by 2030. By early 2020, the same leaders were enacting a social and economic transformation of a similar scale, on just a few months' notice. The pandemic illustrated that sudden change was no longer unrealistic, it had actually happened.

That spirit did not last, and it was far from perfect while it did. But, above all, the government response – from countless governments, of varying levels of prosperity and from many different ideological lanes – was immense, demonstrating at least the possibility that a whole-society effort could be mobilized in response to an imminent threat, with states utterly disregarding what had appeared to be, just months before, the absolute boundaries of political plausibility. The effect was clearest in those nations across East Asia and Oceania that effectively contained the virus through large-scale government intervention matched with widespread social trust and reflexive solidarity (an alignment we might hope to emulate on climate). But as Adam Tooze documents in *Shutdown*, his history of the pandemic crisis, even the botched response of the slow-moving nations of Europe and the Americas showed that, when it came to public spending, every nation in the world was suddenly operating in an entirely new reality, without any of the political and social constraints which had previously set the speed limit on climate action. In the years to come, one lesson drawn from the pandemic response will inevitably be: there is no speed limit but the one we set ourselves. There must not be.

Unfortunately, even as political leaders in the Global North learned a new playbook on the fly, they were reluctant to apply it to the project of decarbonization. In the midst of the pandemic response, you could have easily imagined boundless possibilities for climate action too, and there were those who saw the possibilities already in perfect alignment. 'We don't know what the recovery packages of Covid are going to be,' Christiana Figueres, one of the central architects of the Paris Agreement, told me in the summer of 2020. 'And honestly, the depth of decarbonization is going to largely depend on the characteristics of those recovery packages more than on anything else, because of their scale. We're already at $12 trillion; we could go up to $20 trillion over the next eighteen months. We have never seen – the world has never seen – $20 trillion go into the economy over such a short period of time. That is going to determine the logic, the structures, and certainly the carbon intensity of the global economy at least for a decade, if not more.' If we were going to be spending $20 trillion, in other words, why not spend it on climate?

Instead, the first round of spending was not encouraging for those dreaming of a green recovery. The EU was the gold standard, but still promised only that 30 per cent of its stimulus would be earmarked for climate. The US and China each pledged a fraction of that (and in each case, there was fossil stimulus too). By April 2021, less than a quarter of Covid spending in OECD countries was considered 'environmentally friendly', and 41 per cent of energy stimulus was being spent on fossil fuels. By gift of global tragedy, the possibility of engineering a new world – more stable, more secure, more prosperous and more just – appeared before us all. But rather than embracing that project, the world scurried back to its old order as quickly as it could.

How large a missed opportunity was this? According to a team of researchers including Joeri Rogelj of Imperial College, London, just one tenth of Covid-19 stimulus spending, directed toward decarbonization during each of the next five years, would be sufficient to deliver the goals of the

Paris Agreement and stop global warming at a level well below 2°C. Globally, the total cost of a green transition would be half of what was spent on stimulus in 2020, and yet, even in the midst of all that spending, the world couldn't manage to take the deal. In the US alone, the *Wall Street Journal* noted, a full decarbonization of the power sector would require upfront spending of between $1 trillion and $1.8 trillion – less than one fifth of the cost of the country's pandemic relief. But none of the instalments of American pandemic relief put climate spending at the centre. When President Biden finally got around to it, the total outlay was only in the hundreds of billions of dollars – a far cry from the 5 per cent of GDP suggested by Michael Bloomberg and Hank Paulson, no climate radicals, and even further from proposals championed by Senators Ed Markey and Bernie Sanders.

What was perhaps most striking about this failure was that, for the first time, it was enacted by politicians who were at least speaking the language of climate alarm, and asking to be judged – in forum after forum, conference after conference – by its existential standards. By those standards, of course, they have failed, letting the 1.5°C target slip further and further from reach, watching emissions stockpile in the atmosphere year after year as they give ever more heated speeches about the stakes of inaction. But that rhetoric, still empty, also suggests the possibility that the unprecedented collective action and public intervention made necessary by the pandemic may not mark a one-time shift – and indeed that some of that new willingness may come climate's way soon. 'Anything we can do, we can afford,' John Maynard Keynes declared in the midst of the Second World War. The pandemic reminded us of that principle; with climate change, the world might hope to actually enact it.

The pandemic was humbling, too, teaching those who didn't know it already that crises don't reliably or simply solve rivalries and prejudices and the basic crimes of human indifference. And if Covid-19 also taught us, positively, that people respond when they perceive an imminent and immanent threat, it brought some negative lessons as well.

The first is, the longer we wait, the more we lose. In periods of exponential growth, we learned in the first months of the pandemic, delays of days can be catastrophic, and actions that may prove sufficient in week one are hopelessly inadequate by week three. When it comes to climate, we already know the problem is the same. A project of global decarbonization that began in 1988 when James Hansen, Michael Oppenheimer, Syukuro Manabe and their fellow scientists testified before the US Senate, and which was intended to limit temperature rise to 1.5°C, would have required only modest and relatively undisruptive annual change and could have taken more than a

Next pages:
Young protesters
marching during a
Fridays for Future
demonstration
in Jakarta,
Indonesia, in
September 2019.

hundred years to complete. Having instead chosen to ignore those warnings and let emissions continue to grow, stockpiling each year in the atmosphere a generational burden, the world now faces a far more harrowing task – zeroing out emissions within just a few decades, perhaps even sooner in the absence of negative emissions and carbon removal on a 'planetary scale'. What seemed advisable in 1988 now qualifies almost as climate denial; what counted as ambitious in 2008 is already hopelessly inadequate. And if the curves aren't bent immediately, by 2025 even the dispiriting maths we face today will no longer be workable either.

The second lesson is that success in one country is not enough, and that no one should be satisfied by nationalistic responses to global threats. Already today, climate disparities – in responsibility for the present warming and in the burden imposed by future impacts – represent an immoral horror, though those in the Global North prefer to ignore them. The US is responsible for one fifth of all global historical emissions, while all of sub-Saharan Africa has produced about 1 per cent. The burdens of warming are distributed unequally, too, with much of the developing world already pummelled by climate impacts of the kind those in Europe and North America regard as a still-distant fear; promises of nominal support are yet to be fulfilled and estimates of the true need are many times larger still. (Rich countries asking to be applauded for promises of $100 billion annually to help poor countries address climate change should know that the bill for decarbonizing the Global South may be $5 trillion or more.)

The tragedy of vaccine distribution tells the same story – not just that resources will be hoarded by those who can afford to secure benefits for themselves, but that those same groups will impose scarcity where it isn't, and needn't be, perhaps because they find inequalities reassuring. In July 2021, the International Monetary Fund estimated that a global vaccination programme would cost only $50 billion and would generate $9 trillion in additional revenue by just 2025 – a two-hundred-fold return on public investment in the length of a presidential election. The upfront cost was small enough that it could have been covered not just by the world's largest economies, where it would have disappeared on any government balance sheet, but by just one of the world's largest fortunes. Of course, none of those actors elected to take the bargain, preferring to let the Global South fend for itself against a virus the Global North had at least temporarily decided required a totalizing defence. As a result, the disease festered, mutated and continued to infect and kill, just as unaddressed global warming will. We cannot make these mistakes again. /

5.15

Honesty, solidarity, integrity and climate justice

Greta Thunberg

These were not just the regular Andersson, Petersson or Johansson with a double 's'. These were all real Swedish names: Karlberg, Rönnkvist, Nordgren. But this cemetery did not belong to a Swedish church congregation. This was just a random graveyard in Lindström, Minnesota, in the United States. The size and robustness of the aged gravestones spoke of a bygone era. The roots of the surrounding trees had shifted them slightly out of place, just enough to indicate that sufficient time had passed for the people resting here to start fading from memory.

We – me and my father – were 6,780 kilometres from Sweden, yet from a literary perspective this was the heart of the nation. Chisago County is where Vilhelm Moberg's *The Emigrants* largely takes place, a novel series which occupies a special place in Swedish arts and culture. Many years earlier, when I was too ill to go to school, we had read all those books, and they made a huge impression on me. We stopped by the statue of Kristina and Karl-Oskar and took a picture of a sign painted on a traditional Swedish red *dala* horse from the province of Dalecarlia. It read, 'Life is great on highway 8.' We looked out across South Lindström Lake and saw that the surrounding shores looked exactly like anywhere or anyplace back home. Then we got into the electric car and drove west, late into the night, to make up for the time we had lost to Swedish literary history. We slept at a motel in Sioux Falls and before first light we were back on Interstate 90, where we crossed the Missouri River and drove into the majestic badlands of South Dakota before heading south, to the Pine Ridge Reservation, where I was meeting my friend Tokata Iron Eyes.

Pine Ridge is one of the poorest areas in the US and has huge, poverty-related problems such as alcoholism, high rates of child mortality and suicide, alongside a life expectancy that is among the lowest in the entire western world. Tokata and her father, Chase, drove us around the town and showed us the abandoned churches, the houses with boarded-up windows. It was almost unimaginable that we were right in the middle of the world's richest nation. We stopped at Wounded Knee and walked up the track to the tiny memorial site. The afternoon sun was warm and there was not a cloud in sight. A gentle October breeze was blowing through the high prairie grass. Chase told us there had been an attempt to open a little museum in a nearby house but that the project had failed as there was no money to keep it running.

The monument stands on top of the graves of those who died there on December 29, 1890, when the massacre took place. Or rather *the* grave – singular. There were barely any individual gravestones at Wounded Knee. Just one mass grave marked by a simple memorial stone, enclosed by a fence and with two white-painted concrete pillars marking the entrance. Around 300 people are buried here, all members of the Lakota people, an Indigenous community in the US. Many were women and children. They were massacred – after years of forced migration, broken treaties, and violence – by the 7th US Cavalry Regiment. Twenty of the soldiers who carried out the slaughter were given Medals of Honour.

Countless similar incidents took place during the European colonization of the Americas that started with the arrival of Christopher Columbus in 1492. The beginning of this period is sometimes called 'the Great Dying'. It is estimated that as many as 90 per cent of the Indigenous people of the Americas – or 10 per cent of the world's population – was massacred or lost to infectious disease. These atrocities cannot be described without using words such as 'genocide' or 'ethnic cleansing', yet there are hardly any monuments. The nations responsible have still not atoned for their history. It is hard to imagine how any nation can allow itself to move forward without dealing with the causes and consequences of such social and racial injustice.

Even though the people buried in Lindström and at Wounded Knee lived at the same time and in neighbouring regions, their graveyards were worlds apart. And it was crystal clear that the Swedes in Minnesota were in a superior position to my friend's ancestors in South Dakota. The last twenty-four hours, travelling from Lindström to Wounded Knee, had given me a new perspective on the world. And it was not an easy one to accept.

Between 1850 and 1920 almost one quarter of the Swedish population emigrated to the United States – around 1.2 million people. They were driven by poverty and by dreams of a better life. But their story is also intertwined with the fate of the Indigenous peoples who already inhabited the lands that they claimed in places such as Minnesota, Wisconsin and other recently established US states and territories. And that land acquisition made way for everyone else to follow. Not only was it legal for them to obtain that land, it was encouraged. Just as with the colonization of Africa and other 'blank' spots on European world maps, it was expected that emigrants, trading companies or colonial nations would take over the lands they encountered and treat the previous inhabitants as commodities, property, savages or brutes, as Sven Lindqvist writes in *Exterminate All the Brutes*.

At the same time as the Spanish, French, Portuguese, Dutch and English were expanding their empires across the Americas, Sweden was expanding its borders in a similar way. But apart from attempts to colonize Delaware, Saint Barthélemy and Guadeloupe, we headed north, into Sápmi. This land, stretching across Norway, Sweden, Finland and Russia, is home to the Sámi people, who have lived there for many thousands of years. But the Swedish state claimed it as Swedish territory and began a slow process of expansion and land grabbing. A colonization process that picked up speed as the pursuit of natural resources started gaining traction in the 1800s. Sápmi had vast quantities of iron ore, silver and timber. So the Sámi people were pushed further and further away. Then came the forced movement of entire communities. Families were split up. Children were taken away from their parents. We tried to take away their language, their religion, their traditions, their culture – their entire way of life. Sweden established a State Institution for Racial Biology which measured their skulls. In the twentieth century came the hydroelectric industry, whose dams took away much of the grazing ground for the Sámi's reindeer herds. Then came the timber companies, clear-cutting the forests that supply much of the reindeer's food. Then came the mining companies. Then, in this century, came the wind turbines, chewing away further at the Sámi's ancestors' lands – this time to provide super-discounted 'green' electricity for Facebook servers and Bitcoin mining.

And we got away with nearly all of it. Sweden stole the Sámi's land, we stole their sacred places and artefacts, we stole their religion, we stole their forests and other natural resources. And this theft continues today. As we learned from Elin Anna Labba in Part Three, as climate change makes the conditions for reindeer herding increasingly difficult, it is becoming harder and harder for the Sámi people to maintain their traditional way of life. Many are giving up. New mines are being prospected. Old-growth forests are being clear-cut – forests that cannot be replanted. Any chance of further economic development is always the top priority.

And still Sweden does not in any way consider itself a colonial nation. If you described Sweden using those two words today, most people would probably think that you were completely insane. We tell the story we want to tell. We see what we choose to see. As individuals, we are responsible only for ourselves. But nations and corporations are a different matter altogether. They have accumulated wealth, assets and infrastructure from their past actions. And if that wealth has been created through wrongdoings such as theft, destruction and genocide, then we must find ways towards reconciliation and compensation.

Throughout history we have been adept at keeping these historic atrocities as far away from us as possible. The problem has always been someone else, in some other faraway place. But the climate crisis was created by us, the nations of the Global North. It is a crisis of inequality that dates back to colonialism and beyond. Those who have done the least to cause it are the ones who will suffer the most. And those who have done the most will likely suffer the least. All this is ultimately a symptom of a much larger crisis. A crisis arising from the idea that some people are worth more than others and therefore have the right to exploit and steal other people's land and natural assets – as well as the right to use up the planet's finite resources at an infinitely higher rate than others. A crisis shaped by a mindset that still infects our societies today. A crisis that everyone would benefit from dealing with. But it is naive to think that we can do so without confronting the roots of the problem.

The climate and sustainability crisis is in many ways the perfect saga. Or the ultimate moral test, if you like. Our emissions of carbon dioxide stay in the atmosphere for up to a millennium. And now, science is shining a bright light on all those invisible traces that we have left behind in our pursuit of power, dominance and wealth. Traces that the people on the front lines have been trying to tell us about for centuries. It is like one big unpaid bill that we in the historically responsible part of the world can no longer run away from. Because if we fail that moral test, then we will fail all else too. And all our astonishing achievements will eventually be for nothing.

To solve the climate and ecological crisis and to change everything, we need everyone. And that is never going to happen unless those responsible start to clean up their mess in an equitable way. The financially fortunate nations have already signed up to lead the way, and it is time for us to do this. That means paying for losses, damages and reparations. It means taking full responsibility for historical emissions. It means that polluters pay. It means including all our actual emissions in our statistics, including consumption, imports, exports, shipping, aviation, military and biogenic emissions. It means honesty, solidarity, integrity and climate justice.

A Just Transition

Naomi Klein

Most of us have learned to think about political change in defined compartments: environment in one box; inequality in another; racial and gender justice in a couple more. Education over here. Health over there.

And within each compartment, there are thousands upon thousands of different groups and organizations, often competing with one another for credit, name recognition and, of course, resources. It's not all that different from corporate brands competing for market share. And that shouldn't be surprising: we are all working within the logic of the existing capitalist system.

This compartmentalization is often referred to as the problem of 'silos'. Silos are understandable – they carve up our complex world into manageable chunks. They help us feel less overwhelmed. The trouble is, they also train our brains to tune out when a true crisis needs our help and attention, as we tell ourselves, 'That's someone else's issue.' The deeper problem with silos is they keep us from seeing glaring connections between the various crises tearing apart our world, and they stop us from building the largest and most powerful movements possible.

In practice, what this has meant is that the people focused on the climate emergency rarely talk about war or military occupation – even though we know that the thirst for fossil fuels has long driven armed conflict. The mainstream environmental movement has become a little better at pointing out that the nations getting hit hardest by climate change are populated by Black and brown people. But when Black lives are treated as disposable in prisons, in schools and on the streets, the connections are too rarely made.

Because we don't have a lot of practice working together across silos, the solutions coming out of various movements often seem disconnected from one another. Progressives have long lists of demands – things we all want to change. But what we often are still missing is a holistic picture of the world we're fighting for. What it looks like. What it feels like. And what its core values are.

Fortunately, there are all kinds of conversations and experiments going on to try to overcome these barriers and develop popular platforms that articulate a common vision. These platforms go by many names: the Leap; Green New Deal; the Black, Red and Green New Deal; and more.

What they all share is a recognition that the climate crisis is not the only crisis we face. We face so many overlapping and intersecting emergencies – from surging white supremacy to gender-based violence, to gaping economic inequality – that we simply can't afford to fix them one at a time. We therefore need an integrated approach: policies designed to bring emissions down to zero while creating huge numbers of good, union-ized jobs and delivering meaningful justice to those who have been most abused and excluded under the current extractive economy. We need a *just transition.*

A just transition is about recognizing that the work of confronting the climate emergency at speed and scale opens up a window to build a society that is fairer on every front, and where everyone is valued.

I have been involved in various climate justice coalitions over the past decade and a half, and there is no one definition of a 'just transition'. But there are some core principles that movements have developed and which future work should build upon.

A just transition begins with recognizing that the bottomless quest for profits that forces so many to work upwards of fifty hours a week with no security, fuelling an epidemic of isolation and despair, is the same quest for bottomless profits that has pushed our planet into peril. Once we recognize that, it then becomes clear what we need to do: insist that, as we respond to the climate crisis, we create a broader culture of care-taking in which no one and nowhere is thrown away – in which the inherent value of every person and every ecosystem is foundational.

Science-based climate action means getting our energy, agriculture and transportation systems off fossil fuels as rapidly as humanly possible. Justice-based climate action demands more. It demands that as we make these huge transformations we also build a more equal and democratic economy.

A good place to start is with energy ownership. Right now, a handful of fossil fuel corporations control the global supply and dominate most local markets. One of the great things about renewable power is that, unlike fossil fuels, it's available wherever the sun shines, the wind blows and the water flows. That means we have a chance for more decentralized and diverse ownership structures: green energy co-ops, municipal energy, community-owned microgrids, and more. Under these structures, the profits and benefits of new green industries stay in communities to help pay for services rather than being siphoned off to corporate shareholders.

This just transition principle is often known as **energy democracy**.

But true climate justice requires more than energy democracy – it requires energy justice, and even energy reparations. Because the way energy generation and other dirty industries have developed since the Industrial Revolution has systematically forced the poorest communities to bear a vastly disproportionate share of the environmental burdens while deriving few of the economic benefits.

In North America, where I live, the people who have been forced to bear these unjust burdens have overwhelmingly been in Black, Indigenous and immigrant communities, often referred to as 'front-line communities'. That's why many just transition platforms call for front-line communities to play a leadership role in developing new green infrastructure, in controlling land rehabilitation programmes and in receiving funding for green job creation. Indigenous groups that have had their land rights systematically violated, and whose traditional ecological knowledge systems provide a living alternative to current ecocidal practices, are also calling for greater control over their ancestral territories as part of the response to the climate crisis.

This just transition principle is sometimes called **front lines first** and it is a form of reparations for past and present harm.

One of the great benefits of climate action is that it will create millions of green jobs around the world – in renewables, in public transit, in efficiency, in retrofits, in cleaning up polluted land and water. A truly just transition means making sure that those jobs pay family-supporting wages and benefits and are protected, wherever possible, by trade unions. But there is another aspect to this too.

A just transition also calls for reimagining what a 'green job' is. Environmentalists don't usually mention it, but teaching and caring for kids doesn't burn a lot of carbon. Nor does caring for the sick. Making art is pretty low-carbon too. In a just transition, we would recognize this labour as green and prioritize it because it makes our lives better and our communities stronger. As we reduce our reliance on jobs that are based on encouraging wasteful consumption and dangerous extraction, we can invest in more care-sector jobs and make sure that they pay a living wage.

This just transition principle is sometimes called **care work is climate work** and it will help ensure that women's labour is fully recognized and appreciated in the next economy.

As we make these changes, we must also recognize that there are people who are stuck – through no fault of their own – in regions where polluting industries are virtually the only employer in town. Many of these workers have sacrificed their health in coal mines and oil refineries so that the rest of us can keep the lights on.

These workers, facing the prospect of large-scale job losses as oil and coal infrastructure is decommissioned, cannot be expected to bear the burden of climate action. That's why a just transition calls for massive investments in retraining workers for the post-carbon economy, with workers serving as full and democratic participants in the design of these programmes. A key measure is guaranteeing the income of workers during these periods – all too often, when industries go through massive change, working-class livelihoods and communities have been sacrificed on the altar of 'change' and 'progress'. A just transition would do things differently. It would also create huge numbers of jobs rehabilitating and restoring the lands that have been harmed from extraction, by, for instance capping the countless abandoned oil and gas wells around the world that are currently leaking toxins into the environment. Many workers currently working in high-carbon sectors are already trained for this work. These kinds of programmes and policies are how we make sure that everyone benefits from the transitions required to radically and rapidly lower emissions.

This just transition principle is often called **no worker left behind**.

Of course, creating a new, low-carbon economy is going to cost money. Lots of it. Governments can create some of it, as they did during the Covid-19 pandemic, in the aftermath of the 2008 financial crisis, and as they do during wars. But we live in a time of unprecedented private wealth, and the transition should also be funded by the polluters and overconsumers. The idea that we are too broke to afford to save our one and only home is simply untrue. The money needed for this transition is out there, we just need governments to have the courage to go after it – to cut and redirect fossil fuel subsidies, to increase taxes on the rich, to reduce spending on policing, prisons and wars, and to close down tax havens.

This principle of a just transition is known as **polluter pays** and it's based on a simple idea: the people and institutions that have profited most from pollution should pay the most to repair the harm it has done.

This principle includes not just corporations and wealthy individuals but also the nations of the Global North: we have been putting carbon into the atmosphere for a couple of hundred years, and we did the most to create this crisis, while many of the nations that are most vulnerable to its effects contributed the least. So, as financing is raised for a just transition, there needs to be a transfer of wealth from north to south, to help poorer nations leapfrog over fossil fuels and go straight to renewables. Climate justice also demands far greater support for migrants displaced from their lands by oil wars, bad trade deals, drought and other worsening impacts of climate change, as well as the poisoning of their lands by mining companies, many headquartered in wealthy countries.

The bottom line is this: as we get clean, we have to get fair. More than that, as we get clean, we must begin to redress the founding crimes of our nations. Land theft. Genocide. Slavery. Imperialism. Yes, the hardest stuff. Because we haven't just been procrastinating about climate action all these years. We've been procrastinating and delaying the most basic demands of justice and reparation. And the reckoning is here on every front.

Some find these kinds of connections daunting. Lowering emissions is hard enough, we are told – why weigh it down by trying to fix so much else at the same time? It's a strange question. If we are going to repair our relationship to the land by shifting away from endless resource extraction, why wouldn't we begin to repair our relationships with one another in the process? For a very long time, we have been offered policies that amputate the ecological crises from the economic and social systems that are driving them, searching endlessly for purely technocratic fixes. That is precisely the model that has failed to yield results.

Holistic transformations, on the other hand, have never been tried in the face of the climate crisis. And there is good reason to think that they might yield breakthroughs where technocratic climate policies have failed. The hard truth is that environmentalists can't win the emission reduction fight on our own. It's not a slight against anyone – the load is just too heavy. The transformation that scientists have told us we need represents a revolution in how we live, work and consume.

Winning that kind of change will take powerful alliances with every arm of the progressive coalition: trade unions, migrant rights, Indigenous rights, housing rights, teachers, nurses, doctors, artists. And to build these alliances, our movement needs to hold out the promise of making daily life better by meeting pressing needs that all too often go unmet – for affordable housing, for clean water, for healthy food, for land, for health care, for good public transportation, for time with family and loved ones. For justice. Not as an afterthought – but as an animating principle.

I have laid out five planks for a just transition. Energy democracy; front lines first; care work is climate work; no worker left behind; and polluter pays. This only scratches the surface. Climate justice also requires new kinds of trade deals that move us away from ever-growing levels of consumption; a robust debate on a guaranteed annual income; full rights for immigrant workers; getting corporate money out of politics and fossil fuel companies out of climate negotiations; the right to repair our broken products rather than replace them – and more.

Though the specific responses to the climate crisis will vary from place to place, there is an underlying ethic which connects all this work. As we change our economies and societies to get off fossil fuels, we have a responsibility, and a historic opportunity, to repair many of the injustices and inequalities that scar our world today. The great strength of a just transition framework is that it does not pit important social movements against one another or ask anyone suffering from injustice in the here and now to wait their turn. Instead it offers integrated and intersecting solutions grounded in a clear and compelling vision of our future – one that is ecologically safe, economically fair and socially just. /

The bottomless quest for profits that forces so many to work upwards of fifty hours a week with no security, fuelling an epidemic of isolation and despair, is the same quest for bottomless profits that has pushed our planet into peril.

What Does Equity Mean to You?

Nicki Becker

The first time I marched, it was International Women's Day. I was fourteen years old and I had just discovered that women did not have the same rights as men. I asked my mum to go with me because I was so little. And since then, I haven't missed a single march.

On March 8, 2019, my fifth march, in the middle of the crowd I came across a sign that read, 'Neither Land nor Women are Territories of Conquest'. I took a photo of the sign and continued marching. A week later, together with a group of other young people, we organized the first climate strike in Argentina. More than 5,000 people came and, among the signs that were there, I found the same one I'd seen a week before.

This is what the struggle for equality means to me. We are not fighting for different causes. Whether we are fighting for climate justice, social justice or gender equality, we are fighting together for justice.

I am a climate justice activist because I believe that the environmental movement has the opportunity to break new ground. In a world of increasing uncertainty, environmentalism is one of the engines for questioning the status quo and building a better world. Climate justice is not only about preventing a climate catastrophe, it is about building a world that is just and equal. We do not want to 'conserve' the world as it is now but to create a fairer one.

We refuse to live in an Argentina where a million hectares of land were burned in 2020 and where 10 per cent of the province of Corrientes burned down in 2022 because of the climate crisis, or where a woman is murdered every thirty-two hours, or where six out of ten children live in poverty.

That is why, in a world where many things do not make sense, we have the obligation to redefine and rethink absolutely everything. Equity means believing that another world is possible but also building it with the understanding that the only way to get there is through collective action.

Disha A. Ravi

Every time there is a climate calamity, for example a cyclone, an economic valuation of the losses is immediately published: 'Cyclone Yaas caused damages estimated at estimated ₹610 crore ($83.63 million) in Odisha, India.' This figure is intended to help people understand the severity of the damage the calamity caused. Yet even with monetary values assigned to them, the storms that have robbed people of everything are ignored. These people may have survived the catastrophe, but the rupture caused to their lives is irreparable.

When humankind started asking questions about ownership, these questions were supposed to help us take care of our land, but instead they created more questions. 'Who owns this land? Who owns this tree? Who owns this rock? Who owns the minerals under this rock? Who owns the oceans? And who owns the fish and the oil in the oceans?' We've taken ownership of the land and everything it had to offer, we dug the land until there was nothing left, and once we were done with the land, we moved on to the ocean. The plundering of the Earth by a few has brought all of us to the brink of extinction. The only way to reverse this is to stop laying the Earth to waste and unlearn our extractive ways.

We need a 101 on how to respect the planet. We need to shift our focus from ownership to responsibility. Our questions should be, 'Who is responsible for this land? Who is responsible for this tree? Who is responsible for this rock? Who is responsible for the minerals under this rock? Who is responsible for the oceans? And who is responsible for the fish and the oil in the oceans?' Once we make people accountable for taking care of our planet, they will begin to see the Earth as an extension of themselves, that they are part of the ecosystem. Relearning behaviours around the Earth and climate starts from acknowledging our close proximity to the climate crisis, changing our language to reflect that it's not in the future but here in the present, and acting with urgency; it comes from acknowledging that we are the Earth itself and we are fighting for ourselves and for each other. Solving the climate crisis will require us to change our relationship with the planet and with each other. We need a politics of love; we need people to pick each other first. We need a world where we cannot put a price on the rice we eat, the trees that give us oxygen, the oceans we swim in and the land that gives us our finite and fleeting resources.

Hilda Flavia Nakabuye

Uganda, like many African countries, has faced a number of challenges, especially injustices and social inequalities that stem from slavery and colonialism. The colonial system created marginalized groups in every society, and women were one of the most marginalized groups of all.

The system that created social inequalities gave birth to imperialism and continues to pit poor countries against rich ones. It is shocking that in the twenty-first century people of colour still have to prove that they are human beings! How is racism still happening today?

The climate crisis is without a doubt a global problem that affects us all. Our capacity to respond, however, is different. The best way to address a problem is to first understand the root cause, its drivers, and to ask ourselves tough questions – among them, why don't the developed countries who have historically polluted our planet take responsibility and pay for the damages they have caused?

Equity in Uganda would necessitate building strong, firm foundations for a social justice in which policies are binding. The current system of shelving policies and talking about problems without taking action is instead widening inequality at all levels.

If Uganda and other African countries are to achieve equity in the future, we must start holding big polluters accountable. They need to pay for the damage they have done and support vulnerable countries as they adapt to the changing climate. They need to stop funding fossil fuel projects in Africa.

An equitable future must be a future free from exploitation. Developing countries must not be used as dumping grounds for unwanted products and waste; natural resources must be protected. Children must not be left to die from the effects of pollution or be forced to worry about the climate crisis.

Equity and sustainability go hand in hand. There cannot be sustainability without equity, and there cannot be equity without sustainability. Climate justice must manifest everywhere, for everyone.

Laura Verónica Muñoz

I am a collision of resistance and oppression, of impoverishment and privilege. My roots are Indigenous, but also Spanish. I am the fruit of my peasant grandmothers and the seed my parents planted and nurtured after migrating from the countryside to the city searching for a better future.

I am love and contradiction, and when I look at my reflection I remember who I am and where I come from.

I am privileged. I speak English and have had access to education. However, the most valuable privilege I hold is my identity. Thanks to my peasant heritage, I can still feel and recognize nature in the midst of the superficiality and toxicity of the western world.

I am an ecofeminist climate activist because I understand the power of the Earth and of women. Only these powers will allow us to fight against the racist, patriarchal, capitalist and media-driven systems of exploitation we inhabit and which have produced the social and ecological crises we face today.

I know that a grassroots Latin American and Colombian activism, woven with the voices of those who have had their hands anointed with the soil of the Earth, is much more powerful and transformative than an activism based on individualism and online algorithms. I am certain that to achieve climate justice we must work together, creating safe spaces where diversity is the foundation and decoloniality the path we tread.

I am the product of colonization and exploitation, but I am also a fertile land of brimming resistance. I am the decolonial crop my ancestors sowed.

Ina Maria Shikongo

Namibia has been facing continuous droughts over the past decade. The Kunene region has been hit the hardest, forcing the Indigenous Himba communities to migrate to the cities in search of a better life. In addition to being hit by drought, the Kavango Basin, my ancestral homeland, is currently under threat. ReconAfrica, a gas and oil company from Canada, is hoping to extract 120 billion barrels of oil from the area, leading an oil industry publication to ask if this is 'the largest oil play of the decade'.

For me personally, it feels like déja vu. Having been born in the refugee camps of Angola and having lost a father and four siblings in the war, it pains me to know that my father's death was in vain. Colonialism and apartheid have forced the displacement of so many people in the past, and this is why my father took up arms in the first place. He was killed because he was fighting an oppressive system that did not value our lives as Namibian Black or Indigenous peoples, and this same lack of acknowledgement of other lives is what ReconAfrica represents.

Today, I see that investment and development are no different to the concept of colonialism. Over the past 500 years, African people have been oppressed and have lost their lands to foreigners. ReconAfrica is not only coming to Namibia to pollute our water and destroy our environment and ecosystem, it is threatening to disrupt the ways of the Kavango and the San peoples who live off the lands working as subsistence farmers and hunter gatherers. And the Kavango Basin is also home to the Okavango Delta, the habitat of the largest living population of endangered African elephants and of many more threatened species.

In spite of the intimidation and death threats I continue to face, my calling to protect and fight against this company has become vital, not only in an attempt to save my people and my homeland, but because there is only one Kavango, only one Okavango Delta. What happens in Kavango will not stay in Kavango!

Ayisha Siddiqa

I was born in the northern region of Pakistan and was raised with the belief that, just as your phenotype is composed of DNA from your parents, your spirit is made up of the spirits that came before you. My grandparents don't watch over me, they live within me. That's why the fight for climate justice is, for me, a fight for love. This world is held together by memories of the ones we love and I am trying to preserve them while I still have time.

My work is in equal parts motivated by pain as it is by love. The region I am from, the South West Asian / North African region, has paid with blood for oil for the past thirty years. What is, for the Global North, a conversation about carbon emissions, is a reality of hunger, homelessness, helplessness, and indescribable suffering for us. With so many geopolitical actors at play – militaries, terrorist groups, presidents and dictators – it is no coincidence that the very people who have experienced the threat of extinction by war, imperialism and white supremacy, also understand the pain of the Earth, and the danger of fossil fuels. There is no easy way to say this: moderation, phasing out and false net zeros will eventually kill us.

We need to change our thinking. We cannot allow the same socio-economic systems that are leading us to our own destruction to be the foundations of a new world. We must learn from the people who are still alive after power and greed tried to kill them over and over again. We must learn that gentleness and harmony are not weaknesses; these are the traits of our mothers. These are the things that have kept us alive.

Mitzi Jonelle Tan

On a cloudy afternoon in August 2017, one of the leaders of the Lumad, an Indigenous group in the Philippines, told me something that would change my life. As he was telling us about how the Lumad were being harassed, displaced and killed for protecting their land, he said, with a slight shrug and a chuckle, 'We have no choice but to fight back.'

It was that simple. I had the privilege to choose to be an activist, yet there are people on the front lines, such as the Lumad, whose very existence pushes them to resistance. But at this point, he's right: none of us have any choice but to fight back.

The Philippines is one of the most climate-vulnerable places in the world, despite contributing so little to this global crisis. My country is also one of the most dangerous places on the planet to be an environmental defender. It isn't fair that we have to grow up full of fear. Fear of the next bang of thunder from the storms that will wash away our homes. Fear of the next bang on our door from the police who will whisk us away from our loved ones.

As typhoons destroy our homes and floods rise, the people are rising up to break down systemic oppression. There's a growing movement in my country, led by small farmers, fisherfolk, Indigenous peoples and workers fighting for liberation. Together, we fight for land for the tillers, for reparations for injustices perpetrated under imperialism, for a just transition into a greener society and for a world with a united community full of love and cooperation.

This is what we mean when we talk about equity. Equity is justice. Equity is liberation. Equity is what we need, so there is no choice but to fight back. /

There cannot be sustainability without equity, and there cannot be equity without sustainability. Climate justice must manifest everywhere, for everyone.

5.18

Women and the Climate Crisis

Wanjira Mathai

In Kenya, where I am from, and across much of Africa, women are the backbone of the local community, the family, the small enterprises and the farms. In Africa's urban centres, towns and villages, you can see women walking briskly and with purpose as early as 5 a.m. every day, hugging the edges of unpaved highways and parched roads. Who are they? Many are the heart of an informal economy – the invisible nucleus of a continent straining under the impact of an invisible force.

Africa is one of the most vulnerable continents to climate change because it is so dependent on agriculture, and agriculture, in turn, is highly dependent on the climate. Only 5 per cent of cultivated land is under irrigation, and most agriculture across the continent is rain-fed. African yields are already suppressed relative to other regions, and the IPCC projects that further climate-induced yield reductions are highly likely this century, especially for cereal crops such as maize, which is the continent's most important and widespread staple crop.

We think that women comprise, on average, 43 per cent of the agricultural labour force in developing countries, but we cannot say for sure. Collecting data on these women is difficult because many of them are part of an informal labour force. They often do not own the land they work on. They often do not pay taxes. They do not enjoy any workers' rights. They do not have health insurance. They do not use formal childcare services. They do not have 'data points' to make them visible, despite their enormous contribution to the economies of African countries. But we do know that they are over-represented in unpaid and underpaid, seasonal and part-time work.

As the primary custodians of farms, households, food and water, rural women are disproportionately vulnerable to the effects of climate change. They are the hardest impacted by the dwindling of rural employment opportunities, due to lack of education, traditional perceptions of gender roles, lack of social mobility and a host of other sociocultural factors. But they are also a major part of the climate solution in Africa. They have unique knowledge and skills that can help make the response to climate change more effective and sustainable.

In most of Africa south of the Sahara, since women rarely own land, women farmers usually access it through a male relative, leaving them highly vulnerable to a change in that man's circumstances, or his mind. But when women possess land, and the seeds and tools to work it, they possess agency to adapt to climate change.

The women of the Green Belt Movement are a good example of this. This non-governmental organization was founded in 1977 by Wangari Maathai[1] to empower communities in Kenya, especially rural women and girls, to conserve their environments and protect their livelihoods. More than engaging with tree planting on our landscapes, the Green Belt Movement invests in ensuring that women understand their connection to the land and the degradation it faces. The women work in groups to establish tree nurseries, taking turns to nurture the seedlings and prepare them for the planting season. One of the group leaders, Nyina wa Ciiru, gathers the women in her group once a week under her mango tree to discuss the condition of their nursery and whether the seedlings are ready for planting. Together, they take turns watering the seedlings, often singing in unison as they work. When the seedlings are 2 feet tall, they decide where the seedlings will be planted – on their farms, in their children's school compounds, in marketplaces, along a river, or anywhere they feel is in need of trees. Today, because of the partnership the Green Belt Movement has developed with the Kenya Forest Service, the seedlings are also planted in nearby government-run forests.

When the Green Belt Movement was founded, over forty years ago, the women in the movement always planted the seedlings on their family land first: fruit trees, fodder trees, shade trees and trees that gave them firewood for cooking. They noticed that when they planted trees around their farms, forming belts of green, the birds came back, their families had lots of delicious fruit to eat and their homesteads were cooler even during the hottest time of day. Trees, they believed, were the source of everything good.

After they had planted enough trees on each of their farms, they moved on to planting trees on public land. They taught others to plant trees, and the most beautiful part is the joy it brought them. These women became the

---

1   Wangari Maathai is the founder of the Green Belt Movement and a 2004 Nobel Peace Prize Laureate.

primary suppliers of seedlings in their communities, ensuring that everyone participated in tree planting and that their farms were covered in green vegetation. It is women like these, women who protect the soil and produce food for the community, who are the landscape guardians and climate activists of our time.

These communities have enlisted scores of women to mobilize and begin the important work of planting trees. And now they are everywhere. They are in the homes, on the streets and in the fields, and we need to give them the opportunity to prepare the whole continent for what is coming. As Wangari Maathai so aptly said, 'In the course of history, there comes a time when humanity is called to shift to a new level of consciousness, to reach a higher moral ground . . . A time when we have to shed our fear and give hope to each other. That time is now.' As breadwinners, entrepreneurs and providers of food, shelter and education for their children, women will not surrender their livelihoods to climate change. They will prepare. They will adjust, and they will adapt. They just need the means to do so. It is incumbent upon governments to ensure that policies, laws and financial institutions support the backbone of our societies to the fullest because, if they break, we all will. /

When women possess land, and the seeds and tools to work it, they possess agency to adapt to climate change.

Decarbonization Requires Redistribution

Lucas Chancel and Thomas Piketty

Let's face it: our chances of staying under a 2°C increase in global temperature are not looking good. If we continue business as usual, the world is on track to heat up by at least 3°C by the end of this century. At current global emission rates, the carbon budget that we have left if we are to stay under 1.5°C will be depleted in six years. The paradox is that, globally, popular support for climate action has never been so strong. According to a recent United Nations poll, 64 per cent of people around the world see climate change as a global emergency. So, what have we got wrong so far?

There is a fundamental problem in the contemporary discussion of climate policy: it rarely acknowledges inequality. Poorer households, which are low CO_2 emitters, rightly anticipate that climate policies will limit their purchasing power. In return, policymakers fear a political backlash should they demand faster climate action. The problem with this vicious circle is that it has lost us a lot of time. The good news is that we can end it.

Let's first look at the data. In 2021, an average human being emitted about 6.5 tonnes of greenhouse gases. This average, however, masks huge inequalities. The top 10 per cent of emitters eject on average about 30 tonnes per year per person, while the poorest half of the population emits about 1.5 tonnes per year per person. Put differently, the top 10 per cent of the world's population are responsible for about 50 per cent of all greenhouse gas emissions, while the bottom half of the world contributes just 12 per cent of all emissions (Fig. 1 and Fig. 2).

Over the past three decades, the share of emissions of the global top 1 per cent of emitters (a group fifty times smaller than the global bottom 50 per cent) rose from around 9.5 per cent to 12 per cent. In other words, global carbon inequalities are large, but the gap between the very top and the rest of

Average per capita emissions by income group in 2019

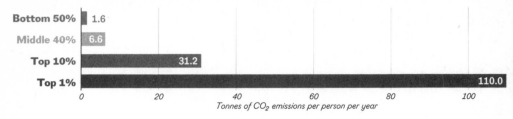

Bottom 50%	1.6
Middle 40%	6.6
Top 10%	31.2
Top 1%	110.0

Tonnes of CO₂ emissions per person per year

Group contribution to world emissions in 2019

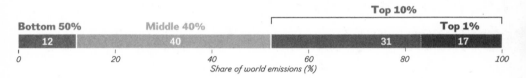

Bottom 50%	Middle 40%	Top 10% / Top 1%
12	40	31 / 17

Share of world emissions (%)

Figures 1, 2 (above) and 3 (overleaf): All figures are expressed in CO2-equivalent: they include CO2 and other greenhouse gases. Personal carbon footprints include emissions from domestic consumption, public and private investments, and imports and exports of carbon embedded in goods and services traded with the rest of the world. Estimates are based on the systematic combination of tax data, household surveys and input–output tables. Emissions are split equally within households.

the population has been growing over time. This is not simply a rich versus poor countries divide: there are huge emitters in poor countries, and low emitters in rich countries.

Consider the US, for instance. Every year, the poorest 50 per cent of the US population emit about 10 tonnes of CO₂ per person, while the richest 10 per cent emit close to 75 tonnes per person. That is a gap of more than seven to one. Similarly, in Europe, the poorest 50 per cent emit about 5 tonnes per person (less than the global average), while the richest 10 per cent emit about 30 tonnes – a gap of six to one. In East Asia, and in China in particular, the richest 10 per cent have a higher carbon footprint than the richest Europeans. Poorer world regions also exhibit significant inequalities, although it is necessary to zoom in on very wealthy groups (i.e. the top 0.1 per cent or above) to observe levels of emissions that are broadly comparable to those observed for wealthy groups in rich countries.

We stress that a lot remains to be done to precisely measure carbon inequalities. Governments should publish up-to-date numbers every year – at least as often as they publish GDP and growth statistics. We provide updated figures on carbon inequalities on the World Inequality Database (wid.world). Such information is necessary to design and evaluate any successful climate transition pathway.

Where do the large carbon inequalities that we have documented come from exactly? The rich emit more carbon through their direct carbon emissions (that is, the oil they put in their cars), but also from the goods and services they buy, as well as from the investments they make. Low-income groups emit carbon when they use their cars or heat their homes, but their indirect emissions – that is, the emissions from the stuff they buy and the investments they make – are significantly lower than those of the rich. As we show in our recent World Inequality Report (2022), the poorest half of the population within each country in the world barely have any wealth, meaning that they have little or no responsibility for emissions associated with investment decisions.

Why do these inequalities matter? After all, shouldn't we all reduce our emissions? Yes, we should, but obviously some groups will have to make a greater effort than others. Intuitively, we might think here of the big emitters – the rich – right? True, and also poorer people have less capacity to decarbonize their consumption. It follows that the rich should contribute the most to curbing emissions, and the poor be given the capacity to cope with the transition to 1.5°C or 2°C. Unfortunately, this is not what is happening – if anything, what is happening is closer to the opposite.

In France, in 2018, the government raised carbon taxes in a way that hit rural, low-income households particularly hard, without much affecting the consumption habits and investment portfolios of the well-off. Many families had no way to reduce their energy consumption. They had no option but to drive their cars to go to work and to pay the higher carbon tax. At the same time, the aviation fuel used by the rich to fly from Paris to the French Riviera was exempted from the tax change. Reactions to this unequal treatment eventually led to the reform being abandoned. These politics of climate action, which demand no significant effort from the rich yet hurt the poor, are not specific to any one country. Fears of job losses in the automobile, fossil or heavy metal industries are regularly used by business groups as an argument to slow climate policies.

Countries have announced plans to cut their emissions significantly by 2030, and most have established plans to reach net zero somewhere around 2050. Let's focus on the first milestone, the 2030 emission reduction target: according to a recent study, as expressed in per capita terms, the poorest half of the population in the US and most European countries have already reached or almost reached the target. This is not the case at all for the middle classes and the wealthy, who are well above – that is to say, behind – the target.

One way to reduce carbon inequalities is to establish individual carbon rights, similar to the schemes that some countries use to manage scarce environmental resources. For instance, in France, in times of stark water shortages, it is possible to prevent all water use that is not strictly essential (that is, for drinking, sanitation, cooking or emergency uses). This approach amounts to equalizing water consumption across the population.

Equal individual carbon quotas set by the authorities of a country would inevitably raise multiple technical issues, but from a social justice standpoint, it is a strategy that deserves attention. There are many ways to reduce the overall emissions of a country, but the bottom line is that anything but a strictly egalitarian strategy inevitably means demanding greater climate mitigation efforts from those who are already at the target level, and less from those who are well above it. This is basic arithmetic.

Arguably, any deviation from an egalitarian strategy, for example quotas, would justify serious redistribution from the wealthy to the worse off to compensate the latter. Many countries will continue to impose carbon and energy taxes on consumption in the years to come. In this context, it is important that we learn from previous experiences. The French example shows what not to do. In contrast, British Columbia's implementation of a carbon tax in 2008 was a success – even though the Canadian province relies heavily on oil and gas – because a large share of the resulting tax revenues goes to compensate low- and middle-income consumers via direct cash payments. In Indonesia, the ending of fossil fuel subsidies a few years

Per capita emissions across the world in 2019

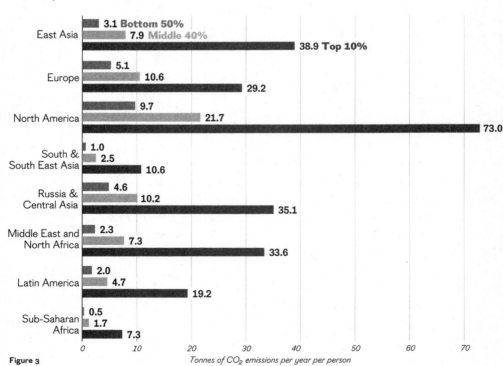

Figure 3 *Tonnes of CO$_2$ emissions per year per person*

ago meant extra resources for its government but also higher energy prices for low-income families. Fiercely opposed at first, the reform was accepted when the government decided to use the revenue to fund universal health insurance and support for the poorest.

To accelerate the energy transition, we must also think outside the box. Consider, for example, a progressive tax on wealth, with a pollution top-up. This would accelerate the shift out of fossil fuels by making access to capital more expensive for the fossil fuel industries. It would also generate potentially large revenues for governments which they could invest in green industries and innovation. Such taxes would be more equitable, since they target a fraction of the population, not the majority. At the world level, a modest wealth tax on multimillionaires with a pollution top-up could generate 1.7 per cent of global income, as has been shown in recent research. This could fund the bulk of extra investments required every year to meet climate mitigation efforts.

Whatever the path chosen by societies to accelerate the transition – and there are many potential paths – it's time for us to acknowledge that there can be no deep decarbonization without profound redistribution of income and wealth. /

The rich should contribute the most to curbing emissions, and the poor be given the capacity to cope with the transition to 1.5°C or 2°C.

5.20

Climate Reparations

Olúfẹ́mi O. Táíwò

The climate crisis is the culmination of centuries of racial injustice –
an injustice that built itself into the very structure of our energy system, our
economic networks and our political institutions. The task of racial justice
and climate justice depends on our meeting this challenge at the scale it
requires: remaking the world.

This is no metaphor. As the activists who challenged the colonial political system in the 1960s and '70s realized, justice requires us to reconstruct
our political and economic systems on a planetary scale. Political theorist
Adom Getachew has called this ethos and ambition 'worldmaking'.

This may seem daunting, for the world is a complex thing made of
many moving parts – but it is a real system that we can, and must, try to
understand. It's often poorly described with the static metaphor of a 'blue-
print', or a chart of institutional hierarchies, while in reality our politics and
economics are in constant motion. Instead, we might think of our trade and
political system as something like a web of aqueducts that spans the globe.

Rather than flowing water through its channels, however, this aque-
duct system produces and distributes social advantages and disadvantages:
wealth and poverty, finished products and pollution, medical knowledge
and ignorance. The resulting distribution of advantages and disadvantages
does not passively reflect the inherent civilizational merit and genius of the
parts of the world where the advantages pool, any more than it reflects the
inherent unworthiness of the parts of the world that drown in disadvantages.
It reflects, instead, centuries of human efforts and decision-making.
Deliberate attempts to create an unjust social structure, failed attempts
to create a just one, and attempts to manage the consequences of both –
all have combined, over time, to form the structure that determines our
present circumstances and constrains future possibilities. These historic
aqueducts allow us to predict where future streams of advantages and disadvantages will naturally run, and where they won't – at least, if the channels
are left where they are now.

The construction of today's world came by way of the global racial
empire: the historically unprecedented colonial conquest and racial slavery

that began in the 1400s. At the beginning of this period, European empires did not stand atop the global political hierarchy – to the contrary, they were essentially middlemen in a vast trade and political network that was centred in Asia. But by the end of it, they had created a planetary system of economic dominance. They built this system using colonies on land that they secured through the domination and elimination of Indigenous peoples, and which they made productive with the labour of enslaved and trafficked Africans, both on a historically unsurpassed scale.

In the 1700s and 1800s, the British empire combined its network of colonies and slave labour with new coal- and steam-powered technologies to massively increase production and mechanize work, resulting in the Industrial Revolution. It is this same Industrial Revolution, and the changes in global energy use and thus carbon emissions it wrought, that scientists treat as the beginning of our era of anthropogenic climate change.

The same history of global racial empire that produced the Industrial Revolution and the climate crisis has given us the networks and channels that direct advantages and disadvantages to different people and different places around the world today. The Global North – which comprises those countries that were atop the hierarchies built in past centuries – contains the lion's share of wealth, political power, research capacity and other social advantages. The Global South, which was disproportionately home to those colonized and otherwise exploited over the same years, contains the lion's share of poverty and pollution. Black and Indigenous peoples, within and across these geographic distinctions, tend to accumulate the least advantages and the most disadvantages when compared to their neighbours.

The injustices that built our present order demand to be addressed. They are not one-off events to be merely apologized for or acknowledged: they have sedimented themselves into the very structure of how we live together on this Earth. We will need to alter that structure if we want to truly address the harms of the past.

This insight underlies the 'constructive' approach to reparations for slavery and colonialism: we must build pipes that will funnel advantages to the previously disempowered, that will make those enriched and empowered by yesterday's injustice take up their fair share of the global burden of responding to the climate crisis and protecting our life on this planet.

What kinds of specific action does this call for? We should start with a goal that has been rightly viewed as central in the long history of Black radical agitation for reparations: giving cold, hard cash to the people most deprived by the aqueducts of history. This includes giving cash unconditionally to people. In the US, a number of strategies have been proposed:

William Darity and A. Kirsten Mullen have endorsed a strategy rooted in direct payments to African Americans descended from those enslaved in the US, to be governed by a National Reparations Board that would empower recipients to research and make decisions about the funds. Scholar and organizer Dorian Warren, of the Economic Security Project, has proposed a universal basic income for all which would add an additional amount for African Americans to account for the owed reparations. Beyond the US, others have also suggested a global universal basic income, which could be weighted along the lines Warren suggests.

Unconditional transfers of cash aren't just for individuals or households. Redirecting the historic currents of capital can – and must – also happen at the level of countries and multinational institutions. Rich nations promised to do exactly this under the framework of the United Nations Green Climate Fund, but both under-promised and under-delivered: the $100 billion target 'does not even come close' to the amount of money needed for developing nations to address the climate crisis, and the world's rich nations have not provided even a fraction of that meagre pledge. Private investors and corporations have offered to bridge the gap, but reliance on the market got us into this mess in the first place.

To the extent that our world-building efforts require financial resources, a better bet is to apply direct political pressure to private institutions rather than putting them in the driver's seat. Such 'divest–invest' strategies would use activism to drive funds out of fossil fuels and other polluting industries and into projects that promote the public good: cash would flow to Black and Indigenous households and communities, to publicly owned renewable energy generation and storage, to rural broadband and urban community orchard projects. We can do this with a planetary scope in mind, pairing this tactic with pressure on the trillions of dollars' worth of wealth hoarded in tax havens across the globe.

But we should not fall into the trap of being overly focused on redistributing money within the same political and economic system. True world-making calls for us to rebuild the system itself, and not just the attempt to compensate for its unequal allocation of resources. That means redistributing power directly by challenging how political decisions are made in the first place.

Within our present system, private corporations have unilateral and authoritarian control over whole swathes of public life: working conditions, the provision of utilities such as energy and water, supply chains for both dirty and clean energy. Fossil fuel companies and other private interests have also wormed their way into democratic processes by way of legal and illegal bribes, turning legislators and regulators alike into accomplices. An important alternative is the idea of 'community control' – an ethos with a long history, but which was championed by radical groups such as the Black Panther Party in the 1960s and '70s as they organized campaigns for community-level democratic decision-making over land, housing, education and even policing.

We already have long-standing examples of this ethos at work in real political systems. The Brazilian Workers' Party pioneered 'participatory budgeting' in Porto Alegre in the 1980s, putting city residents in direct democratic control of the use of public funds. This approach has travelled widely since: it is built into all levels of governance in the Indian state of Kerala and is used to effectively manage public spending in cities such as Maputo and Dondo in Mozambique. In Kenya, the 'Harambee movement' has resulted in formal government funds being dedicated to tens of thousands of 'community self-help' programmes which force legislators to materially serve the people they represent. Even in the Global North, activists fight for 'energy democracy': public ownership and democratic control over energy, to replace investor control over important determinants of people's lives.

We can go even further. We should aim to build and justly distribute freedom-promoting practical affordances: that is, parts of the world that we can use to build safe, meaningful, self-determined lives. Redistributing money and abstract political power are important aspects of doing this, but we should also take this very literally and build physical structures and management systems that will help us create a just and climate-resilient world. We need to make and justly distribute physical drainage systems for flood protection; to build energy-efficient new public housing; to retrofit upgrades to existing housing; and to develop secure, resilient energy transmission and storage infrastructure.

If climate justice and racial justice are about world-making, then ultimately justice is a design project: we are trying to restructure the unjust world. Dollars alone will not redress the problems caused by uranium mining in the Navajo Nation or Niger, nor will they by themselves address the long-term pollution caused by fossil-fuel-extraction zones in the Niger Delta. We need to address environmental problems specifically and directly, while also challenging the power hierarchies that caused them.

As with politics, we have many existing examples to follow. Bangladesh is one of the world's most climate-vulnerable countries, but it is also a leader in climate adaptation, as Saleemul Huq explained in Part Three. Bangladesh's comprehensive system of disaster preparedness includes both physical structures, such as dykes to manage flood risk, and social measures. It has built emergency food distribution programmes and integrated disaster response into education programmes which provide emergency evacuation protocols to make sure the elderly are not left behind in crises. Farmers in Hanoi and Kolkata have devised natural waste management systems that replenish nutrients in agriculture and aquaculture without requiring the use of industrial fertilizers. And, in cities across the US and Canada, community orchard and tree-planting projects are bringing food into the commons, rather than under the control of private profiteers, shoring up food sovereignty and creating future oases from rising urban heat.

Whatever designs we come up with, we will have to literally remake the world – this time for the many, rather than the few. That is a task accomplished with hands, feet and shovels, not with accounting tricks or empty promises. We have nothing to lose and a world to gain. /

If climate justice and racial justice are about world-making, then ultimately justice is a design project: we are trying to restructure the unjust world.

Mending Our Relationship with the Earth

Robin Wall Kimmerer

Where is the snow? It's December and forty degrees warmer than it should be. Glaciers melting, raging wildfires, towns shredded by epic tornados – grief is everywhere. The most I can hold at the moment lies in my hands: a fallen oriole's nest that was blown down from the bare winter branches in an unseasonable thunderstorm. This little pouch woven of roots and bark was a container for incipient birdsong and is now the container for my grief.

My heart breaks at visions of climate refugees fleeing drought and floods, storms and starvation. The world is full of climate migrants – it's estimated that 30 million people in 2020 were displaced by the floods, droughts, wildfires and heatwaves that are increasing in frequency and intensity as a result of climate change. What about the bird people and the forest beings? What of their removal, their uncounted suffering?

My orioles fly back and forth between northern New York and Central America. They are safe when they're here with me but traverse a broken landscape on their way to wintering grounds. Sixty per cent of all songbirds have been lost in my lifetime. The odds are against them returning this spring.

This fallen nest, as every bird nest, beaver lodge, bear den and womb, is in the shape of a bowl. It's a sacred shape, the shape that nurtures life. My Anishinaabe people, as well as the Haudenosaunee people who are my neighbours, have adopted the bowl as the symbol for the nurture and provisioning of the land. We have agreements with one another, known as the One Bowl, One Spoon treaties. The land is understood as the Bowl, filled by Mother Earth with everything that we need. It is our responsibility to share it and keep that bowl full. How we take from the bowl is represented by the spoon. There's just one spoon, the same size for everyone, humans and more-than-humans alike. Not a tiny one for some and a gouging shovel for others. One of the oldest 'conservation policies' on the planet is a statement about sharing, about justice, about reciprocity with the gifts of the land.

After a long winter I welcome every returning bird with delight, from the first raucous red-winged blackbirds to the crescendo of warblers, but none with more joy than my orioles. We greet each other with what feels like mutual joy when they announce their arrival; they with their clear song like musical sunshine and me with an open-arm twirl of love and relief at their safe return. They are loyal to one grandmother maple, where they have raised babies who have faithfully returned for decades now. They join me at dawn in the morning thanksgiving and at twilight as I put the tools away. Hoeing the corn or reading in the shade, my summer is made of oriole song and flittering glimpses of orange and black, like tiger lilies in flight.

Did you know that, according to a recent study of human mental health, psychological well-being is strongly correlated with the presence of birdsong? Of course you did.

The little 7-acre patch of land I care for is songbird paradise. With a combination of intention and benign neglect, the land and I have together fostered thickets, groves, flowery meadows and wetlands that call birds from the sky. My neighbours out here in farm country tend to uniform lawns, pastures, hay and cornfields. It's green and pastoral but designed for human economy. My shared-fence-line neighbour thinks I've ruined my pasture with thickets and brambles, but it's a choir of birdsong, toad song, frog song, bug song and a sparkling lightshow of fireflies in July. My neighbour and I have different definitions of wealth.

The land is a sharp reflection of the worldview of the peoples who care for it, or don't. My orioles fly over hundreds of miles of land degraded by the outcomes of the western worldview, miles of pavements, mines, oil rigs, frack wells flaring methane, industrial sacrifice zones, urban sprawl. Many of the green places are monocrop agriculture – fields or forest plantations toxic with herbicides and nothing to eat. This, the worldview of human exceptionalism, understands land primarily as natural resources, property, capital and ecosystem services; this worldview is not One Bowl, One Spoon, but the land as a warehouse of commodities where the spoon is the property of just a few members of a single species. My beloved singers navigate this wasteland, looking for a place to rest. They must be as grateful as I for the patchwork of protected land, public and private, refuges, parks and forests. These intact places are ever more critical, not only for shelter for other species but for purifying air, sequestering carbon and making it rain.

These are places, islands in a sea of loss, where the birdsong still rises and insects stitch together the fabric of the land, pawprints follow ancient trails, fish still guard the water as they were asked to do and where human people have not forgotten their gifts and their responsibilities.

From a bird's-eye view they are full bowls of life, forest islands beckoning them to safety. If you look at a map of 'biodiversity hotspots', those places remaining on the planet where ecological integrity is intact, where species richness is highest – they overlap to a very high degree with hotspots of cultural diversity, with the homelands of Indigenous peoples.

Estimates indicate that 80 per cent of the world's remaining biodiversity is sheltered in lands under the care of Indigenous peoples. A 2019 report from the United Nations found that biodiversity is declining perilously all over the planet, but the rates of loss are dramatically lower in areas under Indigenous control. After centuries of colonial land dispossession, genocide, forced assimilation and attempted erasure of the Indigenous worldview, the dominant society is now waking up to the understanding that what it once sought to exterminate is essential to survival today. My elders spoke of this time. Against all odds, they protected our knowledge, our philosophy, our sacred One Bowl, One Spoon worldview against the colonial onslaught, because, they said – with prophetic clarity – there would come a time when the whole world would need it. The humans, the waters and the orioles too.

Many orioles spend the winter in the tropics of Mexico. The Yucatán Peninsula, housing the great Mayan Forest, is a vast biodiversity hotspot, where Indigenous land care nurtures the well-being of people and their more-than-human relatives. I'm told that orioles are beloved there as well, and people slice open oranges from their gardens to greet them, like delicate orange bowls of welcome. I fantasize that my orioles migrate from my little patch of Indigenous Potawatomi land care to a verdant patch stewarded by a Mayan family on the Yucatán Peninsula. Traditional Mayan communities use sophisticated practices of silviculture, working with the successional processes of the land – its cyclical developmental changes – to continually renew the forest. We grow corn, beans and squash for our families and forests and thickets and berry-rich hedgerows for the other species, because we recognize that the world does not belong to us, that we are all fed from the Bowl with One Spoon, human people, salamander people, tree people and oriole people. We are all related, woven together in webs of reciprocal connection, where what happens to one happens to all. I like to think that my orioles join the morning prayers of their Mayan family just as they join mine. But all around them, the forces of warehouse thinking, the worldview of human exceptionalism, threaten their homelands.

Why are Indigenous homelands also biodiversity hotspots? At the superficial level of geography, it is undeniable that remnant tribal lands are often in remote places deemed inhospitable to the colonizing forces of development. These lands and peoples survive due to fierce defence by Indigenous land protectors, from the Arctic to the tropical rainforest. But the causes of the teeming species richness go far deeper than geography and governance. Biodiversity thrives in Indigenous homelands because of the land's response to traditional land care practices, which are grounded in Indigenous science or Traditional Ecological Knowledge. It is the way we keep the Bowl with One Spoon full. Such practices, which western conservation terms 'land management', are myriad; they represent locally evolved adaptive strategies which enhance biodiversity. Some of these practices are becoming well known to mainstream conservation, like the skilful application of prescribed fire, methods for carbon sequestration, intentional habitat creation, agroforestry. For centuries, such practices of Indigenous science were dismissed as unscientific and destructive. My ancestors could have been jailed for using their fire knowledge for the good of the land. Currently, western science is beginning to take off its colonial blinders and there are glimmers of understanding of the brilliance of Indigenous science. These carefully tended cultural landscapes offer western science a glimpse of how people and land can be a source of mutual thriving. They are libraries of ancient knowledge and every single one is threatened.

Our work is clear. It's not enough to revere Indigenous wisdom. We have to fiercely defend land rights for Native peoples. It's not enough to uphold teachings like the Honourable Harvest as paragons of virtue and sustainability. Everyone must become a humble student and learn to live as if they were native to place, as if the Earth were One Bowl and One Spoon, to live as if the future were in our hands. Because it is.

Indigenous homelands are the finger in the dyke holding back a flood of extinction. Yet only 10 per cent of those lands are legally protected with Indigenous title. And all are being encroached upon by corporate, private and governmental interests all over the world. This crisis demands that the world's governing bodies, nations and states prohibit any further loss of Indigenous homelands and strengthen protections. They must uphold the hard-won provisions of the United Nations Declaration on the Rights of Indigenous Peoples and ensure that climate mitigation practices do not displace Indigenous peoples from their homelands in a new, green colonialism. Indigenous peoples have been at the forefront of climate warnings, have suffered disproportionate climate impacts and have created visionary approaches to climate justice, mitigation and adaptation.

Collectively, dominant society has a responsibility to elevate Indigenous voices in climate justice leadership.

Climate action must prioritize nature-based solutions for mitigation, supporting the plants in doing what they do best: absorb carbon, store it away, regulate microclimates, cool the planet, generate oxygen, rebuild soil and make rain. The movements that call for protection of half the world's lands from development are essential for reducing climate impacts. However, Indigenous homelands show us clearly that people and nature can coexist and even promote mutual flourishing. It's not a matter of locking up 'nature' in one place and being granted permission to wreck it elsewhere. The call for land protection cannot be one of removing Indigenous and local people from land, but of harmonizing people and land, of aligning economies with the laws of nature. Let's remember that ecology and economy share the same root word, *oikos*, the Greek word for home.

Our work is not just to protect the remnants of biodiversity but to restore them with a combination of the tools of environmental science and the philosophy and know-how of Indigenous knowledge. Restoration must also include restoration of an honourable relationship with land, of re-storyation, the adoption of a new narrative for the relationship between people and place. One that asks not 'What more can we take from the Earth' but 'What does the Earth ask of us?'

Meaningful climate action rests on many changes. We have to change tax structures, laws, policies, industries, governance, technologies, ethics, but at root the most important thing we have to change is ourselves. Transformation of worldview from Warehouse to Bowl is a spiritual change. David Suzuki has written that 'spirituality may be our chiefest adaptation – the means by which we touch the sacred, hold together against disintegration. The forms and varieties of spiritual belief and ritual among cultures on Earth may be another example of evolution's incredible, extravagant invention of ways for life to survive.'

When I listen to that song from the orioles and all their feathered kinfolk, to me it sounds a wake-up call. I am heartened by how many have risen to a new consciousness and are giving heart and soul to the transformation we need. There are breathtaking examples of Indigenous and ally leadership in the protection of land and waters, of restoration and healing, in seeding old/new ideas into law through the application of Indigenous principles, from the UNDRIP to Rights of Nature. We have to celebrate it, even while acknowledging that it's not yet enough to stem the tide of climate disruption.

Why is it not enough? Because even with alarms sounding everywhere, so many have not woken up. I've come to think that the sleepers are drugged with a powerfully addictive narcotic of material wealth and spiritual poverty. I almost can't blame them. If you wake to a world in which all that is asked of you is to be a good consumer and a passive bystander, wouldn't you put the blanket back over your head too? It is fear and powerlessness that keep people from awakening, intentionally produced by a worldview which understands Mother Earth to be nothing more than stuff to be consumed. Instead of living in a world blessed with a wealth of species richness, sacred water and living mountains, they live in a world of rapidly dwindling stuff. What is it that makes someone wake up, put their feet on the ground and get to work? For too long, fear has been that tool, and here we are, still floundering as the climate clock ticks. I don't think it's fear that we need.

I'm often asked: where do you find hope in these dark times? I'm not sure I really know what we mean by hope. A source of optimism? Wishful thinking? Evidence of a turning towards life and away from destruction? I don't know about hope, but I do know about love. I think we are in this perilous moment because we have not loved the Earth enough, and it is love that will lead us to safety. I'm dreaming of a time when we are propelled not by fear of what is coming towards us, fearsome as it is, but by love for a beautiful vision of a world whole and healed. One of the great gifts of Indigenous environmental philosophy is that it provides that expansive vision of what it means to be a human: it is an invitation to be a member of the sacred web of life, to belong. As we join the oriole in singing thanks to the Earth, we can live in such a way that the Earth will be grateful for us.

As I have travelled and listened, I have been deeply moved by the myriad manifestations of people's love for the land, by their deep longing for a different way of being which celebrates the joy of reciprocity, of giving back to the Earth in return for all we've been given.

In my culture, a warrior is not someone who is motivated by fear or power, but someone called by love. Not the sentimental, pink-hearts kind of love, but the kind of love that makes sacrifices for the well-being of the other, who puts the beloveds ahead of themselves. Let us ask each other, what do you love too much to lose?

For me, my acts of love for the land are teaching and writing and science and voting, raising good children, raising a garden and raising a ruckus when needed. This is how that love calls to me: I will do the big things and the small things, even though I don't know which is which. I will work for system change. I will write for cultural change. I will tend my patch of berry-full ground with science and with love, so that the One Bowl is filled, for my grandchildren and for the grandchildren of orioles.

Listen. How does love call to you? /

Hope is something you have to earn

Greta Thunberg

Right now, we are in desperate need of hope. But hope is not about pretending that everything will be fine. It is not about sticking your head in the sand or listening to fairy tales about non-existent technological solutions. It's not about loopholes or clever accounting.

To me, hope is not something that is given to you, it is something you have to earn, to create. It cannot be gained passively, through standing by and waiting for someone else to do something. Hope is taking action. It is stepping outside your comfort zone. And if a bunch of weird schoolkids were able to get millions of people to start changing their lives, just imagine what we could all do together if we really tried.

The transformation we need in order to stay below 1.5°C or even 2°C of warming may not be politically possible today. But we are the ones who determine what will be politically possible tomorrow. We now live on a planet where technology has allowed nearly all of us to be connected to each other. In some nations, the political regime does not allow this. But still, if something big enough happens somewhere around the globe, then nearly everyone will instantly know about it. This opens up a whole new realm of possibility. No one yet knows what we are capable of once we collectively decide to respond to change. I am convinced that there are social tipping points that will start to work in our favour the minute enough of us choose to take action. The possibilities that follow are infinite.

The destruction of the biosphere, the destabilization of the climate and the wrecking of our common future living conditions are in no way predestined or unavoidable. Nor is it human nature – we are not the problem. This is all happening because we, the people, haven't yet been made fully aware of our situation, or of the consequences of what is about to happen. We have been lied to. We have been deprived of our rights as democratic citizens and left unaware. This is one of our biggest problems, but it is also our greatest source of hope – because humans are not evil, and once we understand the nature of the crisis we will surely act. Given the right circumstances, there are no limits to what we can do. We are capable of the most incredible things – the ability to change our minds, to invent, to forgive. Once we have been given the full story – and not something that has been conjured up to benefit certain short-term economic interests – we will know what to do. There is still time to undo our mistakes, to step back from the edge of the cliff and choose a new path, a sustainable path, a just path. A path which leads to a future for everyone. Not just for those who think their money can buy them a way of adapting to dying ecosystems and mass extinctions. And no matter how dark things may become, giving up will never be an option. Because every fraction of a degree and every tonne of carbon dioxide will always matter. It will never be too late for us to save as much as we can possibly save.

Some of the people with the strongest voices in the climate movement today were barely even aware of this crisis a few years ago and now they are a key part of changing the fate of humanity. I believe that in the years to come this phenomenon will keep repeating itself – and this is where you come in. You see, this is the end of the book. It is where I am supposed to round up my thoughts and write some inspirational words worthy of last sentences. But I will not do that. Instead, I will leave that to you. Because some of the best ways of igniting the changes we need have not yet been discovered. It is my belief that the best ideas, tactics and methods are still out there, yet to be thought of. Some have been tried, and some have failed because the timing was wrong – because the level of public awareness was not high enough at the time. So we must try them again.

Things are changing, faster and faster. And all those changes have been made possible by the people who pioneered the climate and environmental movement. The scientists, the activists, the journalists, the writers. Without them we would not stand a chance. This time, we need everyone on board – especially the most affected people in the most affected areas. This is a moral issue, and you have the moral high ground. Use it.

Everyone is needed, everyone is welcome, no matter where you live, no matter where you come from, no matter your age or your background. You must take it from here and carry on connecting the dots yourself because, right there, between the lines, you will find the answers – the solutions that need to be shared with the rest of humanity. And when the time comes for you to share them, I would give you just one piece of advice. Simply: tell it like it is. /

What Next?

If you, for example, live in Warsaw and want to buy the most sustainable tomatoes from your local food shop, which ones should you buy? The organic ones from Spain or the non-organic ones grown in Poland? A likely answer is that none of them are sustainable. But perhaps an even better reply would be: who cares?.

Of course it is important to support and develop organic farming methods, and if we had a hundred years to solve this crisis, then these choices would really matter. But if we keep focusing solely on small, separate issues concerning our individual consumption we will not stand a chance of reaching our international climate targets. We don't need to keep telling people to change their light bulbs, to vote, or to stop throwing away food. Not because these things aren't important – they are – but because we can safely assume that the people who read books, watch TV documentaries or attend seminars about the climate crisis are already well aware of the importance of the democratic process and the fact that people in the Global North should use fewer resources.

In fact, these narratives might even risk doing more harm than good, as they send a message that we can solve this within our current systems – and we no longer can. Voting is the most essential duty for all democratic citizens. But who do you vote for when the politics needed are nowhere in sight? And what do we do as democratic citizens when not even the universal compromise of voting for the best available candidate will bring us closer to finding a solution to our greatest problems?

In 2021 the container ship *Ever Given* ran aground in the Suez Canal, generating a feast for creators of social media memes. There it was, an enormous dark green ship stuck in the desert, the word 'EVERGREEN' painted in gigantic white letters across the hull, while a lonely excavator chipped away at the vast shoreline. It was the perfect image for our modern world: the 400-metre-long vessel, leased by a Taiwanese shipping company and registered in Panama for tax reasons, which single-handedly brought global supply chains and large parts of the world's trade to a one-week standstill.

The *Ever Given* was on its way from China and Malaysia to the Netherlands, carrying some 18,000 containers filled with whatever stuff containers are filled with – electronics, household products, footwear, fast fashion, mountain bikes, outdoor furniture, barbecues, and so on. There are over 5,000 ships criss-crossing the oceans today, just like the *Ever Given*. Many run on bunker fuel – an extremely dirty oil refining residual product which also happens to be extremely cheap. So cheap that few shipping companies can afford not to use it. But since emissions from international shipping have been negotiated out of our national frameworks – for the greater good of economic growth – we do not have to worry about them. They only exist in real life and, as we have learned in this book, reality does not always count in the world of climate statistics.

Take a minute to picture the circle of consumption. A plastic toy is being manufactured in China by an American toy company taking advantage of the cheaper labour, fewer restrictions and weaker environmental legislation there. When finished and packaged, the plastic toy is shipped to Europe on vessels like the *Ever Given*. Once the toy has arrived, it is loaded on to a lorry and driven across Europe to reach the shelves of a local shop, where someone buys it, puts it into a plastic bag, then hops into their petrol car and drives home. Perhaps, after they have unpacked the toy, they recycle the wrapping and packaging. Then, years later, when the toy is broken or forgotten, the consumer recycles the toy, to make room for new ones. The recycled materials go off in different directions. A small portion might be used to produce new plastic toys, bottles or packages. But that portion is tiny. Even in a progressive recycling nation such as my home country, Sweden, only about 10 per cent of the plastic we recycle actually ends up being recycled. The rest is often burned for energy. Another very likely fate for our recycled waste products is that they are driven back to ports such as Rotterdam and once again loaded on to ships like the *Ever Given*. This time their destination is one of the countless landfills in Southeast Asia or Africa where a huge proportion of our recycled materials end up, contaminating communities, soil, coastlines and fresh water. Unless, of course, they are burned in various unregulated ways on sites close to those same landfills, causing further pollution.

The idea of these gigantic container ships carrying all our recycled plastic waste is controversial and provocative, to say the least. But perhaps not as disturbing as the fact that those mighty vessels often sail back completely empty to ports halfway around the world, where they are once again filled up with our stuff. And so the circle of consumption goes on and on and on and on.

- **Every year,** an estimated 8 million tonnes of plastic waste are dumped into our oceans.

- **Every day,** we use around 100 million barrels of oil.

- **Every minute,** we subsidize the production and burning of coal, oil and gas by $11 million.

- **Every second,** an area the size of a football field of forest is cut down.

No number of individual actions can make up for that. We cannot live sustainably in an unsustainable world, no matter how hard we try. The truth is, many of us exceed the planetary boundaries just by paying our taxes, as so many of our collective assets go into fossil fuel subsidies.

The world will of course not end if we exceed 1.5°C or 2°C of global average temperature rise. But for a lot of people who lack the privilege of being able to adapt to the initial consequences of such a climate destabilization it will be the end of many things – food security, safety, stability, education, livelihoods and ultimately, an ever-growing number of human lives. Let us not forget that in a 1.2°C hotter world, people are already losing their lives and livelihoods. This may be acceptable to some people in the Global North. But from a moral perspective it is as far from acceptable as you can possibly get. Not least because billions of people who are already on the front lines of the climate emergency have done almost nothing to cause the problem in the first place.

Also, there are tipping points. Some we have already passed; others might be waiting just around the corner. There is a reason for the figure 1.5°C. And that reason is to minimize the risk of causing further irreparable damage to our life-supporting systems.

If you are looking for answers to how we can fix the climate crisis without changing our behaviours, then you will be forever disappointed, because our leaders have left it far too late for that. However, that does not mean we do not have solutions, because we do. We have lots of them. We just have to change our perspective about them – just as we need to redefine hope and progress so that those words are no longer synonymous with destruction. A solution is not just something that automatically replaces whatever is no longer working. A solution can also be to simply stop doing something.

Some of these ways forward might be very different depending on who you are and where you are. For instance, if you live in Angola, Peru or Pakistan, you might already be suffering the consequences of the climate crisis. Then maybe the best thing you can possibly do is jump on a plane to a climate conference in Europe or North America and tell your story to try to create change – if you get the opportunity to do so. Whereas if you live in the US, Belgium or the UK, one of the most effective ways of communicating that same crisis might be to surrender your privilege to fly.

But it is important that we do not blame anyone for what they do or don't do. Life is complicated enough as it is. In no way can we expect that we as individuals should compensate for the wrongdoings of governments, media, multinational corporations and billionaires. That is an absurd idea. As individuals we can do many things, but this is not a crisis that can be solved by one person acting alone.

In order to create the necessary changes, we need a series of different layers of actions. We need both structural system changes and individual changes. And on top of that we need a cultural transformation when it comes to norms and discourse. All of this is entirely possible. If we are prepared to change, then we can still avoid the worst consequences. There is still time. So yes, we can still fix this.

Fundamentally changing an unsustainable society is not such a bad thing to do. On the contrary. Replacing unsustainable habits with sustainable ones will likely give us a greater sense of purpose and meaning. Once we stop pretending that we can fix this without treating the crisis like a crisis and without fundamentally changing our societies, then action begins. And a new hope is born. A better hope. A real hope.

We have little to fear because all the best things in life will still be there: friends, culture, sports, entertainment, family, nature, food, drink, arts, travel, adventure, people. None of those things will go away, even if we will need to approach some of them in different ways.

The climate crisis cannot be solved within today's systems. But that must not stop us from taking whatever action we can, right now. Not only are these changes needed, they themselves will create positive feedback loops and tipping points that will lead us away from our current path of planetary destruction.

Throughout this book, and in this section in particular, I talk about 'solutions' to the climate crisis. It is important to remember that while we can (and must) put solutions in place that will reduce carbon emissions, protect biodiversity and rid our skies of toxic air pollution, we cannot 'solve' the climate crisis for everyone.

The UN Secretary-General António Guterres called the recent IPCC Sixth Assessment report 'an atlas of suffering'. The climate crisis is already impacting people around the world with devastating consequences —

particularly those living in poor economies. Even if we could stop all greenhouse gas emissions today, we have already inflicted irreparable damage to the planet and to the people whose livelihoods and lives have been destroyed by floods, droughts, wildfires and storms. And the best available science is clear that temperatures are continuing to increase and that these impacts will certainly get worse.

Our leaders have failed to take action, and that has turned the changing climate into a crisis which can no longer be avoided. They have failed us up until now, but that doesn't mean we can give up. Far from it.

As Guterres said, 'Now is the time to turn rage into action. Every fraction of a degree matters. Every voice can make a difference. And every second counts.'

I am not telling anyone what to do but, based on the information given by the scientists and experts in this book, here is a list of actions that some of us can take, if we want to. /

The climate crisis cannot be solved within today's systems. But that must not stop us from taking whatever action we can, right now.

What needs to be done

Start treating the crisis like a crisis

The longer we pretend that we can solve the climate and ecological crisis without treating it as such, the more precious time we will lose. /

Face the emergency

Thanks to our leaders' complete failure to address any of the issues connected to sustainability, this is no longer about what we want to do, it is about what we have to do. We do not just need to lower our emissions or become a low-carbon society. We need to get as close to zero as is physically possible. There is no middle ground left for small steps in the right direction. We must start getting our priorities straight. /

Admit failure

Even if we were to stop our destruction of nature this very minute, we have already inflicted irreparable damage to our life-supporting systems. Therefore, we have failed. Our political ideologies have failed. Our economic systems have failed. And we keep on failing, because we have not begun to slow down. We are speeding up the process. Unless we admit this failure we will not be able to learn from our mistakes. Nor will we be able to fix them. /

Include all the figures

One of our first priorities must be to include all of our actual emissions in our statistics. How else can we get an overview of the situation in order to start making the necessary changes? The fact that this has not been done tells us all we need to know about the efforts of our societies up until this point. Until we start including all of our emissions – the consumption of imported goods, international aviation and shipping, military, exports, investments of pension funds, biogenic emissions, etc. – then the fact remains: our emperors are naked. /

Connect the dots

The ability of our ecosystems to absorb carbon is rapidly deteriorating, thanks to deforestation, pollution, over-exploitation, and so on. Industrial farming is ruining our soils, rivers and coastlines. The ongoing destruction of our biosphere has initiated a potential mass extinction and the destabilization of our entire climate. And as we keep encroaching on nature, we are creating the conditions for new pandemics to emerge. But the environment is not the only thing that is suffering. Social inequality is growing and the imbalance between the world's richest and poorest is completely absurd. These crises are interlinked, and we cannot address one without also addressing the others. /

Choose justice and historic reparations

The climate and ecological crisis is a crisis of inequality and social injustice. Those who are being affected the most are the ones who have done the least to create the problem. That makes it a moral issue, an issue of social, racial and intergenerational injustice which involves almost 8 billion people. To find mutual ways forward we need to get as many people as possible on board. It must be done because failure is simply not an option. And none of this will be possible in the long run unless the nations accountable for having drained 90 per cent of the carbon budget already spent face up to the consequences of their actions and pay for the damage they have caused. Paying for these damages is the least they can do – no price tag can be put on lives. We cannot move towards a better future without taking action to heal the wounds of the past. /

What we can do together as a society

Educate ourselves

Decades of information that should have reshaped our entire society have failed to reach the general public. Unless we rapidly undo this violation of democracy and basic human rights, none of the necessary changes will be even remotely possible. Because why would we transform our societies entirely unless we understood that we had to? /

Leave no one behind

We must reshape our current system into one that protects workers and the most vulnerable, in order to reduce all forms of inequality and eliminate discrimination. /

Establish binding commitments

Starting today, establish annual, binding carbon budgets based on the current best available science and the IPCC's budget, which gives us at least a 67 per cent chance of limiting the global temperature rise to below 1.5°C of warming. We must make sure that these budgets include the aspect of global equity, the consumption of imported goods, global shipping and aviation and biogenic emissions, and they must not depend on future negative emissions technologies that do not already exist at scale – and perhaps never will. /

Rewild nature

This is one of the most effective tools we have at hand. And all we have to do is to step back and leave nature to heal itself. /

Restore nature

In places where nature is not able to heal itself we must help it along and restore what has been wrecked by human activities or extreme weather events. Mangroves, forests, wetlands, peat bogs, ocean floors, rivers and grasslands have a huge potential to sequester carbon – much more than any current technological alternative. /

Plant trees

If it is suitable for the soil and local biodiversity, then afforestation is a great solution. This should not be confused with industrial monoculture tree plantations intended to be cut down the moment it is financially profitable to do so. /

Maximize all possible carbon sinks

Our emissions need to be in unprecedented decline. And since we do not have the techno-logical solutions to make that happen, we must stop doing things or do things significantly less. It also means that we must use whatever means we have to catch and store carbon. One of the most effective ways of doing this is to leave large areas of our remaining forests alone. A living tree must be valued more than a dead one, and we must develop a system where we pay for carbon storage instead of paying for deforestation. Such a system must, however, be developed with a just and equitable perspective, where Indigenous rights and knowledge are at the forefront. /

Abandon phrases like 'carbon offsetting' and 'climate compensation'

The idea that in the foreseeable future we will be able to compensate for present or even future emissions is very misleading. None of the above – afforestation or rewilding and restoring nature – should be mistaken for carbon offsetting, which leads people to believe that we can compensate for emissions yet to be made. We have decades of past emissions to compensate for, and with our current capacity – as well as our current emission levels – we can barely scratch the surface of our historical pollution. /

Divest from fossil fuels

Banks, private investors, equity funds, pension funds, governments, and so on, must admit responsibility and completely halt all investments in fossil fuels, including their exploration and extraction. /

End all fossil fuel subsidies

Annually, we are spending $5.9 trillion to fund the destruction of our life-supporting systems. That is the ultimate definition of insanity. It must – and can – be brought to an immediate stop. /

Make local public transport free of charge

I often avoid advocating specific individual solutions as it risks stealing attention away from the larger systemic changes needed. I do not want to send a signal that we can fix this problem within our current system. However, if we are at least remotely interested in lowering our emissions of greenhouse gases, then improving, repairing and expanding our local public transport – while also making it free of charge – is one of the lowest-hanging fruit we have. /

Rethink transportation

There is no such thing as a sustainable car. Nor will there be one until we learn to grow them on trees or develop magic wands. There are currently around 1.4 billion motor vehicles in the world. A recent study estimates that figure will reach 2 billion by 2035. The thought of replacing them with new electric ones while staying within the planetary boundaries is far from realistic. So we have to rethink the entire concept of private road transportation. In many cases it is possible to retrofit electric engines into existing cars; another solution is car pools or car sharing. But overall, public transport must become more accessible and dominate our transportation systems. Restore, develop and expand low-carbon public transport – trains, trams, buses and ferries. In many regions, we have a huge network of infrastructure already in place. Long-distance electric buses can serve as an alternative to trains. Bring back night trains. And instead of subsidizing air travel, subsidize trains. The low-emitting option should always be made the cheapest by far. /

Make ecocide a crime

Widespread destruction of the environment must be made an international crime so that we are able to hold accountable those responsible for the destruction of nature. /

Leapfrog to renewable energy

If the Global South was given the opportunity to skip over fossil-fuel-based energy infrastructure and leapfrog straight to renewables, then everyone would benefit. But it must be paid for by those who have built their own wealth and infrastructure by polluting the atmosphere to the point where our CO_2 budgets have been used up. Doing this cannot be used as an excuse for richer countries to 'compensate' for failing to cut their own emissions. The idea that some countries can buy their way out of transforming their societies is deeply flawed. 'Like paying poor people to diet for you,' as Kevin Anderson has put it. /

Leapfrog social norms

We must move the public discourse on and skip over the mindset behind phrases such as 'taking small steps in the right direction'. The changes necessary can no longer be achieved within today's systems, and the ongoing attempts to 'slowly bring the public along' risk doing more harm than good. /

Avoid false solutions

In order for biofuels and the burning of biomass for energy to be sustainable we first of all need sustainable forestry and agriculture. And that does not exist at scale anywhere on this planet. We cannot continue sacrificing nature and biodiversity for the benefit of maintaining a loophole which enables nations and regions in the Global North to go on with business as usual. /

Invest in wind and solar power

In many cases the miracle has already happened. There are no perfect solutions, but when wind and solar infrastructure are built in the right places, and proper consideration is given to the local environment, they are a global game-changer. /

Avoid both-sidesism

Both-sidesism is when you treat two sides of an issue as equally important. In past decades, this phenomenon was evident in how the media gave climate deniers and delayers undue attention in order to appear impartial, as George Monbiot explained in Part Five. It has fuelled an existential crisis and the initiation of a mass extinction. Now both-sidesism has shifted to the media giving economic interests – in the best case – equal status to ecological ones, as in: 'Yes, this mine will contaminate the drinking water and pollute the air across the entire region, but it will also create 250 new jobs.' Survival is not a story with two sides. Extinction is not something that should be up for debate.

Or, actually . . . hey! Come to think of it – let's have both-sidesism! It's a chance to get things right. Since the media have spent the last seventy years or so reporting on economics and economic progress without any reference to their effects on nature, now they can make up for it by spending the next seventy years reporting only on ecological interests. Then they will have proven their impartiality. Bring it on. /

Ban high-carbon advertising

The idea that you can legally promote the destruction of our present living conditions and our future is ridiculous. If we are to have even an outside chance of meeting our climate targets, it must be banned. But since we no longer have the luxury of implementing non-holistic solutions, this ban must cover all high-emitting sectors. Otherwise, a ban just on advertising fossil fuels will mean an indirect stamp of approval for things such as unsustainable biofuels, burning wood for energy, and so on. /

Invest in science, research and technology

Technology alone will not save us. We have left it way too late for that. Nevertheless, we desperately need it – our lives depend on a scientific understanding of our situation. For instance, farm-free food production – food made from ingredients grown in laboratories – is on the verge of revolutionizing the way we feed ourselves. This – along with perennial crops and no-till farming practices – could open up a game-changing series of positive feedback loops, potentially returning huge amounts of carbon to our soils and forests. /

Heed the principles of safety

In 2021, global wildfires caused an estimated 6.45 gigatonnes of CO_2 emissions. That is around 15 per cent of our global emissions of carbon dioxide. In any other situation, an unaccounted-for 15 per cent increase to a serious crisis would likely have most of us reaching for the emergency brake. But when it comes to the climate crisis it does not even amount to major news. This ignorance must stop and the same principles of safety that apply to the rest of our society must be valid for the climate and ecological crisis. /

Sue carbon-polluting governments and companies

Take them to court. Make them pay for loss and damage and force them to act. But make sure you also communicate that we do not currently have the laws to put things right. Before the pandemic we were using around 100 million barrels of oil every single day. Prognoses say we are on track to surpass that in 2023. There are no laws to keep that oil in the ground. There are no laws against forest companies clear-cutting forests and burning them for energy. There are no laws protecting us in the long term against destroying our biosphere. It is perfectly legal to saw off the branch we all live on. So yes, we should sue them with all the means we have at our disposal. But also make sure to inform everyone that it will not be enough, especially in the unlikely event that we win in court. /

Create new laws

Make polluters pay for the damage they have done. Oil companies and fossil-fuel-producing nations must be held accountable for the irreparable damage they have caused and are causing. /

What can you do as an individual

Educate yourself

The moment you understand the full extent of the situation, then you will know what to do. Start study groups and share your knowledge with your friends and colleagues – use books, articles and movies and share them widely. /

Become an activist

This is by far the most effective way to fight the climate and ecological emergency. Advocate for change. Speed up the process of democracy. Shift social norms. Shine a light on justice and equity. Pass the mic to those who need to be heard. Take action. March. Boycott. Strike. Use non-violence, civil disobedience. We need billions. We need you. /

Defend democracy

There is no way for us to safeguard our future living conditions without democracy. Democracy is the most important tool we have. So defend it. Fight for it. Develop it. Expand it. Help others to register to vote. Stand up against all anti-democratic forces such as authoritarianism and xenophobic prejudice and all oppression against human rights and freedom of expression. Democracy must always be in motion, so we must find new ways of using it, for example citizens' assemblies. Vote, but also remember it is public opinion that runs the free world – and that is created every hour of every day. Not just on election day. /

Become politically active

This crisis cannot be solved with today's party politics, but that could be changed if enough people on the inside of political parties became aware of the situation. /

Talk about it

All the time. Be a nuisance. Be disruptive. There are very few pleasant things associated with the climate and sustainability crisis, so it is not easy to be nice about it. But we must always try. Look for common ground. Never use hate, especially against individuals. /

Amplify the voices on the front lines

The most affected people in the most affected areas (MAPA) are on the front lines of the climate crisis. But they are not on the front pages of our newspapers. Their voices must be heard, and we can all help. Share their stories, spread their names. /

Avoid culture wars

The moment we start treating the climate crisis like a crisis and implement annual, binding carbon budgets, and the moment we start including all our actual emissions in our statistics and face the climate and ecological emergency, we will no doubt start discussing all specific, individual solutions from a holistic point of view. But, until that day, try to avoid getting caught up in culture wars – the endless debates whose primary purpose is to stall the conversation, to create division and to delay the changes needed. There is no single solution that by itself will come close to making a significant dent in our emissions curve. So let's focus on the full picture. /

Shift towards a plant-based diet

As Michael Clark noted in Part Four, even if we were to reduce all other emissions to zero, those from our food systems alone would still push us beyond 1.5°C of global warming. Shifting towards a plant-based diet could save us up to

8 billion tonnes of CO_2 every year. The land requirements of meat and dairy production are equivalent to an area the size of North and South America combined. If we continue making food the way we do, we will destroy the habitats of most wild plants and animals, driving countless species to extinction. If we lose them, we will be lost too. If we changed to a plant-based diet, we could feed ourselves using 76 per cent less land. And if that is not enough you could do it for health reasons. Or moral ones. We currently kill more than 70 billion animals every year, excluding fish, whose numbers are so great that we measure their lives only by weight. Keep in mind that veganism is a privilege that is mainly available for the affluent citizens of the Global North. Many parts of the world maintain sustainable small-scale food production that includes fish, meat and dairy, particularly Indigenous communities and areas of the Global South. /

Be sceptical

According to Scientists for Global Responsibility, the combined CO_2 emissions from the world's military – and the industries that manufacture their equipment – are estimated to be around 6 per cent of our global total. But these numbers are often either unaccounted for, or subject to 'very significant under-reporting'. This is due to the fact that a large amount of our emissions have been negotiated out of our climate frameworks, and therefore they don't exist in our national statistics.

So whenever you hear someone say, 'Our emissions have decreased by so and so many per cent' – ask if that number includes consumption of imported goods, biogenic emissions, exports, methane leaks, military as well as international aviation and shipping. /

Stay on the ground

Flying is in many ways about privilege. Our remaining carbon budget is rapidly disappearing and – within the timeframe of us staying below 1.5° or 2°C of warming – there are no solutions in sight for air travel. Air travel is a growing sector. Today it accounts for about 4 per cent of our total climate impact, but that is expected to grow rapidly in the future. A recent study showed that the emissions for the entire tourist industry make up around 8 per cent of our global emissions. Around 80 per cent of the world's population have never set foot on an aeroplane, whereas the richest 1 per cent are responsible for 50 per cent of all aviation emissions, as Jillian Anable and Christian Brand explained in Part Four. So, if you live in the Global North, giving up your privilege to fly has been proven to be a very effective way of highlighting these inequalities. It is not remotely enough to solve the crisis, but it sends a clear message to people that we are in one. /

Buy less and use less

The chapters in this book clearly show that we are living way above our planetary means. That is not true for everyone, though. Some people need to raise their standard of living. Electricity, clean water and non-polluting cooking facilities are examples of things we need more of in many places around the world. Even in the Global North the situation varies enormously between different income groups. However, there is no way around the fact that we need a drastic overall reduction in resource usage. The three main problems are that our economy is dependent on growth, our politicians are ignoring the problem and we have a small group of high-income earners who are using up our common resources at an incredible speed. You and I can stop buying new things, we can use less stuff, we can repair, swap and borrow – but we should keep in mind that we do it as a form of activism, or as a moral choice, or a way to amplify our voices. We do it as citizens, not as consumers. This problem cannot be solved solely by individuals – nor can it be solved without a system change. /

Some of us can do more than others

Politicians

Being an elected official at this time in history means having responsibilities and opportunities beyond all imagination. Use them wisely. Be bold and brave. Lead by example. Change the narrative. Dare to risk your own popularity – as often as you can. Democracy is in your hands. You have to make sure that the solutions needed are available in today's politics. We need new policies, new economics, new frameworks, new legislations, new protection plans for workers. But above all we need to wake people up and inform them about our current whereabouts – about the fact that we are facing an existential crisis and that the time we have remaining to avert the worst consequences of that crisis is rapidly running out. So one of your top priorities must be communicating the urgency of our situation. There are many ways of doing that. One is to stand up and leave your place at the table, to say, 'This is clearly not working and I will not be a part of it.' /

Media and TV producers

If you are a media producer in search of new programmes, formats or stories, then you are probably already carrying around a loose idea of creating a new, optimistic series on the climate which will educate people while also giving them a feeling of hope. Before you decide to go ahead with that idea, ask yourself whom you wish to create hope for. The ones who are causing the problem, or the ones who are already being affected by it? All those young people who show up in the statistics as 'worried' or 'extremely worried' about the climate crisis are well aware of the problem. To them, news about the climate crisis is nowhere near as depressing as the fact that the news is being ignored. They do not need game shows with high-emitting celebrities talking about avocados being bad for the environment. They do not find it the least bit hopeful to be told that people can lower their carbon footprint by trying to go vegetarian once a week. In fact, your past and present failures are often one of the reasons why they feel hopeless. So unless the reason you became whatever you are today was to silently support the destruction of the living planet, then I suggest you start doing your job. /

Journalists

The responsibility to tell the stories, write the articles about this crisis and to hold accountable those responsible ultimately belongs to the media. If your editors are not taking these matters seriously, then it should be your duty as a reporter to make them change their minds. This is not too complicated for you to understand – even your own children often get it. The time when you, as an individual reporter, could blame this on ignorance or the fact that you were not aware has come and gone. Without the media, there is simply no way for us to reach our international climate targets. /

Celebrities and influencers

If you worry about the climate and happen to be a celebrity, an influencer or just someone with a lot of friends and followers on social media, then I have fantastic news for you. You have a unique opportunity to create crucial change at a crucial time in history. We humans are social animals, we copy the behaviour of others and we follow our leaders. You are some of those leaders. People aspire to be like you. When you got your Covid vaccination you probably posted something about it on social media. You might even have been a part of some official vaccination campaign. I was. Why did we do that? Because we know it works. It has a positive effect on a majority of the population. The climate is no different; it matters what we say but what we do matters even more. If you post a picture of yourself dressed in expensive fashion at a luxury resort on the other side of the world, then many of your followers and many of your friends will aspire to do what you are doing. That is how we work as a species. But if you, on the other hand, decide to try adopting a lifestyle closer to the planetary boundaries and become an activist, then those choices will have a huge impact on your surroundings. It might even push us beyond social tipping points.

Speaking out about the climate crisis while at the same time living it up like there is no tomorrow probably does more harm than good as it sends a clear signal that you can live an extreme way of life while still being someone who cares about ending our climate destruction. The time for 'little steps in the right direction' is over. We are in a crisis, and in a crisis you adapt and change your behaviour. We are all responsible for solving this situation. But we are not equally responsible. The bigger your platform, the bigger your responsibility; the bigger your carbon footprint, the bigger your moral duty. So this is not about what you write on social media. This is not about the money you give away to whatever charities or carbon-offsetting programmes. This is not a crisis we can buy our way out of. This is about what we do. /

The most affected people in the most affected areas

The voices in this world which have the most power belong to those who are destroying it: high-income nations, global leaders, corporations, oil companies, car manufacturers, high-emitting celebrities and billionaires with individual carbon footprints the size of entire villages or towns. These are the ones the world is mostly listening to, the ones expected to solve our problems. Not the Indigenous peoples who take care of the nature that up until now has been spared from the onslaughts of modernity. Not the scientists. Not those who have already been affected by the destruction. Not the children who one day will have to clean up the mess of all those powerful voices – or whatever will still be possible to clean up. It has to be the other way round.

We say that we need hope to survive – and yet we focus on giving hope only to those who are causing the problem rather than to those who are already suffering the consequences.

'We can still do this,' the powerful voices of the Global North say in their tremendous struggle to maintain a system that has been proved flawed, incapable and doomed in more ways than we can possibly imagine. 'We pledge to be climate neutral by 2050,' they say, sending everyone back to sleep. If they were honest about hope being something we need, then they would immediately reduce their emissions for the benefit of the billions of people who are already being affected, and for their own children. But they are not being honest. Instead they use hope as a powerful weapon to delay all necessary changes and prolong their business as usual.

Climate justice is not about the Global North saving the world in some act of white saviourism. That idea belongs to the same colonial mindset that got us into this mess in the first place – the idea that some people are worth more than others and therefore have the right to determine the world order. Climate justice is about the Global North acknowledging its past and present wrongdoings and starting the process of repairs by paying for loss and damage. Because our history is very much alive today. Just look at global economic inequality, vaccine inequality, pollution, or the rate at which some of us are using up our remaining natural resources – such as our rapidly disappearing carbon budgets.

The climate crisis is the greatest challenge humanity has ever faced. But it is also a historic opportunity to undo some of our past mistakes. We cannot solve this crisis with the same methods and mindsets that got us into it. Truth is on the side of those of you who are most affected by the crisis. Morality is on your side. Justice is on your side. I urge you to raise your voices and demand what you are owed. /

Index

Page references in **bold** indicate images

Illustration Credits

i 'Global Average Temperature 1850–2020' adapted for 2017–21 from 'Changes over time of the global sea surface temperature as well as air temperature over land' by Robert Rohde, Berkeley Earth Surface Temperature project, http://berkeleyearth.org/global-temperature-report-for-2020. Reproduced with permission

ii (top) 'Atmospheric CO2 Concentration' from Global average long-term atmospheric concentration of CO2. Measured in parts per million (ppm) by Hannah Ritchie and Max Roser, Our World in Data. Data source: EPICA Dome C CO2 record, 2015, and NOAA, 2018. Creative Commons license

ii (bottom) 'Annual Global CO2 Emissions (1750–2021)' by Bartosz Brzezinski and Thorfinn Stainforth, The Institute for European Environmental Policy, 2020, https://ieep.eu/news/more-than-half-of-all-co2-emissions-since-1751-emitted-in-the-last-30-years. Data sources: Carbon Budget Project, 2017, Global Carbon Budget, 2019, Peter Frumoff, 2014. Reproduced with permission of IEEP; and 'The 10 largest contributors to cumulative CO2 emissions, by billions of tonnes, broken down into subtotals from fossil fuels and cement' by Hansis et al., 2015. Carbon Brief using Highcharts, Global Carbon Project, Our World in Data, Carbon Monitor, Houghton and Nassikas

iii 'The Countries with the largest cumulative emissions 1850–2021' from 'The 10 largest contributors to cumulative CO2 emissions, by billions of tonnes, broken down into subtotals from fossil fuels and cement', Carbon Brief analysis of figures from the Global Carbon Project, CDIAC, Our World in Data, Carbon Monitor, Houghton & Nassikas, 2017, and Hansis et al., 2015. Reproduced with permission of Carbon Brief

xvi–xvii © Streluk/istock/Getty Images

4 'Global income and associated lifestyle emissions' from Extreme Carbon Inequality, Oxfam Media Briefing, 2015, https://www-cdn.oxfam.org/s3fs-public/file_attachments/mb-extreme-carbon-inequality-021215-en.pdf, Figure 1, updated with data from 'Confronting carbon Inequality', Oxfam, 2020, https://www.oxfam.org/en/research/confronting-carbon-inequality and 'Carbon inequality in 2030', Oxfam, 2021, 3–4, https://www.oxfam.org/en/research/carbon-inequality-2030. Reproduced with permission of Oxfam

16–17 © Johnny Gaskell

28 Composite graph of 'Atmospheric CO2 at Mauna Loa Observatory', Dec 2021, Scripps Institution of Oceanography; NOAA Global Monitoring Laboratory; #ShowYourStripes – Graphis & lead Scientist: Ed Hawkins, National Centre for Atmospheric Science, University of Reading; Data: UK Met Office. Design by sustention [PG]. Creative Commons License

34–5 Adapted from 'Socio-economic trends' and 'Earth System Trends' from 'The trajectory of the Anthropocene: The Great Acceleration' by Will Steffen, Wendy Broadgate, Lisa Deutsch, et al., The Anthropocene Review, 01/04/2015, Vol 2(1), 81–98, SAGE Publications, copyright © 2015, SAGE Publication. Reprinted with permission of SAGE Publications

36 © Johan Rockström. Reproduced with permission

38 (top) Adapted from 'Tipping elements in the Earth's climate system' by T. M. Lenton et al., PNAS, 12/02/2008, Vol 105(6), 1786–1793, https://www.pnas.org/content/105/6/1786

38 (bottom) Adapted from 'Climate tipping points – too risky to bet against' by T. M. Lenton et al., Nature, 27/11/2019, Vol 575, 592–595, https://www.nature.com/articles/d41586-019-03595-0

39 © Johan Rockström, with data from Global Warming of 1.5 ºC, IPCC, 2018, SPM.2; Climate Change 2014, IPCC, 2014, SPM10; and TAR Climate Change 2001, IPCC, 2001, copyright © IPCC, https://www.ipcc.ch/. Reproduced with permission

46–47 © Steffen Olsen, Danish Meteorological Institute

55 Adapted from Climate Change 2021: The Physical Science Basis. Contribution of Working Group I to the Sixth Assessment Report of the Intergovernmental Panel on Climate Change, Summary for Policymakers, IPCC, 2021, Figure SPM.2, copyright © IPCC, https://www.ipcc.ch/

63 (top) 'Near-surface air temperature change in the Arctic and the globe as a whole since 1995 for all months' ERA-5 reanalysis, NOAA, https://psl.noaa.gov/cgi-bin/data/testdap/timeseries.pl

63 (bottom) Aerial Superhighway, NASA 07/02/2012: https://svs.gsfc.nasa.gov/10902. copyright © NASA. Reproduced with permission

64 'Comparison of conditions with a cold Arctic and relatively straight jet stream and conditions with a relatively warm Arctic and wavy jet stream', NOAA, https://www.climate.gov/news-features/event-tracker/wobbly-polar-vortex-triggers-extreme-cold-air-outbreak

70–71 © Pat Brown/Panos Pictures

79 Adapted from 'Changes over time of the global sea surface temperature as well as air temperature over land' by Robert Rohde, Berkeley Earth Surface Temperature project, http://berkeleyearth.org/global-temperature-report-for-2020. Reproduced with permission

81 © Stefan Rahmstorf, CC by-SA 4.0. With data from 'Persistent acceleration in global sea-level rise since the 1960s' by Sönke Dangendorf et al., in 'Nature Climate Change', Springer Nature, 05/08/2019, 705–710, https://www.nature.com/articles/s41558-019-0531-8, copyright © The Authors, 2019, under exclusive licence to Springer Nature Limited

82 'The observed sea surface temperature change since 1870' from Observed fingerprint of a weakening Atlantic Ocean overturning circulation by Levke Caesar, Nature, Vol 556, 11/04/2018, 191–196, https://www.nature.com/articles/s41586-018-0006-5. Reproduced with permission

94–95 © Katie Orlinsky/National Geographic

103 'The global distribution of forests, by climatic domain' from Global Forest Resources Assessment 2020, FAO, 2020, https://www.fao.org/documents/card/en/c/ca9825en, with data from 'Proportion of global forest area by climatic domain, 2020', XI, 14, adapted from United Nations World Map, 2020. Reproduced with permission of FAO

104 © Beverly E. Law, with data from 'British Columbia Managed Forests (MMT CO2e)' in Provincial greenhouse gas emissions inventory, British Columbia, https://www2.gov.bc.ca/gov/content/environment/climate-change/data/provincial-inventory, copyright © 2021, Province of British Columbia

105 Data from 'Strategic Forest Reserves can protect biodiversity and mitigate climate change in the western United States' by Beverly E. Law, Logan T. Berner, Polly C. Buotte, David J. Mildrexler and William J. Ripple, Nature Communications Earth & Environment, 2021, Vol 2(254); and 'Land use strategies to mitigate climate change in carbon dense temperate forests' by Beverly E. Law, Tara W. Hudiburg and Logan T. Berner, PNAS, 03/04/2018, Vol

115(14), 3663–3668, copyright © 2018 the Authors

108 'The number of severe threats to biodiversity around the world' from 'Mapping human pressures on biodiversity across the planet uncovers anthropogenic threat complexes' by D. E. Bowler, A. D. Bjorkman, M. Dornelas, et al., People & Nature, 27/02/2020, 380–394, figure 6. Creative Commons Attribution 4.0 License

120 'Permafrost in the Northern Hemisphere', copyright © GRID-Arendal/Nunataryuk, https://www.grida.no/resources/13519

123 (top and bottom) Summary for Policymakers, IPCC, 2021, Figure SPM.5 (b&C), copyright © IPCC, https://www.ipcc.ch/. Reproduced with permission

125 Adapted from 'Historical and projected future concentrations of CO2, CH4 and N2O and global mean surface temperatures (GMST)', Climate Change 2021 The Physical Science Basis, IPCC, 2021, Figure 1.26; and 'Selected indicators of global climate change under the five illustrative scenarios used in this Report', SPM.8(e), copyright © IPCC, https://www.ipcc.ch/

126–27 © Dmitry Kokh

130–31 © Josh Edelson/AFP via Getty

138 Data from 'Heat-related deaths (2000–19)' in 'Global, regional, and national burden of mortality associated with non-optimal ambient temperatures from 2000 to 2019: a three-stage modelling study' by Q. Zhao et al., The Lancet PH3, July 2021, https://www.thelancet.com/journals/lanplh/article/PIIS2542-5196(21)00081-4/fulltext; and GBD Compare 'Global annual mortality in 2019 attrib-

uted to a selection of causes of death or due to specific risk factors', 15/10/2020 https://www.healthdata.org/data-visualization/gbd-compare, Institute for Health Metrics Evaluation. Used with permission. All rights reserved

141 Adapted from 'Temporal and spatial distribution of health, labor, and crop benefits of climate change mitigation in the United States' by Drew Shindell et al., PNAS, 16/11/2021, Vol. 118(46), Figure 7.C, copyright © 2021 the Authors

144 Data from 'Projecting the risk of mosquito-borne diseases in a warmer and more populated world: a multi-model, multi-scenario intercomparison modelling study' by Felipe J. Colón-Gonzàlez et al., The Lancet Planetary Health, 01/07/2021, Vol 5(7), E404–E414, https://www.thelancet.com/journals/lanplh/article/PIIS2542-51962100132-7/fulltext

145 Data from 'Projecting the risk of mosquito-borne diseases in a warmer and more populated world: a multi-model, multi-scenario intercomparison modelling study' by Felipe J. Colón-Gonzàlez et al., The Lancet Planetary Health, 01/07/2021, Vol 5(7), E404–E414, https://www.thelancet.com/journals/lanplh/article/PIIS2542-51962100132-7/fulltext

152–53 © Rakesh Pulapa

155 'Cumulative emissions (1850–2021) per current population, selected countries' from 'The 10 largest contributors to cumulative CO2 emissions, by billions of tonnes, broken down into subtotals from fossil fuels and cement', Carbon Brief analysis of figures from the Global Carbon Project, CDIAC, Our World in Data,

Carbon Monitor, Houghton & Nassikas, 2017, and Hansis et al., 2015. Reproduced with permission of Carbon Brief

178–79 © Ami Vitale

183 © Solomon M. Hsiang

184 Data from 'GDP per capita in 2019', World Bank, 2021; 'Valuing the Global Mortality Consequences of Climate Change Accounting for Adaptation Costs and Benefit' Working paper 27599, NBER July 2020, revised August 2021, https://www.nber.org/system/files/working_papers/w27599/w27599.pdf; and 'Global non-linear effect of temperature on economic production' by Marshall Burke, Solomon M. Hsiang & Edward Miguel, Nature, 2015, Vol 527, 235–239, https://www.nature.com/articles/nature15725

188 Data from Uppsala Conflict Data Program. Retrieved January 2022, UCDP Conflict Encyclopedia: https://www.pcr.uu.se/research/ucdp/, Uppsala University

190 'Quantifying the Influence of Climate on Human Conflict' by Solomon M. Hsiang, Marshall Burke & Edward Miguel, Science, 2013, 341, Figure 2

194–95 © Richard Carson/REUTERS

198–99 © Daniel Beltrá

202 Graph by Robbie M. Andrew based on mitigation curves from Raupach et. al. 2014, using data from Global Carbon Project, Creative Commons Attribution 4.0 International. Emissions budget from IPCC AR 6. curves

207 Graph by Kevin Anderson based on data from IPCC AR6, headline carbon budget for a 67% chance of staying below 1.5°C, 2020, updated to the start of 2022 based on data from Global Carbon Project, by Robbie M.

Andrew and Glen Peters, et al. https://www.globalcarbonproject.org

212 'Utsläpp från Sveriges ekonomi' by Maria Westholm, https://www.dn.se/sverige/sverige-ska-ga-fore-anda-ar-klimat-malen-langt-ifran-till-rackliga/, copyright © Dagens Nyheter. Translated and reproduced with permission

214–15 © Pierpaolo Mittica/INSTITUTE

225 Data from 'High Strain-rate Dynamic Compressive Behavior and Energy Absorption of Distiller's Dried Grains and Soluble Composites with Paulownia and Pine Wood Using a Split Hopkinson Pressure Bar Technique' by Stoddard et al., Bioresources, Dec 2020, 15(4). 9444–9461; and 'Global Carbon Budget 2021' by Friedlingstein et al., 2021 Creative Commons Attribution 4.0 License

242–43 © Wang Jiang/VCG via Getty Images

246 Data from 'Harmonization of global land use change and management for the period 1600–2015 (LUH2) for CMIP6' by G.C. Hurtt et al., Geoscientific Model Development, 2020, Vol 13(11), 5425–5464, copyright © Authors 2020. Creative Commons Attribution 4.0 License

249 'Multiple health and environmental impacts of foods' by Michael A. Clark et al., PNAS, 12/11/2019, Vol 116(46) 23357–23362, copyright © 2019 the Authors. Creative Commons Attribution License 4.0

250 'Comparative analysis of environmental impacts of agricultural production systems, agricultural input efficiency, and food choice' by Michael A. Clark & David Tilman, Environmental Research Letters, 2017, Vol 12(6),

Creative Commons Attribution 3.0 license

251 'Global food system emissions could preclude achieving the 1.5° and 2°C climate change targets' by Michael A. Clark et al., Science, 06/11/2020, Vol 370(6517), 705–709, American Association for the Advancement of Science. Reproduced with permission

257 'Long-term model-based projections of energy use and CO2 emissions from the global steel and cement industries' by Van Ruijven et al., Resources, Conservation and Recycling, September 2016, Vol 112, 15–36, Figure 9, copyright © 2016 The Authors. Published by Elsevier B.V. Reproduced under Creative Commons CC-BY license

258 Data from The Global Carbon Project's fossil CO2 emissions dataset by Robbie M. Andrew and Glen P. Peters, Zenodo, 2021. Creative Commons Attribution 4.0 International license

259 Data from Net Zero by 2050, Data product, IEA, chapter 3, https://www.iea.org/data-and-statistics/data-product/net-zero-by-2050-scenario, Figure 3.15. Reproduced with permission of IEA

263 Graph by Ketan Joshi with data from 'Historical' and 'Planned 2020 report', Appendices 6.1 and 6.2, https://www.globalccsinstitute.com/wp-content/uploads/2021/03/Global-Status-of-CCS-Report-English.pdf; and 'planned 2021 report' from https://www.globalccsinstitute.com/wp-content/uploads/2021/11/Global-Status-2021-Global-CCS-Institute-1121.pdf; The Global Status of CCS, 2020; and 2021, copyright © Global CCS Institute, Australia.

A Note on the Cover

Ed Hawkins

No words. No numbers. No graphs. Just a series of vertical, coloured stripes showing the progressive rise in global temperatures in a single, striking image.

Each stripe on the cover of this book represents the average global temperature for a single year. Shades of blue indicate cooler years, while red shows years that were hotter. The stark band of deep red stripes on the front demonstrates the unmistakable rapid heating of our planet in recent decades.

The Warming Stripes were designed to start essential conversations about climate change – and they do. They have been downloaded and shared by millions of people – from politicians and artists to weather presenters and rock stars – spreading the message that no corner of the globe is immune from the effects of climate change.

Similar images for almost every country can be downloaded for free from **showyourstripes.info**.